Silicate Glass Technology Methods

WILEY SERIES IN PURE AND APPLIED OPTICS

Advisory Editor

Stanley S. Ballard, University of Florida

ALLEN AND EBERLY · *Optical Resonance and Two-Level Atoms*

BABCOCK · *Silicate Glass Technology Methods*

BOND · *Crystal Technology*

CATHEY · *Optical Information Processing and Holography*

CAULFIELD AND LU · *The Applications of Holography*

GERRARD AND BURCH · *Introduction of Matrix Methods in Optics*

HUDSON · *Infrared System Engineering*

JUDD AND WYSZECKI · *Color in Business, Science, and Industry,* Third Edition

KNITTL · *Optics of Thin Films*

LENGYEL · *Introduction to Laser Physics*

LENGYEL · *Lasers,* Second Edition

LEVI · *Applied Optics, A Guide to Optical System Design,* Volume I

LOUISELL · *Quantum Statistical Properties of Radiation*

McCARTNEY · *Optics of the Atmosphere; Scattering by Molecules and Particles*

MOLLER AND ROTHSCHILD · *Far-Infrared Spectroscopy*

PRATT · *Laser Communications Systems*

SHULMAN · *Optical Data Processing*

WILLIAMS AND BECKLUND · *Optics*

ZERNIKE AND MIDWINTER · *Applied Nonlinear Optics*

Silicate Glass Technology Methods

CLARENCE L. BABCOCK

Optical Sciences Center
University of Arizona, Tucson

John Wiley & Sons, New York / London / Sydney / Toronto

Library of Congress Cataloging in Publication Data:

Babcock, Clarence Lloyd, 1904-
Silicate glass technology methods.

(Wiley series in pure and applied optics)
Bibliography: p.
Includes index.

1. Glass manufacture. I. Title.
TP857.B2   666'.1   76-30716
ISBN 0-471-03965-9

Printed in the United States of America

10 9 8 7 6 5 4 3 2 1

Preface

Confined to his spinning satellite, man is dependent on his physical environment as year by year he continues his trips around the sun. He has advanced from searches for the means of a primitive existence to ever-increasing uses of available materials and energy to provide added comforts to his life. Rapid advancements in technology and accelerating population growth, particularly during this century, have made him aware that sources of materials and energy are not inexhaustible. He is in much the same position as the astronauts who carefully planned their material and energy needs on their trips to the moon. Glass has played a major role in man's history since Stone Age people first used the natural glass obsidian to make hunting tools. The uses of man-made glass have become increasingly diversified and necessitate the development of a systematic technology of this versatile material.

This book was written for those interested in methods of materials research on inorganic glasses. This interest may range from research on glass per se to the need for general knowledge of glasses as it may be involved in related research efforts. The interests of technologists engaged in the making and uses of glasses have also been kept in mind. It is believed that the author's new approach to glass technology methods will be useful to those engaged in the teaching of glass science and technology.

This book deals with only one type of the inorganic glasses; namely, those known as silicate glasses, whose major constituents are silicon and oxygen. There are compelling reasons for this choice. (1) The crust of the earth consists essentially of raw materials used in the making of silicate glasses. (2) The history of glassmaking, from earliest times, reveals that the making and use of silicate glasses has consequently been predominant. (3) Most of the published glass literature deals with compositions, properties, and postulated ionic structures of silicate glasses.

Silicate glasses in common with all types of oxide glass are the products of fusion and can be cooled to room temperature without crystallizing. They can all be characterized as supercooled materials and have somewhat similar time-temperature characteristics. Thus knowledge of the behavior of silicate glasses, and research techniques used for their study, can serve as a general introduction to all types of oxide glass.

This book discusses methods of silicate glass technology and is not intended as a comprehensive review of all facets of this now extensive subject. The author has used information selected from published literature, conditioned by his experience in commercial glass research and technology, to outline the basis of a systematic technology of silicate glasses. The method is based on the postulated presence of ionic substructures in silicate glasses, which are similar in kind to the substructures in their crystalline counterparts. The merits of this method are demonstrated in terms of quantitative representations of experimental data on a number of electromagnetic, mechanical, and thermal properties of glasses. Comparisons are drawn between this method and the one in general use, which is based on the assumption that constituent cations and anions are arranged at random in the glass.

It is a pleasure to acknowledge help from a number of sources. The author was introduced to the fascinating subject of glass during the time he held the Corning Glass Works Fellowship in the Physics Department at Purdue University. Cooperative research and technology efforts with colleagues in Owens-Illinois, Inc., have been most helpful. Important guidance has been received through correspondence and discussions with colleagues in the National Bureau of Standards, the American Ceramic Society, and the International Commission on Glass. Dr. Peter Franken, Director of the Optical Sciences Center, generously provided time for preparation of this book and has been a source of constant encouragement. Finally the author is indebted to his son, Dr. Larry L. Babcock, for many clarifying discussions on the mineralogy of crystalline silicates.

CLARENCE L. BABCOCK

Tucson, Arizona
February 1977

Contents

PART II ELECTROMAGNETIC PROPERTIES

PART IV   THERMAL PROPERTIES

PART V   GLASSMAKING PROCESSES

I

Principles of Glass Technology

I often say ... that when you can measure what you are speaking about, and can express it in numbers, you know something about it: but when you cannot measure it, when you cannot express it in numbers, your knowledge is of meager and unsatisfactory kind; it may be the beginning of knowledge, but you have scarcely, in your thoughts, advanced to the stage of science, whatever the matter may be.

Lord Kelvin (1824-1907)

The purpose of glass technology is to develop quantitative relations between compositions, properties, and their ionic and electronic structures and to apply them in the making and use of glasses. Part I reviews existing information on silicate glasses from this viewpoint and makes critical judgments on its relevance to the purpose stated. Evaluation of these data indicates that the ionic structures of silicate glasses are similar to their crystalline silicate source materials. Glasses are unique materials deriving their properties from ionic structures coupled with their nature as supercooled liquids. Sample preparation and property measurement methods are outlined for use in glass research.

The development of methods of relating glass properties to their oxide compositions is related to advancements in our knowledge of glass structures. Schott and Abbe expressed their property data in terms of the amounts of oxides in the glasses. Thus they implied that glasses were simple solutions of the constituent oxides. This method seemed to be justified after the development of the random network theory, initiated in the 1930s, according to which the constituent cations and anions were distributed at random in glass structures. The author departs from this concept of silicate glass structures and postulates the presence in glasses of substructures that are somewhat similar to the substructures in their crystalline primary phase counterparts. Uses of this method are demonstrated in quantitative property calculations, the calculation of primary phase diagrams, and the classification of silicate glasses.

Part I provides background information on silicate glasses as an introduction to the understanding of their electromagnetic, mechanical, and thermal properties.

1

Structures of Minerals and Glasses

Physical properties of all materials are dependent on their constituent atomic nuclei and attendant electrons, the interatomic distances involved, and the spatial arrangements of atoms. Minerals and manufactured materials used as raw materials for making glasses are ionized in the solid state. Their crystal structures, in terms of ionic size, bond types, and spatial arrangements, are known in considerable detail.

Relatively few data, in the foregoing terms, are available on the structures of glasses. The structures of glasses continue to be important subjects of research. Minerals are characterized by the presence of periodicity and symmetry, but these features are absent in glasses. However the ionized atoms in glasses are linked by forces essentially the same as those found in crystalline minerals.

1.1 GLASSMAKING ELEMENTS

Geological research has provided information on the distribution of the elements in the earth's crust. Clarke and Washington (1924) made 5159 analyses of materials in the continental areas of the crust. These materials are far from being equally distributed, and objections can be raised against the use of such averages. Poldervaart (1955) included the submarine crust and arrived at somewhat different figures. The figures of Clarke and Washington, used here as general guidelines, give average amounts of the 12 most abundant elements, in weights percent, as follows;

Oxygen	46.60	Potassium	2.59
Silicon	27.72	Magnesium	2.09
Aluminum	8.13	Titanium	0.44
Iron	5.00	Hydrogen	0.14
Calcium	3.63	Phosphorus	0.12
Sodium	2.83	Manganese	0.10

These elements make up about 99.39% of the total, and all except hydrogen are useful constituents of silicate glasses. Oxygen is predominant; making up about

47% by weight, more than 60% of the total number of ions, and it occupies more than 90% of the total volume. Thus the continental crust is essentially a packing of oxygen anions bonded principally to the cations of silicon, aluminum, iron, and so on. Elements are also found in halides, phosphates, carbonates, sulfates, sulfides, and hydroxides. The same data expressed as weights percent of the oxides on a water-free basis are as follows:

SiO_2	60.1	MgO	3.6	TiO_2	1.1
Al_2O_3	15.5	CaO	5.2	P_2O_5	0.2
Fe_2O_3	3.1	Na_2O	3.9	MnO	0.2
FeO	3.9	K_2O	3.2		

Thus the earth's crust is an abundant source of materials for making silicate glasses. That they exist principally in the form of oxides attests to their stability and in turn to that of oxide glasses made from them.

We will confine our attention to one type of oxide glass, namely, the silicates whose major constituents are silicon and oxygen. References to Caley (1962), Morey (1954), Hovestadt (1902), and to catalogs of glass manufacturers, show increasingly diversified uses of silicate glasses in new products. It is noted however that several other types of inorganic glass have found important uses in modern technology. Rawson (1967) gives comprehensive review of inorganic glass-forming systems. The technologies of these systems are presently in various stages of development. The elements sulfur and selenium and the oxides SiO_2, GeO_2, B_2O_3, P_2O_5, and As_2O_3 readily form glasses by themselves. Several other oxides including TeO_2, SeO_2, and V_2O_5 form glasses in combination with other oxides. The list of inorganic glasses includes the halides BeF_2 and $ZnCl_2$, the nitrates such as KNO_3-$Ca(NO_3)_2$, and a variety of chalcogenide glasses made from combinations of the elements Se, Ge, As, Sb, Bi, S, Si, and Te.

1.2 PRINCIPLES OF CRYSTAL STRUCTURES

Max von Laue and his colleagues made the important discovery that x-rays were diffracted by crystals to produce interference patterns, the crystals acting as three-dimensional gratings. The initial research was described by von Laue (1913) and by Friedrich, Knipping, and von Laue (1913). X-ray analysis depended largely on mineralogy, especially in the beginning. Mineralogists supplied most of the crystalline materials for x-ray analyses and introduced physicists to the subject of crystallography. The first crystals to be analyzed were the most simple natural minerals such as rock salt and calcite. X-ray analyses were later extended to more complex structures, principally the silicates. Bragg (1937), Evans (1946), Pauling (1960), Winkler (1955), Mason (1960), and others have described the principles of crystal structures that have resulted from this extensive research.

The cohesion of materials is seldom caused by the action of a single type of chemical bond, and minerals are no exceptions. Two categorical bond types, ionic and covalent, have been found to exist simultaneously in minerals. The concept of the ionic bond is most useful in describing mineral structures and their related properties. This is the case when a sodium atom loses its one M electron and becomes a singly charged positive cation. Much less is known about the covalent bond that is formed when adjacent atoms share pairs of electrons. This is the case when silicon shares its four M electrons with another atom such as oxygen and becomes a positive cation having an electrical charge of 4. Single, double, and triple covalent bonds involve the sharing of two, four, or six electrons. Despite detailed differences between these two bond types, minerals can be accurately described in terms of cations and anions. Following Pauling (1960), the dual character of bonds in oxide minerals can be roughly evaluated as follows:

| Element Bonding | Cation-Oxygen Bonds | |
with Oxygen	Ionic (%)	Covalent (%)
Si	50	50
Al	63	37
Mg	70	30
Ca	75	25
Sr	75	25
Li	75	25
Na	80	20
Ba	80	20
K	82	18

These particular numbers should not be taken too seriously. Simply remember that all bonds in minerals are intermediate between the two types and that they are all somewhat covalent in character. The Coulombic attractions in an ionic bond arise from electrical fields in all direction, whereas a covalent bond is stereospecific; it has a specific direction in space. Our interest is in silicates wherein cations form bonds with oxygen. The bonds in crystalline silicates have appreciable covalent character, which suggests that the constituent ions have somewhat ordered positions and are not randomly distributed in the structure. This suggestion is followed up in our later discussions on crystalline silicates and silicate glasses.

Distances between ions are such that the total attractive and repulsive forces are equalized as illustrated in Figure 1.1. The electrostatic potential energy at this distance is minimal, and the thermodynamic equilibrium is achieved under prevailing temperature and pressure conditions. Electrical stability does not come about by pairing off individual cations and anions. The positive charge of a cation is divided equally between the surrounding oxygen anions, whose

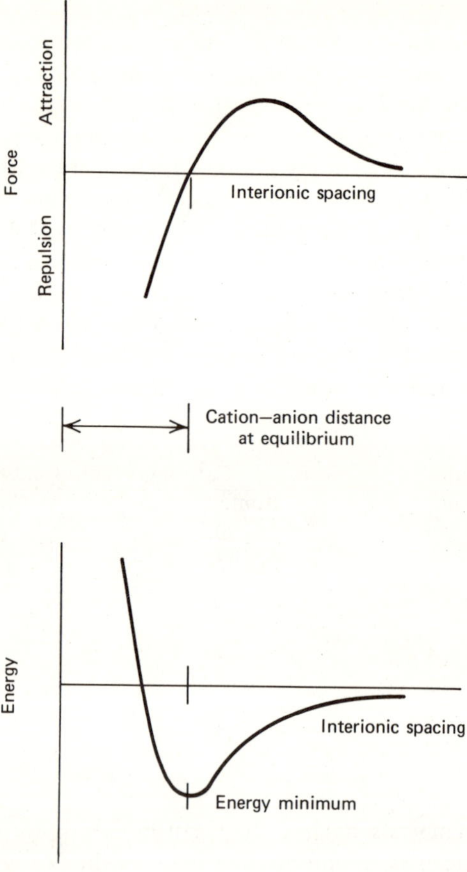

Figure 1.1 Interionic forces and potential energy minimum.

number is determined by the relative sizes of cations and oxygen anions. The number of surrounding oxygens is called the coordination number (CN) and is related to the ratio of r/R, where r and R are the cation and anion radii, respectively. Table 1.1 lists ionic radii and CN for elements commonly found in crystalline silicates and silicate glasses. Atomic radii of neutral atoms are shown for comparisons. The observed CN are those obtained from research on crystals. The predicted CN come from considering the ions as electrically charged spheres in the arrangements indicated in Table 1.2.

Table 1.1 reveals generally good agreement between CN predicted on the basis of the idealized scheme just given and those observed in structure studies on minerals. It is noted that agreement is best for smaller cations such as B and Si. If the cations are nearly the same size or larger than the oxygen anion, the CN becomes less well defined. Pauling (1929) observed in this connection that anion groups around the smaller and more highly charged cations tend to share corners only, whereas edges and even corners are shared by larger groups around weak cations. When this idealized scheme is applied to mineral structures, the need for certain modifications becomes apparent. Crystals found in the natural minerals are seldom, if ever, perfect. They are subject to structural disorders arising from temperature and pressure conditions during the time of their formation. Crystallographers have categorized structural disorders in terms of point defects, dislocations, boundaries, and so on. Mason (1960) points out that high temperatures and low pressures favor low CN, and low temperatures and high pressures favor high CN. For example, Al tends to fourfold coordination and to substitute for Si in high temperature minerals, but it tends to sixfold coordination in minerals formed at low temperatures. Mason (1960) states that ionic substitutions are the rule rather than the exception in minerals; research indicates that such substitutions can occur if their radii do not differ by more than 15%. Thus O, OH, and F with radii of 1.40, 1.40, and 1.36 Å can replace each other.

The spatial configurations given in Table 1.1 imply that distances between ionic centers can be obtained by adding the cation radius to that of the anion. Thus the distance between Si and O would be 0.42 + 1.40 = 1.82 Å. However x-ray diffraction studies on silicate minerals show this distance to be close to 1.60 Å. Likewise the predicted 0-0 distance is 2.80 Å, whereas the observed distance is near 2.60 Å in SiO_2. Strong directional covalent bonds distort spherical symmetry of the electron clouds, and the ions are said to be polarized. Fajans (1923) suggested that such mutual polarizations are related to departures from additivity of ionic radii. The polarizing power of an ion may be defined as its ability to influence the distribution of electrons around nuclei of neighboring ions. When an Si atom becomes an ion, its radius is reduced from 1.18 to 0.42 Å. When this occurs the remaining electrons are pulled closer to its highly charged nucleus and become less responsive to external electrical fields. When an O atom becomes an ion, its radius is increased from 0.60 to 1.40 Å. Thus its electron cloud is greatly expanded and its electrons then become more responsive to external electrical fields. When Si and O interact to form SiO_2, the strong charge of Si attracts and deforms the otherwise spherical electron cloud around O. The electrical centers of the electron cloud of the O and that of its nucleus are no longer identical; the O anion is then said to be polarized. The polarization of the Si cation is this interaction is quite small. In silicates the polarizing power of Si is high and its ability to be polarized is low. Likewise the polarizing power of O is low and its ability to be polarized is high. The polarizing powers and polarizabilities of the other ions are intermediate between these two extremes. Departures from additivity are greatest for small, high valence cations and least

TABLE 1.1 Atomic Radii of Neutral Atoms, Ionic Radii in Crystals, and Oxygen Coordination of Cations in Crystals

| Element | Symbol | Radii (Å) | | Oxygen coordination CN | |
		Atomic	Ionic	Predicted	Observed
Aluminum	Al	1.43	0.51	4	4, 5, 6
Arsenic	As	1.25	As^{3+}, 0.58; As^{5+}, 0.46	4	4
Barium	Ba	2.17	1.34	12	6-12
Beryllium	Be	1.12	0.35	4	4
Boron	B	0.46	0.23	3	3, 4
Cadmium	Cd	1.49	0.97	6	6, 8
Calcium	Ca	1.97	0.99	8	6-10
Chromium	Cr	1.25	Cr^{3+}, 0.63; Cr^{6+}, 0.52	6	6
Cobalt	Co	1.25	0.72	6	4, 6
Copper	Cu	1.28	0.72	6	4, 6
Fluorine	F	0.60	1.36		
Germanium	Ge	1.22	0.53	4	4, 6
Iron	Fe	1.24	Fe^{2+}, 0.74; Fe^{3+}, 0.64	6	4, 6
Lead	Pb	1.75	Pb^{2+}, 1.20; Pb^{4+}, 0.84	8	8-12
Lithium	Li	1.52	0.68	6	4, 6
Magnesium	Mg	1.59	0.66	6	6-10
Manganese	Mn	1.12	Mn^{2+}, 0.80; Mn^{4+}, 0.60	6	4, 6
Nickel	Ni	1.25	0.69	6	4, 6
Oxygen	O	0.60	1.40		
Phosphorus	P		0.35	4	6
Potassium	K	2.31	1.33	8	8-12
Silicon	Si	1.18	0.42	4	4
Sodium	Na	1.86	0.97	6	6-12
Strontium	Sr	2.15	1.12	8	6-12
Tantalum	Ta	1.43	0.68	6	6
Tellurium	Te	1.43	Te^{4+}, 0.70; Te^{6+}, 0.56	6	6
Titanium	Ti	1.46	Ti^{3+}, 0.76; Ti^{4+}, 0.68	6	6
Uranium	U	1.38	0.97	6	6, 8
Vanadium	V	1.32	0.59	6	4
Zinc	Zn	1.33	0.74	6	4, 6
Zirconium	Zr	1.58	0.79	6	6, 8

Sources. Atomic radii from *Metals Handbook,* American Society of Metals, Cleveland, 1961.

Ionic radii from Ahrens (1952).

Oxygen coordination CN from Zarzycki (1965).

TABLE 1.2 Oxygen Coordination Numbers (CN)

Radii ratios, r/R	CN	Arrangement of oxygens around cations
0.155/0.225	3	Corners of equilateral triangle
0.225/0.414	4	Corners of a tetrahedron
0.414/0.732	6	Corners of an octahedron
0.732/1.000	8	Corners of a cube
1.000	12	Closest packing

TABLE 1.3 Geometry of Crystal Systems

System	Space lattices	Axes	Axial angles
Cubic	3	$a_1 = a_2 = a_3$	All angles = $90°$
Orthorhombic	4	$a \neq b \neq c$	All angles = $90°$
Tetragonal	2	$a_1 = a_2 \neq c$	All angles = $90°$
Monoclinic	2	$a \neq b \neq c$	2 angles = $90°$; 1 angle $\neq 90°$
Triclinic	1	$a \neq b \neq c$	All angles different; none = $90°$
Hexagonal	1	$a_1 = a_2 = a_3 \neq c$	Angles = $90°$ and $120°$
Rhombohedral	1	$a_1 = a_2 = a_3$	All angles equal, but none = $90°$

for large, low valence cations. For example, the predicted Na-O distance of 2.37 Å is almost the same as the observed distance. The concept of charge polarization of ions clarifies the understanding of crystal structures and their characteristics. Charge polarization of ions is qualitatively related to the stability of silicate minerals and to the formation and properties of silicate glasses.

Research on crystalline materials shows that the ions are arranged in repetitive patterns following a limited number of geometric schemes. The seven systems presented in Table 1.3 can be expanded to include only 14 possible space or Bravais lattices. The properties of crystalline materials derive from the specific ions involved and from their ordered arrangements in space. Reference is made in this connection to Van Vlack (1970). Those interested in the use of detailed crystallographic data are referred to Buerger (1956) and similar publications.

1.3 STRUCTURES OF SILICATE MINERALS

The arrangement of four oxygens about one silicon in tetrahedral coordination is universal in the more than 800 known natural silicate minerals. The Si-O bond is so strong that the four oxygens are always found at the corners of tetrahedra of nearly constant dimensions and regular shape regardless of the rest of the structure. The various types of crystal lattice result from the ways in which the ionic $(SiO_4)^{4-}$ groups are linked together. These arrangements are somewhat similar to polymerization in organic compounds. The five principal types of tetrahedral linkages are given in Table 1.4. Silicate minerals with these linkage types further crystallize into the seven crystal systems and 14 space lattices in Table 1.3. Deer, Howie, and Zussman (1966) provide details on structures and properties of silicate minerals arranged under the five principal types.

Type 1

In the separate $(SiO_4)^{4-}$ groups each oxygen has a valence of -1 and is linked to cations other than Si. Examples are forsterite Mg_2SiO_4, and fayalite Fe_2SiO_4. The ionic radii of Fe^{2+} and Mg^{2+} are the same size within the 15% limit and substitute for each other as in the mineral olivine, $(Fe, Mg)_2SiO_4$. The Mg and Fe cations are located in orderly positions in relation to the ordered positions of SiO_4 tetrahedra. Deer notes that "half the (Mg, Fe) atoms are located at centers of symmetry and half on reflection planes."

Type 2

In the $(SiO_7)^{6-}$ complexes of type 2, two SiO_4 tetrahedra are linked by one inert oxygen. The three oxygens at either end of the double tetrahedra have single links to silicon, hence valencies of -1 for interaction with other cations. The mineral akermanite $Ca_2MgSi_2O_7$ has Mg cations located at the corners and Si cations at the centers of SiO_4 tetrahedra joined in pairs to form Si_2O_7 groups. The structure consists of SiO_4, AlO_4, and MgO_4 tetrahedra arranged in a sheet-like pattern, the sheets being held together by Ca-O linkages.

Type 3

The $(Si_2O_7)^{6-}$ complexes may combine into rings or chains. Three double tetrahedra form into a ring in the rare mineral benitoite $BaTiSi_3O_9$; each ring is attended by a barium cation and a titanium cation. Six such complexes form rings in beryl $Be_3Al_2Si_6O_{18}$ which accommodate three beryllium cations and two aluminum cations.

Type 4

Sheets of tetrahedra formed by linking three corners of each tetrahedron to neighbors result as a continuation of the process that converts chains into hexagonal sheets. These sheets, made up of $(Si, Al)O_4$ tetrahedra, are held together

TABLE 1.4 Tetrahedral Linkages in Silicate Minerals

Type	Structure in plane	Oxygens shared with next tetrahedron	Oxygen/silicon ratio	Description
1	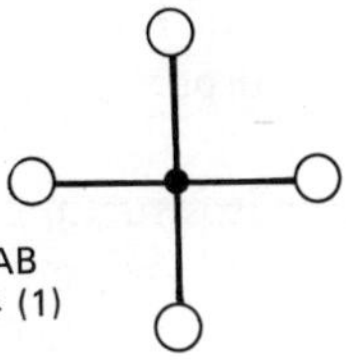 TAB 1–4 (1)	0	4	Isolated tetrahedra
2	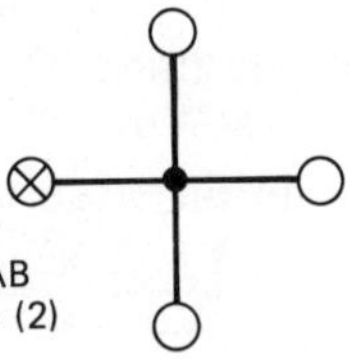 TAB 1–4 (2)	1	3.5	Two linked tetrahedra
3	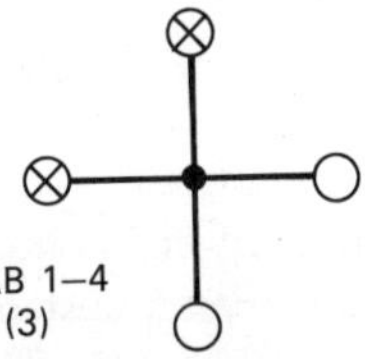 TAB 1–4 (3)	2	3	Rings or chains of tetrahedra
4	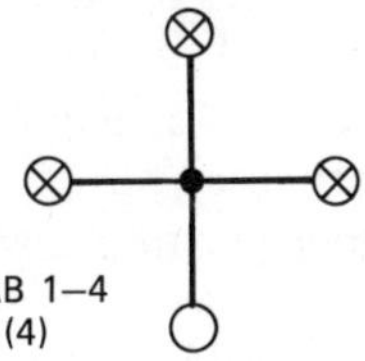 TAB 1–4 (4)	3	2.5	Two-dimensional sheets
5	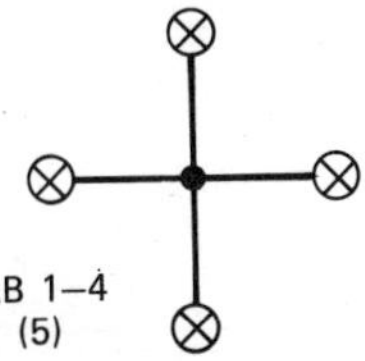 TAB 1–4 (5)	4	2	Three-dimensional networks

by layers of octahedrally coordinated cations such as Na, K, Ca, and Ba. Talc $Mg_6(Si_8O_{20})(OH)_4$ is an example.

Type 5

Continuous three-dimensional frameworks are formed when each SiO_4 tetrahedron shares all its oxygens with other tetrahedra. Quartz, tridymite, and cristobalite, the mineral forms of silica, have this arrangement. Framework silicates from SiO_2 to more complex species, such as the feldspars, have lattices made up of combinations of SiO_4 and AlO_4 tetrahedra.

Structural data on minerals reveal that isomorphous or isostructural substitution of one element for another is the rule rather than the exception. Ionic size, as illustrated under type 1 silicates, is the governing factor in such substitutions. It is not essential that substituting ions have the same valency or charge, provided electrical neutrality is maintained by simultaneous substitution of other ions. Such substitutions occur in the feldspars that are three-dimensional networks of SiO_4 and AlO_4 tetrahedra. The substitution of Al^{3+} for Si^{4+} requires additional cations to maintain local electrical neutrality. In the alkali feldspars $(Ca^{2+} + Al^{3+})$ substitutes for $(Na^{1+} + Si^{4+})$ and in the barium feldspars $(Ba^{2+} + Al^{3+})$ substitutes for $(K^{1+} + Si^{4+})$. Reference to Table 1.1 demonstrates that substitution requirements are satisfied. The monovalent and divalent cations are in ordered positions in relation to the orderly positions of SiO_4 and AlO_3 tetrahedra, thus maintaining local electrical neutrality. The same principles hold for ionic arrangements in all crystalline silicates and follow because the bonds, being from 20 to 50% convalent, are therefore somewhat directional.

Quartz, tridymite, and cristobalite, the three polymorphs of crystalline SiO_2, have their SiO_4 tetrahedra linked according to different crystal systems, and each has high and low temperature forms. Quartz is a common mineral and, next to the feldspars, is the most abundant mineral in the earth's crust. Tridymite and cristobalite are found in smaller amounts in volcanic rocks. These minerals usually contain small amounts of impurities such as Fe, Al, Ca, and Ti, which reduce their compositions to less than 100% SiO_2. Quartz used for glassmaking contains impurities ranging from a few parts per million up to a few tenths percent.

The temperature ranges of stability and metastability of crystalline quartz, tridymite, and cristobalite have controlling influences on the properties of 100% SiO_2 or vitreous silica glass. This subject is covered in Chapter 4.

1.4 STRUCTURES OF SILICATE GLASSES

The amount of structural data on glasses is meager in comparison with the reliable and extensive information on crystalline silicates. Interpretations of structural data on glasses, in terms of glass properties and characteristics, are still open to question, in marked contrast to the well-established principles of crystalline materials. Review of the glass literature reveals that the lack of data on glass structure has not prevented the publication of hundreds of papers giving theories of glass structures. In the author's opinion the word "theory" imples considerable evidence in support of a formulated general principle. The lack of sufficient evidence on glass structures suggests that "postulate," "hypothesis," or "assumption" should be used in this connection.

X-ray diffraction studies provided the first data on glass structures, followed by newer experimental techniques such as neutron diffraction, infrared and Raman spectroscopy, spectral fluorenscence, and nuclear magnetic resonance. Crystals exhibit sharp line x-ray diffraction patterns, whereas the patterns from liquids and glasses are diffuse bands. If the particle sizes of powdered crystals are gradually made smaller and smaller, the x-ray linewidths broaden and finally become diffuse. This fact led early investigators, particularly Randall, Rooksby, and Cooper (1930), to conclude that glasses contained very small crystals or crystallites. Porai-Koshits (1965), one of the early advocates of the crystallite postulate, states that "Modern views on the nature of the glassy state have long abandoned the suggestion that glass is an ordinary melt consisting of highly disperse crystallites." Earlier papers by Zachariasen (1932) and Warren (1937) had advanced reasons for the untenability of the crystallite postulate.

The classic paper by Zachariasen (1932), an eminent crystal structure specialist, outlined postulates on glass formation that greatly motivated further x-ray diffraction studies, particularly those of Warren and his colleagues. Zachariasen assumed that the atoms in glass are linked by forces essentially the same as those in crystals, that the atoms are oscillating about definite equilibrium positions, and that they form three-dimensional networks. He concluded from x-ray diffraction experiments that this network is not periodic and symmetrical as in crystals, but that it is not entirely random, since the internuclear distances do not sink below a given minimum value. He dealt only with oxide glasses and particularly those which by themselves form glasses—SiO_2. B_2O_3, GeO_2, P_2O_5, and others. He designated Si^{4+}, B^{3+}, Ge^{4+}, and P^{5+} as glass-forming cations and suggested three simple rules governing their roles in glass formation. (1) An oxygen atom is linked to not more than two of these cations; (2) the number of oxygens surrounding these cations must be small; (3) the oxygen polyhedra share corners with one another, not edges or faces. Although subsequent research has shown that a three-dimensional network is not essential for glass formation, Zachariasen's rules are still valid for the most stable silicate glasses used in glass products.

Warren and Gingrich (1934) applied a Fourier synthesis method for interpreting their x-ray diffraction data on a number of liquids and glasses. Mozzi and Warren (1969) improved on earlier methods by using fluorescence excitation, which largely eliminated background scattering. This new technique in terms of "pair distribution" functions allowed calculation of interatomic distances directly

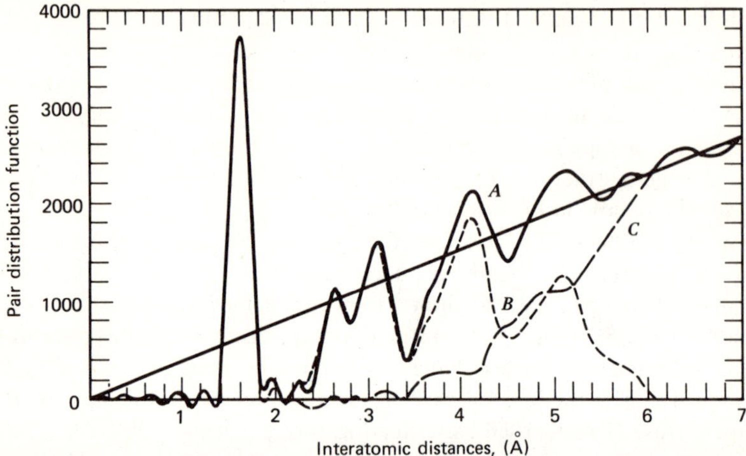

Figure 1.2   The pair distribution function for vitreous silica: *A* is measured curve, *B* is sum of the calculated curve for the first six contributions, *C* is difference between *A* and *B*. From Mozzi and Warren (1969).

from peak positions on a distribution curve. The areas under the peaks indicate the number of neighboring atoms. The Mozzi-Warren analysis of x-ray diffraction pair distributions for vitreous silica appears in Figure 1.2; peaks are identified as follows:

Peak	Ion-Ion Distance (Å)
First	Silicon-oxygen = 1.62
Second	Oxygen-oxygen = 2.65
Third	Silicon-silicon = 3.12
Fourth	Silicon-2nd oxygen = 4.15
Fifth	A combination of oxygen-2nd oxygen and silicon-2nd silicon = 5.1

Table 1.5, complied by Zarzycki (1965), summarizes data on oxide glass structures obtained by different investigators using diffraction methods. These data should be compared with data on crystals given in Table 1.1. The first neighbor cation-anion distances and coordination numbers of cations for glasses are the same as those in crystals within experimental error. Data in Table 1.5 indicate that the CN of glasses at higher temperatures is less than that for glasses at room temperature. This is also the case for crystals, as noted earlier. A further comparison of the data shows the CN of glasses to be about 10% greater than those in crystals. Zarzycki (1965) ascribes this to cumulative errors in the Fourier method, with resultant lack of precision in determining CN. A principal reason involves the rapid overlapping of coordination spheres.

TABLE 1.5 Cation-Anion Distances and CN of Cations in Oxide Glasses Determined by Diffraction Methods

Glass	Cation-anion distance (Å)	CN	Source and temperature*
SiO_2	Si $=1.62$	Si $=4.3$	(1)
SiO_2	Si $=1.60 \pm 0.05$	Si $=4.4$	(2)
SiO_2	Si $=1.60 +0.05$	Si $=4.4$	(3) -1600°C
SiO_2	Si $=1.60$	Si $=4.3$	(4)[†]
SiO_2-Na_2O	Si $=1.60$-1.65	Si $=4.2$-4.5	(5)
	Na$=2.35$	Na$=6$	
SiO_2-K_2O	Si $=1.62$	Si $=4.4$-4.7	(6)
	K $=2.6$-2.7	K $=10$	
SiO_2-CaO-Na_2O	Si $=1.62$	Si $=4.6$-4.8	(7)
	Ca$=2.4$	Ca$=8$	
		Na$=6$	
GeO_2	Ge$=1.70 \pm 0.05$	Ge$=4.4$	(2)
GeO_2	Ge$=1.70 \pm 0.05$	Ge$=3.4$	(3) -1200°C
P_2O_5-CaO	P $=1.57$	P $=4.2$	(8)
	Ca$=2.4$	Ca$=7$	
B_2O_3	B $=1.39$	B $=3.1$	(1)
B_2O_3	B $=1.37$	B $=2.7$	(9)
B_2O_3	B $=1.30$	B $=3.3$	(2)
B_2O_3		B $=2.3$	(2) -1200°C
B_2O_3		B $=2.2$	(2) -1600°C
B_2O_3-Na_2O	B $=1.37$-1.48	B $=3.2$-3.9	(10)
		Na$=6$	
B_2O_3-Na_2O	B $=1.4$	B $=3.1$-3.9	(11)
B_2O_3-CaO	B $=1.45$	B $=4.3$	(8)
B_2O_3-PbO		B $=3.9$	(12)
B_2O_3-PbO		B <3.9	(12) -580-770°C

(1) Warren, Krutter, and Morningstar (1936).
(2) Zarzycki (1956).
(3) Zarzycki (1957).
(4) Breen, Delaney, Persiani, and Weber (1957).
(5) Warren and Biscoe (1938).
(6) Biscoe, Druesne, and Warren (1941).
(7) Biscoe (1941).
(8) Biscoe, Pincus, Smith, and Warren (1941).
(9) Meller and Milberg (1960).
(10) Biscoe and Warren (1938).
(11) Becherer, Brummer, and Herms (1962).
(12) Furukawa (1960).

*Measurements at room temperature unless noted otherwise.
[†]Neutron diffraction measurement.

Source. Zarzycki (1965).

Particular attention is directed to data on B_2O_3 and borate glasses in Table 1.5. The CN of B in B_2O_3 glass is near 3 as predicted, and the B-O first neighbor distance is near 1.35 Å. When other oxides are substituted for B_2O_3 in increasing amounts, the CN approaches 4 and the B-O distance approaches 1.5 Å. Thus B_2O_3 is the only glass-forming oxide having a variable CN. This variability of boron, referred to as the boron anomaly, is encountered in many widely used glasses and is strongly reflected in their properties. Many papers have addressed this anomaly, but the function of boron in glasses remains unclear and continues to invite active research.

The publications of Zachariasen and Warren et al., at almost the same time, made a tremendous impact on glass technologists. Although these authors dealt only with oxide glasses, principally silicates, a rapidly emerging random network theory was soon being applied not only to oxide glasses, but to other types of inorganic glasses. This came about even though information from their publications could not reasonably lead to a valid theory, even for oxide glasses. In essence this theory considered, in the case of silicates, that coherence in glasses derived from strong Si-O bonds in SiO_4 tetrahedra arranged at random in the structure; SiO_2, along with B_2O_3, P_2O_5, and so on, were designated as glass formers. Other oxides such as Al_2O_3, TiO_2, and ZnO came to be called intermediates. Oxides such as Na_2O, CaO, and BaO were called modifiers and were considered to be arranged entirely at random in the holes or interstices between the SiO_4 tetrahedra and other network-forming oxides.

In retrospect it is suggested that early advocates of the random network theory misinterpreted the papers by Zachariasen and Warren. Zachariasen had pointed out that the network in oxide glasses could not be entirely random. Warren had stated that his studies gave information on average quantities only and that the details would have to be filled in from other kinds of measurement. Zarzycki (1965) has commented that a statistical description, of the type used by Warren, does not necessarily mean that the glass structure itself need also be statistically disordered. During the development of the random network theory, many papers on glass properties indicated that glass structures could not be entirely randomly arranged. These property studies suggested the presence of somewhat orderly arrangements of the constituent ions. Terms such as "complex molecules," "complex ions," "aggregates," and "coexisting structures," were used to describe them.

It has been noted that the bonds in crystalline silicates were considered by Pauling (1960) and others to be from 20 to 50% covalent and that covalent bonds are directionaly oriented. This bond type leads to ordered structures in minerals, and cations such as Na, Ca, and Mg are found in somewhat ordered positions in these materials. Papers by a number of authors also indicate that alkali and alkaline earth ions occupy somewhat ordered positions in glass structures. Brosset (1962), taking advantage of the high x-ray scattering power by ions containing a large number of electrons, published results on SiO_2-CsO and SiO_2-BaO glasses demonstrating that Cs and Ba occupied ordered positions. Ohlberg and Parsons (1965) completely replaced Na by Ag in an SiO_2-CaO-Na_2O glass and concluded from their data that Na occupied ordered positions in the glass structure. Other studies by Vogel and Gerth (1958), Zarzycki and Mezard (1962), and Belov (1961) also show that non-glass-forming cations occupy somewhat ordered positions in glass structures. For example, Hartleif (1938) made x-ray diffraction studies on glasses in the system K_2O-SiO_2 and found a higher degree of order than Warren and Biscoe (1938) had found in Na_2O-SiO_2 glasses. Hartleif stated that his research did not support the concept of a random distribution of K cations in the network of SiO_4 tetrahedra. He suggested that a second structure arises as increasing amounts of K_2O are introduced.

The above-mentioned researches indicate that glass-forming cations such as Si are not entirely responsible for the types of structure existing in glasses, but that medium and large size cations impose their own requirements on glass structures. This structural picture seems to be in agreement with the thermodynamic requirement that all ions arrange themselves in positions of minimal potential energy in terms of ionic size and electrical charge. The most comprehensive reviews of this subject are by Weyl and Marboe (1962, 1964, 1967). The publications, which include 1883 references, are recommended to those interested in materials research on glasses and glass technology.

2

The Nature of Silicate Glasses

Silicate glasses are unique with respect to their strong cohesive bonds and the resulting high viscosities in the molten condition, which contribute to the super-cooled nature of glasses. These characteristics exert strong controls on areas of glass formation and glass properties. Depending on glass compositions and temperature regions, properties and characteristics are also functions of time. The property values of glasses at room temperature are characteristic of the temperature at which the structure has reached an equilibrium condition. Glasses can be brought down to room temperature as isotropic solids by following established procedures. Glasses equilibrated at temperatures corresponding to a viscosity of 10^{13} poises will have stabilized property values that will not change after the materials are brought to room temperature and used normally.

2.1 PRINCIPLES OF GLASS FORMATION

Crystalline silicates can be melted into liquids by raising them to temperatures above their melting points as illustrated in Figure 2.1. They are converted into molten liquids as they are made to follow the path *ABCD*. The fusion process entails a sharp gain in heat content as the crystal melts. If the liquid is cooled along the reverse path *DCBA*, it will devitrify or crystallize, and primary phase crystals characteristic of its composition will form. However if the molten liquid is cooled along the path *DCE*, it will form into a glass. The glass exhibits the behavior of a supercooled liquid and remains unstable to the extent of its heat of fusion or heat of crystallization. In the form of glass it has no melting point but, rather, a liquidus temperature that is defined as the maximum temperature at which thermodynamic equilibrium exists between the molten glass and its crystalline primary phase. Gibbs' phase rule defines the conditions for systems in thermodynamic equilibrium. It requires homogeneous equilibrium within each phase and heterogeneous equilibrium between coexisting phases. Our interest is in single liquid homogeneous glass phases. Thus the phase rule does not apply to glass per se. It is most important, however, in establishing equilibrium liquidus

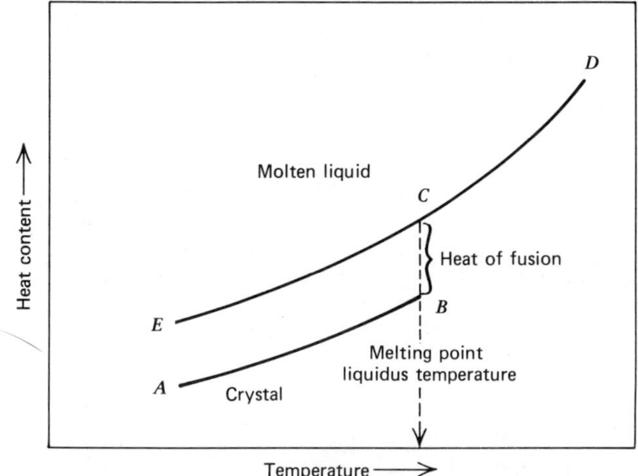

Figure 2.1 Heat contents of crystals and glasses.

temperatures and crystalline species that are in equilibrium with glasses. Chapter 4 proposes that primary phase diagrams form the basis of classification of silicate glasses.

Levin, Robbins, and McMurdie (1964, 1969) have compiled existing phase diagrams of materials, singly and in combination. Phase diagrams in this book are from these sources. Figure 2.2 is the diagram of the binary Na_2O-SiO_2 system. The liquidus temperatures decrease sharply as Na_2O replaces SiO_2 until they reach the binary eutectic between SiO_2 (quartz) and $Na_2O \cdot 2SiO_2$ (sodium disilicate) at 790°C, where the composition is 74% SiO_2 and 26% Na_2O. A eutectic composition is that combination of components of a single system having the lowest melting temperature of any ratio of the components. It is located at the intersection of two solubility curves in a binary system and at the intersection of three solubility surfaces in a ternary system. Figure 2.2 shows two other eutectics at SiO_2 contents of 62 and 43%, at temperatures of 840 and 1025°C, respectively. Glasses in the SiO_2 range from 62 to 74% have $Na_2O \cdot 2SiO_2$ as their primary phases. If such glasses are devitrified, the samples will contain crystals of $Na_2O \cdot 2SiO_2$ embedded in a glassy matrix. If glass of the composition $Na_2O \cdot 2SiO_2$ is devitrified, the sample will consist of these crystals and will contain little or no glass.

Figure 2.3 is a ternary phase diagram of a part of the Na_2O-CaO-SiO_2 system with indicated invariant points, phase boundaries, and liquidus temperatures. These relations can best be visualized by means of a solid model in which the solubility surfaces and liquidus temperatures are plotted vertically above a hori-

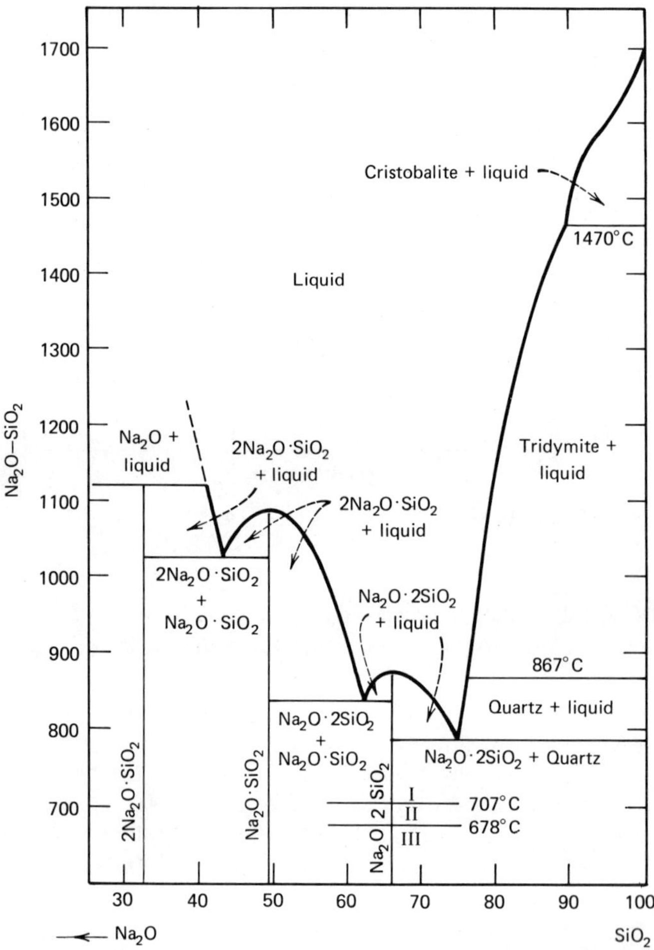

Figure 2.2 Primary phase of Na_2O-SiO_2 system; compositions in weights percent. *Source.* Levin, Robbins, and McMurdie (1964). Originally from F. C. Kracek, *J. Phys. Chem.*, **34**, 1588 (1930) and *J. Am. Chem. Soc.*, **61**, 2869 (1939).

zontal composition plane. It can be seen that glasses having $Na_2O \cdot 2SiO_2$ as primary phases are located in one area. Glasses having SiO_2 as their primary phases may be located in the three separate areas of quartz, tridymite, and cristobalite. Compositions in all the indicated phase fields except $Na_2O \cdot 2SiO_2$ and $Na_2O \cdot 2CaO \cdot 3SiO_2$ form the bases for commercial compositions used in making ordinary glass products such as window glass and containers.

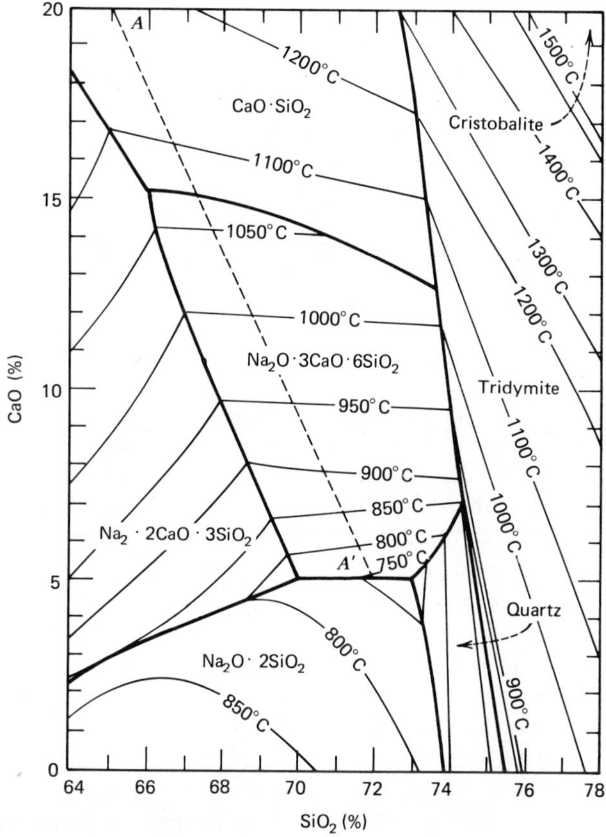

Figure 2.3 Primary phase diagram of Na$_2$O-CaO-SiO$_2$ system. Part of system of interest to glass technology; weight percentage of Na$_2$O obtained by subtracting sum of CaO and SiO$_2$ from 100. *Source.* Levin, Robbins, and McMurdie (1964). Originally from G. W. Morey, *J. Am. Ceram. Soc.,* **13** (10), 700 (1930).

The liquidus temperature of a given glass at atmospheric pressure depends only on its oxide composition. The times required to achieve equilibrium vary widely among glasses. The tendency toward crystallization involves rates of formation of centers of crystallization and rates of crystal growth that are also dependent on composition. These driving forces toward crystallization are opposed by the viscosity of the glass at its liquidus. The question of whether a given molten silicate can be formed into a glass is further dependent on the size or volume of the sample desired. This question is an arbitrary one and is resolved in terms

TABLE 2.1   The Nature of $Na_2O\text{-}SiO_2$ Glasses

Oxides; properties	Composition		
	SiO_2	$Na_2O\cdot2SiO_2$	$Na_2O\cdot SiO_2$
SiO_2 (mole fraction)	1.00	0.67	0.50
Na_2O (mole fraction)	–	0.33	.50
SiO_2 (wt %)	100.0	66.0	49.2
Na_2O (wt %)	–	34.0	50.8
Liquidus ($^\circ$C)	1723	874	1088
Viscosity (poises*)	$10^{7.5}$	10^4	10
Tetrahedrally shared oxygens	4	3	2

*A unit of viscosity for which the viscosity coefficient is equal to 1 dyne-sec/cm^2.

of the winner of the race between its capability to crystallize and the cooling rate of the sample. The information in Table 2.1 is pertinent to glass formation in the $Na_2O\text{-}SiO_2$ system. Oxide compositions are expressed in weights percent and mole fractions. Commercial compositions and those appearing in most glass publications are expressed in weights percent. Following practices of physical chemistry and thermodynamics, the author prefers to use mole fractions in relating the compositions of glasses to their properties and structures. Weights percent compositions are given for easy reference to phase diagrams and other publications. With reference to Table 2.1, molten SiO_2, commonly known as fused or vitreous silica, can easily be converted into glass because of its four tetrahedrally shared oxygens and its very high viscosity at the liquidus temperature. Its very good physical properties and chemical stability recommend this material for many glass uses. Its high manufacturing costs, however, limit its uses in products. There is no reasonable limit on sample sizes of fused silica, and they may range up to tons with careful manufacturing controls. As Na_2O is substituted for SiO_2, the number of shared oxygens becomes smaller, the viscosities sharply decrease, and glass-forming capabilities diminish. As a consequence, only small samples can be obtained, and these require rapid quenching. Such samples are hygroscopic and must be kept in a dessicator if property measurements are to be made. Although glass formation of binary glasses can be rationalized in terms of tetrahedrally shared oxygens, the problem becomes more complex for ternary, quaternary, and other systems containing several oxides. Depending on the kinds and numbers of oxides substituted for SiO_2, glass formation is usually possible when shared oxygens range from two to four. This is an approximate classification.

2.2 SILICATE GLASS SYSTEMS

Silicate glasses are used in more than 95% of the total tonnage of manufactured glasses. These glasses, which contain SiO_2 plus other added oxides, can be made into an almost unlimited number of compositions. It is in order to outline the most commonly used types designated as glass systems in the literature.

The most extensively used glasses, those containing major amounts of Na_2O, CaO, and SiO_2, are known as soda-lime-silica glasses. Table 2.1 and Figure 2.2 show sharp reductions of viscosities and liquidus temperatures as Na_2O replaces SiO_2 in binary Na_2O-SiO_2 glasses. These phenomena make possible the widespread uses of commercial products such as containers, window glass, and plate glass. Levin, Robbins, and McMurdie (1964, Figs. 167, 182) demonstrated that K_2O and Li_2O perform similar functions in binary K_2O-SiO_2 and Li_2O-SiO_2 glasses. Uses of the latter monovalent oxides are generally limited to specialty glasses because of their higher costs. Once viscosities and liquidus temperatures are reduced in this manner, it becomes necessary to add stabilizing oxides to achieve property requirements of the various commercial products. The oxides CaO, MgO, and Al_2O_3 are most commonly added for this purpose in high tonnage products.

The unique properties contributed by B_2O_3 to glasses in the borosilicate system derive from the fact that the oxygen coordination of boron may vary continuously from 3 to 4. In other words, boron may be surrounded by three oxygens in a plane or by four in tetrahedral coordination, as indicated in Chapter 1. This oxide is extensively used in specialty glasses requiring low coefficients of thermal expansion, high chemical resistance, low dielectric losses, and so on. The best coverage of commercial borosilicate glasses is that given by Volf (1961). This chameleonlike characteristic of B_2O_3 in glasses has led to numerous publications discussing its unique properties. A number of methods have been suggested for evaluating the proportions of 3- and 4- coordinated boron existing in borosilicate glasses. None is completely satisfactory, and basic principles are needed to quantitatively predict its effects on glass properties. The oxide effectively lowers both viscosities and liquidus temperature in borosilicate glasses and is comparable in this respect to Na_2O in soda-lime-silica glasses. The problem of property-structure relations becomes more complex for glasses based on the B_2O_3-Al_2O_3-SiO_2 system because both boron and aluminum have varying coordination numbers. In this case one must rely on experimental data for implications of the separate roles of boron and aluminum because there is no reliable way of calculating them.

Aluminosilicate glasses are widely used in specialty glass products. Two general types may be distinguished in terms of monovalent and divalent oxide constituents.

1. Glasses in the R_2O-Al_2O_3-SiO_2 systems, such as Na_2O-Al_2O_3-SiO_2, are made principally from feldspar raw materials and were early discovered to have useful properties such as high chemical and thermal stability.

2. Glasses in the RO-Al_2O_3-SiO_2 systems, such as CaO-Al_2O_3-SiO_2, are of more recent development, in addition to having the properties just named, they are useful in other applications requiring high electrical resistivity and low dielectric losses. The contributions of the aluminum cation to the properties of these glasses can be related qualitatively to its oxygen coordination, which may be either 4 or 6. There is no reliable method for calculating these contributions a priori.

Glasses containing PbO as a major constituent have long had important uses in optical glasses because of their high refractive indices and dispersions. From a glassmaking standpoint, PbO effects reductions in viscosities and liquidus temperatures in these glasses. In this respect its function is similar to Na_2O and B_2O_3 in the above-mentioned systems. Lead orthosilicate $2PbO \cdot SiO_2$, having the composition 78.8% PbO and 21.2% SiO_2, can be made into a glass and can be used in commercial optical products. This glass departs from the scheme for silicate glass formation (Table 1.4). Fajans and Kreidl (1948) attribute the high stability and usefulness of PbO-SiO_2 glasses to the high polarizability of PbO. Compositions based on the PbO-B_2O_3 glass system find important commercial uses for low temperature sealing glasses and for coating metals and glasses.

How does the ionic structure of a glass-forming liquid silicate compare with that of its crystalline primary phase? The answer to this question cannot be given in direct experimental terms. It presupposes that microscopic structures can be related to macroscopic observations on the properties and characteristics of a material. These relations are more definite for crystalline silicates than for their glass counterparts. The structures of glasses are less well defined, and one must infer postulated structures from their macroscopic properties.

The following thermodynamic equation can be used in connection with the question just posed:

$$H = F + TS \qquad (2.1)$$

where H = enthalpy, F = free energy, T = absolute temperature, and S = entropy. The heat content indicated in Figure 2.1 is loosely the same as enthalpy. Entropy of disorder is that part of entropy caused by a disordered arrangement, as opposed to that of a similar but ordered arrangement. The entropy change on fusion of a solid is largely due to entropy of disorder.

Barth and Rosenquist (1949) observe that the entropy of fusion of silicates is not high and that no major changes in structures are to be expected on melting. Simon and Lange (1926) measured the zero point entropy of vitreous silica to be 0.9 ±0.3 cal/($^\circ$C)(mole). Gutzow (1962) places the configurational entropy of vitreous SiO_2 at 1.08 cal/($^\circ$C)(mole). These studies indicate that the entropies of disorder of silicates at their melting temperatures are low and that the degree of order of ions in molten silicates may not differ greatly from the degree of order found in crystalline silicates.

Properties of glasses change in smoothly continuous ways in going from melting temperatures to room temperature. The property-temperature curves exhibit no abrupt changes in slope as they pass through the liquidus temperatures. This suggests that no abrupt changes in ionic structures occur. Riebling (1967, 1968) examined in detail a number of properties of glasses at room temperature, as well as the high temperature liquids of the materials. He concluded that "The structures, or distribution of different polyhedral species, of an oxide melt and its corresponding glass are not very different." Thus present information indicates that structural constituents in the molten glass arrange themselves in positions of minimal energy, kinetic plus potential, and that these structures persist down to room temperature.

It has been noted that the principles of crystal structures are subject to some modifications when they are applied to silicate minerials. In developing these principles, crystallographers had to base them on natural minerals, which were variable in composition, subject to effects of temperature and pressure, and so on. The glass technologist trying to develop schemes of glass structure and glass formation faces equally difficult problems. Ideally glass formation requires complete homogenization of all the ions involved in the melting process. Actually this homegenization is opposed by the viscosities of molten liquids of the given composition. The homogeneity required depends on the intended use of the glass. Optical glass must be very homogeneous, and few impurities are permitted. At the other end of the scale, amber beer bottles are much less homogeneous and may contain sizable amounts of impurities. In fact examination of glass used in containers may reveal visible stria or cords of inhomogeneous glass. In brief, postulated principles of glass structure and glass-formation schemes must be based on published data on a wide variety of commercial and experimental glasses that vary in homogeneity, purity, heat treatment, and precisions of measurements. Thus those interested in glass research and technology must carefully scrutinize such data.

2.3 EFFECTS OF TEMPERATURE ON PROPERTIES

The characteristics of glasses undergo significant modifications as the temperature is varied, changing gradually from viscous liquids at high temperatures to elastic solids at room temperature. In the same range glasses change gradually from electrolytic conductors to insulators or dielectrics. For purposes of discussion, these changes can be arbitrarily categorized in terms of three temperature zones: properties vary continuously from high to low temperatures, and indicated temperatures are dependent on glass composition.

Range from 1550°C to 700°C

Glass is essentially a liquid and a conductor. However it can be fractured by impact forces, and dielectric characteristics can be demonstrated. The properties are essentially functions of composition and temperature, only.

Range from 700°C to 500°C.

Glass may be fractured or made to flow depending on the type of applied mechanical force. Correspondingly, the glass may exhibit conductive or dielectric properties depending on the type of applied electrical stress. The properties are definitely functions of composition, temperature, and time. This is the annealing temperature zone where internal strains are equalized and properties are stabilized.

Range from 500°C to Room Temperature.

As room temperature is approached, the effects of time become less pronounced and are near zero for practical purposes. Time effects can be demonstrated at room temperature and cannot be completely eliminated until the glass reaches absolute zero. The properties are essentially functions of composition and temperature alone.

It is well recognized that the viscosity of glass is a major factor in all stages of glassmaking and glass manufacture. It is not the only factor involved, but it exerts controlling influence on glassmaking processes that involve flow. Glass technologists in cooperation with the American Society of Testing and Materials (ASTM) have related glassmaking processes to viscosity-temperature relations. The viscosity values, which have been defined in terms of glassmaking processes, have been standardized and are widely used in the industry. These controls are also used in the making of experimental glasses. These definitions and standards are published in Part 17 of the *Annual Book of ASTM Standards* (1976). Table 2.2 shows viscosity-temperature data on experimental glass number 70-3, with indicated viscosity levels used in glass processing. The melting of glass batch can be accomplished at a commercially desirable rate when the viscosity of the glass is near 10^2 poises. The working range is also dependent on the type of product; the viscosity level for glass fibers is much lower than that for pressing large shapes such as television tubes. The annealing point corresponds to the viscosity level, where internal strains can be relieved in about 15 min. The strain point viscosity level relates to relief of strain in about 4 hours. Generally speaking internal strain cannot be introduced at temperatures below this viscosity level.

Published information is available on the effects of temperature on most glass

TABLE 2.2   Viscosity-Temperature Data on
Na_2O-CaO-SiO_2 Glass 70-3

Temperature ($^\circ$C)	Viscosity (poises)	Glassmaking process
1318	10^2	Melting of glass batch
1072	10^3	
950	$10^{3.8}$	Liquidus temperature
916	10^4	
805	10^5	Working range
734	10^6	
686	10^7	Glass can be blown
637	10^8	or pressed to shape
603	10^9	
577	10^{10}	
556	10^{11}	
538	10^{12}	
524	10^{13}	Annealing point
513	10^{14}	
507	$10^{14.5}$	Strain point
501	10^{15}	

properties. Our knowledge of these effects is presently based on experimental data because information on electronic and ionic structures of glasses is not sufficient to derive property-temperature relations from basic principles. Chapter 1 stated that the ions in glasses are mutually polarized and that electrons surrounding the ions are unsymmetrically distributed. Subsequent discussions indicate that various glass properties can be qualitatively related to polarizations of ions. The viscosities of glasses increase very rapidly when temperatures are lowered, as illustrated in Table 2.2. Fajans and Kreidl (1948) have reasonably suggested that unsymmetrical polarizations of ions in glasses can qualitatively account for their stabilities and rapid changes of viscosity with temperature.

The annealing of glass accomplishes two purposes; it lowers internal glass strains and stabilizes glass property values. The property values used at room temperature are representative of equilibrium glass structures established at the annealing temperatures. Temperature-equilibrated structures can be achieved rather rapidly at high temperatures. Time durations for their establishment increase exponentially as temperatures are lowered.

It is in order to amplify the meanings of our terms and to relate them to those commonly found in the glass literature. The intermediate temperature range wherein properties are functions of composition, temperature, and time has come to be referred to as the "transformation range." Our term "temperature-equilibrated structure" implies that the glass has been held at a constant temperature long enough for the glass structure to attain an equilibrium configuration that is characteristic of that temperature. This temperature, commonly referred to as the "fictive temperature," has controlling influence on glass properties, particularly those which are dependent on ionic structures. For example, glass conditioned at a lower fictive temperature has a density higher than that of glass conditioned at a higher fictive temperature.

Tool (1945) introduced the concept of fictive temperature and mathematically related property changes in terms of time and temperature. Tool and his colleagues at the National Bureau of Standards did extensive research on time and temperature effects on several properties, including density, refractive index, thermal expansion, and heat absorption. Similar work by others on various properties has added to our extensive experimental knowledge of the subject. Relating time-temperature effects on properties to microstructural changes in glasses has posed difficult problems that have not been satisfactorily resolved. Our discussions here are generally limited to glass structures and properties that have been equilibrated at given annealing temperatures.

2.4 ANNEALING OF GLASSES

The specific volume of silicate glasses decreases of the order of 10% in going from melting temperatures to room temperature. If the molten sample is cooled very rapidly its outer surface layers will become rigid and will continue to contract while the center is still molten. If the sample is not too large, it may reach room temperature in one piece. In this case the outer layers will be under compressive forces that balance the internal tensile forces. Such samples are mechanically unstable and anisotropic. If the sample is cooled slowly enough to allow primary phase crystals to form, it will be heterogeneous; the composite sample will consist of large crystals embedded in a glassy matrix. Glasses made to follow either of these time-temperature paths are of experimental interest only and are not suitable for glass products.

If molten glass is to be brought to room temperature as an isotropic solid, it must be cooled following prescribed time-temperature conditions as illustrated in Figure 2.4. Following the path $ABCD$, the glass is cooled rapidly to its annealing temperature at point B and is held at this temperature long enough to reduce the birefringence, a measure of its anisotropy, to one-half the desired value at room temperature. It is then cooled from C to D the strain point temperature, at a rate such that the resultant added birefringence is limited to one-half the desired room temperature value. The glass can be cooled more rapidly from the strain point to room temperature without adding appreciable birefringence. If the

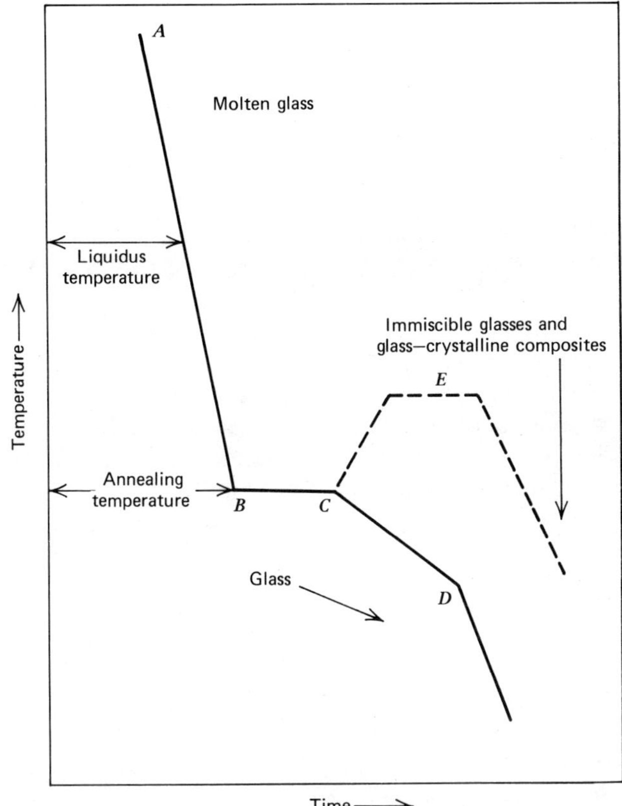

Figure 2.4 Time-temperature conditions of glass formation.

critical stages of the process from B to C and from C to D are carried out correctly, the glass will have the desired room temperature birefringence. The allowable anisotropy in terms of measured birefringence varies with the type of glass product. Optical glasses must have very low birefringence, but ordinary glass products can have much higher levels. The measurement of stress-optical relations and their composition dependence is covered when we discuss individual glass properties.

Although the methods discussed in this book are limited to glass per se, it should be noted that glass at its annealing temperature can be converted into other types of useful products. Special glass compositions can be given additional heat treatments along the path CE in Figure 2.4, to change the glass from a single homogeneous liquid to a material containing immiscible liquids. Alternatively other special compositions, if made to follow the path *CE*, can be developed into useful composite glass-crystalline materials. The temperatures of treatment for both types of material are kept within the range from the annealing to the liquidus temperatures. Neither type of product can tolerate the presence of large primary phase crystals. It is usual for both types to be heated well above their annealing temperatures, permitting the formation of immiscible liquids or very small crystals in reasonable times. Materials of both types are then cooled to room temperature following prescribed time-temperature paths in terms of desired birefringence levels. There is active worldwide research in progress on these useful products. The technologies of these two general types of material have led to their widespread manufacture, and related information is being published in the glass and ceramic journals listed in the Appendix. Résumés of progress in this area can be found in McMillan (1964), Rawson (1967), and Berezhnoi (1970). These new materials are first made into glasses. Thus their properties and manufacturing processes are the same as for glasses from their molten condition down to their annealing temperatures.

3

Sample Preparation and Property Measurements

Information on glass properties and their chemical compositions are available in references cited in this book, in the supplementary publications listed in the Appendix, and in the files of the United States Patent Office. Data for use in relating glass properties and compositions should include information about how the samples were prepared. As noted earlier, glass is a supercooled liquid, and its properties depend on its thermal history. The individual investigator has control of all steps in sample preparation and property measurements and is in the best position to evaluate the worth of property and composition data.

The preparation of glasses for experimental purposes involves adherence to carefully controlled analytical procedures from glass batch formulation to the finished sample. Sample preparation procedures used in the Optical Sciences Center at the University of Arizona form the basis of discussion in Chapter 3. Although all oxide glasses have supercooled characteristics somewhat similar to those of silicates, the experimenter will find it necessary to modify these procedures to fit his needs. For example, glasses containing P_2O_5 should not be melted in platinum crucibles; refractory crucibles are recommended, instead. The preparation of acceptable glass samples requires appreciable time and effort, and the samples should be treated like money in the bank. This suggests the need for careful storage, as well as the keeping of complete records on sample preparation and property measurements.

Measurements of glass properties at room temperatures can generally be carried out in the same manner employed for other solids. Glasses are ionized materials and become more and more reactive to ambient atmospheres and materials in contact with them as the temperature is increased. Glasses vary widely in this characteristic; the experimenter must judge whether such reactions are detrimental to his particular research. The control of such boundary conditions is especially important when glass surfaces are being studied. High temperature property measurements such as viscosity and surface tension involve careful measurement and control of temperature. Noble metal thermocouples are used for most temperature measurements.

3.1 SAMPLE PREPARATION EQUIPMENT

The apparatus needed for making experimental glass melts depends on the scope of the planned research. The following equipment has been found to be satisfactory for preparing samples for measurements on microindentation hardness, stress-optical coefficient, moduli of elasticity, and spectral studies including absorption, luminescence, and vacuum ultraviolet excitation.

Weighing and Batch Mixing.

Major amounts of batch ingredients are weighed on a torsion balance having a capacity of 2000 g and a precision of 0.01 g. Minor amounts of constituents are weighed on an analytical balance having a capacity of 100 g and a precision of 0.0001 g. Batches are mixed in a porcelain jar laboratory blender.

Melting Crucibles.

Glass melts ranging in size from 500 to 2000 g are made in crucibles of either 100% platinum or 90% platinum and 10% rhodium. The pure platinum crucibles are used to prepare glasses for spectral studies, to prevent the retention of small amounts of rhodium in the samples. The platinum and platinum alloy crucibles are placed inside refractory crucibles to prevent contact between the melting crucibles and iron handling tongs. The refractory crucibles also serve to hold the heat of the melt during pouring and casting of samples.

Calcining Furnace.

The calcining furnace, with automatic temperature controls, is used to drive water and other possible volatiles from the batch before it is introduced into the melting furnace. This process minimizes losses of raw materials during early stages of the melting.

Melting Furnace.

The bottom-loading Pereny furnace has a heating space of about 3 ft^3 and a maximum temperature capability near 1600°C. It is equipped with "Kanthal" molybdenum disilicide heating elements. Melts can be made under controlled furnace atmospheres by bleeding in neutral or reducing gases. Melts are usually made in air, which of course contains about 20% oxygen. Temperatures are automatically controlled and recorded with Honeywell equipment in conjunction with a radiation temperature-sensing device. Periodic checks of the radiation head are made with a Pyro optical pyrometer.

Homogenization of Samples.

Molten glasses are viscous liquids and must be homogenized either by stirring or by quenching and crushing. Figure 3.1 is a functional diagram of water-cooled aluminum rollers used for this purpose. They are 6 in. in diameter and 12 in. long and are kept in adjustable close contact. When poured between the revolv-

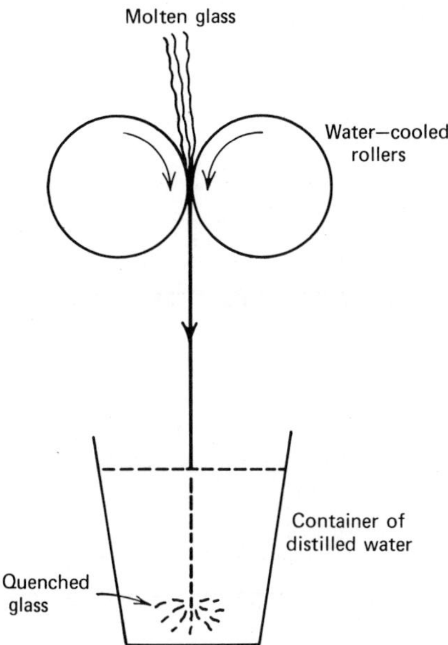

Figure 3.1 Homogenization of molten glass sample.

ing rollers, molten glass is drawn out into a thin sheet that breaks up into very small pieces when it is quenched. Pieces of the quenched glass are then dried and put back in the furnace for further conditioning. This process is repeated as needed. Experience shows that stirring with a platinum rod is much less effective in achieving good mixing.

Annealing Furnace.

The Lindberg Heavy Duty annealing furnace has a heating space of $10 \times 12 \times 24$ in. and a top temperature capability of about $1100°C$. The temperature is measured and controlled by recording Honeywell equipment in conjunction with thermocouples. This equipment incorporates a "Data-Trak" programmer, which permits heating or cooling at desired rates.

Leitz Polarizing Microscope.

Refractive index or density measurements can be used as composition controls where x-ray fluorescence or other means of analysis are not available. The polarizing microscope is employed to measure refractive index using the central illumination method (Becke line) as described by Bloss (1961). Density measurements are made by the method of Archimedes, using the analytical balance.

Glass Saw and Polishing Equipment.

This equipment is used to saw glass samples to size and prepare glass surfaces for property measurements.

Polarizing Apparatus.

The effectiveness of annealing and the evaluation of homogeneity can be determined with polarizing apparatus or with the polarizing microscope.

3.2   RAW MATERIALS AND BATCH FORMULATION

Raw materials come from two sources: (1) naturally occurring minerals, which are purified and beneficiated with regard to grain size after they have been mined from the earth's crust, and (2) manufactured chemicals. Raw materials seldom can be represented exactly by chemical formulas because of the presence of small amounts of impurities. Commercial vendors can furnish glassmaking materials having different purity and grain size ranges. The raw materials used, the size of melts needed, and so on, depend on the nature of the planned research. Glasses to be used for optical absorption measurements in fiber optics transmission lines may require impurity levels in parts per million. Purity requirements of samples for mechanical property measurements are much less restrictive.

The batch materials chosen determine minimum impurity levels, but additional impurities may result from the melting process. The use of platinum melting crucibles assures minimum pickup of impurities, but added impurities may result if bits of furnace refractories or other undesired materials are allowed to get into the melt. All the raw materials in a given batch should be of uniformly fine grain sizes to aid in good batch mixing. Grain sizes should be in the range 50-100 mesh (300-150 microns).

The calculation of a glass batch from analyzed raw materials is a simple procedure and requires no explanation. Incorporation of some raw materials that give off gases during melting is important and calls for clarification. Carbonates are useful constituents in batches and aid in completion of the melting process. The CO_2 is lost during melting and does not change the oxide composition in any way. It is, however, an essential part of a number of chemical reactions and decompositions involved in glass melting. The reader is referred to publications on physical chemistry and differential thermal analysis for details. The CO_2 further aids in agitation and mixing as it leaves the melt. Other batch materials containing sulfates, sulfides, chlorides, and other impurities, may be used in the batch. In these cases small amounts of SO_2 or chlorine may be retained in the finished glass: they also aid in the melting process, but their presence in some experimental glasses may be objectionable. These materials are known in the industry as "refining agents" having the principal function of assisting in making

TABLE 3.1   Molecular Weights of Oxides Commonly Used in Silicate Glasses
($O = 16$)

Oxide	Molecular weight	Oxide	Molecular weight
Al_2O_3	101.96	Li_2O	29.88
B_2O_3	69.62	MgO	40.31
BaO	153.34	MnO_2	86.94
CaO	56.08	MnO	70.94
CoO	74.94	Na_2O	61.98
Cr_2O_3	151.99	NiO	74.71
CrO_3	100.00	PbO	223.19
Fe_2O_3	159.69	SiO_2	60.09
FeO	71.85	TiO_2	79.90
K_2O	94.20	ZnO	81.37

bubble- or seed-free glass. Commercial practice indicates that the choice of a refining agent depends on glass type. Table 3.1 shows molecular weights of oxides commonly used in silicate glasses. The atomic weight of oxygen is 16.

Experimental glass number 70-3 used by Georoff and Babcock (1973) for microindentation hardness measurements illustrates the steps in sample preparation. Oxide compositions of this glass are as follows:

Oxide	Mole Fraction	Weight Percent
SiO_2	0.70	70.03
CaO	0.10	9.34
Na_2O	0.20	20.63

Table 3.2 lists raw materials used and batch calculations. The figures in this table are based on the amounts of batch needed to make 100 g of finished glass. The 21.97 g of CO_2 will be lost during melting, as noted earlier. The use of sulfates would have allowed completion of melting in a shorter time, but the presence of sulfur in the finished sample was not desirable.

3.3   TYPICAL PROCEDURE

Batch sufficient to make 500 g of finished glass was weighed on the torsion balance. The batch was mixed in a laboratory blender for about 30 min, examined visually for homogeneity, placed in fused silica crucibles, and put into the calcining furnace. The time and temperature of heating depends on the batch composition. High temperature sintering at this stage is not desirable. After calcining,

TABLE 3.2 Raw Materials and Batch Calculations for Glass 70-3

Material	Oxides remaining in glass	
	Oxide	Percentage
Quartz	SiO_2	100.0
Reagent $CaCO_3$	CaO	56.0
Reagent Na_2CO_3	Na_2O	58.5

		Batch calculations			
Material	Weight (g)	SiO_2	CaO	Na_2O	CO_2
Quartz	70.03	70.03			
$CaCo_3$	16.68		9.34		7.34
Na_2CO_3	35.26			20.63	14.63
	121.97	70.03	9.34	20.63	21.97

the batch should consist of separate grains that can be easily put into the melting crucible. The batch for glass 70-3 was heated for about 15 hours at approximately 800°C to remove water and volatiles.

The viscosity-temperature relations for glass 70-3 given in Table 2.2 were used as guides in the melting and annealing procedures. The melting crucible should have a volume twice that of the finished glass, to allow for gases given off during melting. Part of the calcined 70-3 batch was poured into a 600 cc capacity platinum alloy crucible and placed in the melting furnace, which was stabilized at a temperature near 1318°C. The remaining batch was added in increments, and after removal of the CO_2, the molten material was poured between revolving aluminum rollers (Figure 3.1). The resulting thin sheet of glass was immediately quenched in a container of distilled water, to minimize contamination by soluble salts, which cause the small pieces of glass to adhere and hinder further mixing. The crushed glass was then washed with alcohol, dried on a hot plate, mixed, and put back in the alloy crucible for further melting. This operation may need to be repeated to thoroughly homogenize the melt. In some cases it is necessary to pour the final melt into a clean platinum crucible because the glass adhering to the first crucible has not been well mixed in the quenching process. After about one hour the final melt is stirred and later poured into stainless steel molds of the proper size and shape. Samples for microindentation hardness were cast in the form of cylinders 3/4 in. in diameter and about 5/8 in. high.

The samples were immediately placed in the annealing furnace, which had been stabilized at about 524°C as indicated in Table 2.2. They were held for one hour at this temperature, then cooled slowly to the strain point at 507°C, after which they were cooled to room temperature at a more rapid rate. The cooling rates were controlled by the "Data-Trak" programmer.

The polarizing equipment was used to ensure that the samples were well annealed and free of inhomogeneous streaks. The refractive index n_D of glass 70-3, calculated using the equation of Babcock (1968), was found to be 1.5213. Refractive indices measured on several pieces from the samples ranged from 1.521 to 1.522. It was concluded that the actual composition was therefore reasonably close to the theoretical one given in Table 3.2. The samples were then sawed into slices about 1/8 in. thick, and the surfaces were polished for measurement. It is noted that the annealing process is reversible; if samples are not properly annealed, they can be satisfactorily reannealed.

3.4 MACROSCOPIC PROPERTIES AND STRUCTURES

Well-established methods for macroscopic property observations on materials in the condensed state are described in many publications. Books by Estermann (1959) and Lark-Horovitz and Johnson (1959) are representative. Glasses differ from other materials because they are supercooled; thus their properties are strongly dependent on thermal history. The properties and atomic microstructures of glasses are determined by the termperature at which these two characteristics are equilibrated before measurement. Thus the thermal history of a given sample must be known in reproducible terms if property values are to be meaningful. The experimenter may not know the microstructure of the glass he is measuring, but he can be reasonably sure that a given thermal history will reproduce the same structure. This poses a major problem for the glass technologist engaged in comparing his property data and their structural implications with those of other investigators. People doing experimental research on glasses are well advised to follow the operational procedure recommended by Bridgman (1936). This approach calls for describing all the operations involved in the measurement of a given material property. In the case of glasses these include purity of raw materials, batch formulation, melting procedure, heat treatment, and the measurement conditions imposed on the sample. Glass research requires accurate measurements of temperature over a wide range. Kingery (1959), Bockris, White, and Mackenzie (1959), and Herzfeld (1962) describe measurements of temperatures and of high temperature properties. In this connection it is emphasized that glasses react readily with other materials at elevated temperatures because of the ionized nature of glasses. There is minimal reaction between hot glasses and the noble metals, but reactions are appreciable when hot glasses are in contact with base metals and refractories.

Glass research consists essentially of subjecting a sample of known characteristics to specified experimental conditions and observing how the glass reacts.

The measurements of refractive index and specific volume at room temperature leave the glass unchanged from its original state; it reacts passively to test conditions. Little or no energy is absorbed in such measurements. Subjecting the sample to other conditions may result in the absorption of energy. Such samples may revert to their original condition after the measurement, or they may undergo some observable change. They are said to react actively to test conditions. We are speaking of samples that are large enough to see and touch and can be manipulated in the measuring apparatus. In such samples the numbers of atoms or other structural constituents is very large compared to the number on or near the surface. Thus such properties as refractive index, specific volume, and moduli of elasticity, are representative of the glass mass; these characteristics are also referred to as bulk or volume properties. The macroscopic properties mentioned derive their meaning from macrophysics, which deals with masses large enough to permit directly observable measurements.

The observations on macroscopic properties may result from the use of x-ray diffraction, spectral absorption, and other techniques that give information on the constituent electrons, ions, and their spatial arrangements. The correlation of macroscopic properties with microstructures is the major concern of investigators working on the condensed states of materials. Lacking a microscopic description of physical properties, we must rely on structural models that relate to experimental observations. In the final analysis we resort to educated guesses in bridging this gap in terms of reasonable postulated relations. Quantum mechanics, which consists principally of rules of computation, has been successful in giving empirical descriptions of a number of physical phenomena on a microscopic scale.

Our goal in this discussion is to postulate reasonable relations between macroscopic properties and microstructures in silicate glasses. Insofar as possible, glass properties will be related to the presence of electrons, ions, and ionic groups in the glasses. These three categorical structural constituents are considered to execute forced oscillations under the actions of electrical, magnetic, mechanical, and thermal force fields. These oscillators are coupled, and the energy supplied is distributed among them. Energy absorptions occur when the oscillator frequencies are in resonance with frequencies of the imposed force fields. For example, electromagnetic radiation with frequencies in the visible range affects principally the electrons, (although electrostatic fields of neighboring ions and ionic groups strongly influence electronic resonance.) When the frequency range of imposed force fields embraces all the frequencies of the structural oscillators, there results a wide spectrum of energy absorptions. In such cases the separate effects of the individual oscillators cannot be clearly isolated, and empirical relations must be developed between glass properties and their microstructures. Examples of such postulated relationships are given in Parts II to IV.

The bulk properties and microstructures of glasses differ from those on or near glass surfaces. Glass usage, be it in set experiments or in product usage, always involves surroundings of the environment. The reactions of glass surfaces with electromagnetic, mechanical, thermal, and chemical forces constitute an extensive and important subject, but one outside our present interest. The reader is referred to the excellent review of glass surfaces by Holland (1964) and to separate investigations described in glass journals.

4

Property-Composition-Structure Relations

The fabrication of glasses and metals probably began soon after man's discovery of fire. These two processes have much in common both historically and technically, and technical development along essentially parallel paths is understandable in that commonly used glasses consist of metallic oxides. The prediction of properties of glasses and metals from their compositions and atomic structures is an important subject of research. Increased usefulness of products made from these materials is closely related to advancements in basic science and technology. This chapter outlines methods leading to property predictions in silicate glasses in terms of their oxide compositions and postulated microstructures. Although these methods are demonstrated specifically for silicate glasses, it is suggested that they may be adaptable for other types of oxide glass. Primary phase diagrams of silicate glasses can be approximated by using glass property data, and this method leads to the classification of silicate glasses in terms of their composition locations in primary phase fields.

4.1 HISTORICAL DEVELOPMENT OF COMMERCIAL GLASSES

It is not definitely known when and where glass was first made, but the long and interesting history of glass has been well documented by many authors. Early history has been related by Petrie (1924-1925), Lucas (1926), Caley (1962), and others. Modern methods are described by Hovestadt (1902), Morey (1954), Volf (1916), Tooley (1974), Weyl and Marboe (1964), and references cited in these publications. Glassmaking grew as a handcraft and for centuries was attended by superstitution and secretiveness, attitudes that have not yet entirely disappeared. The history of the craft can be traced in terms of gradual improvements in available raw materials for making glasses and glass melting vessels, methods for attaining high temperatures, and glass production processes. One must distinguish between the still flourishing handcraft of making glass objects of art and the manufacturing of glass for utilitarian purposes. Our interest here is in the latter—specifically, in replacing some of the art still remaining in glass manufacturing by methods of objective technology. The conversion from hand to machine methods, which began at about the beginning of the present century, compelled manufacturers to pay increasing attention to more scientific methods. The glass hand worker can adjust his efforts to fit glassmaking conditions from experience, but a cycled high speed glassmaking machine cannot. Therein lies the reason for use of objective methods in glass manufacturing. Major advances in the glass industry continue to come about because of engineering know-how and empirical inventions relating to glassmaking processes and glass usage. These developments have been augmented in recent years by advances in the physical sciences, experimental techniques, and electronic computers.

The first comprehensive research on relations between glass properties and compositions was done by Ernst Abbe and Otto Schott in the last quarter of the nineteenth century. Their pioneering work, motivated by the need for new and better glasses for optical purposes, thermometers, and products requiring high chemical and thermal resistance, has been described by Hovestadt (1902). The research of these men has served as a model for subsequent glass studies, which have revealed further diversified uses for glasses. Efficient machine production requires continuous control of glass compositions, properties, and temperatures at levels needed for the various glass products. Straightforward methods now exist for keeping these factors under control. Glassmaking also requires control of heat transfer processes. As we see later, the role of transient heat transfer in glassmaking machines is very difficult to evaluate and control.

4.2 GLASSES CONSIDERED AS SOLUTIONS OF OXIDES

The paper by Winklemann and Schott (1894) is typical of those described by Hovestadt (1902), indicating that some glass properties are linearly related to oxide compositions . Such properties as specific heat, specific volume, elasticity, and thermal expansion were represented by equations of the form

$$P = a_1 b_1 + a_2 b_2 + a_3 b_3 + \cdots \tag{4.1}$$

where P = glass property, a_1, a_2, and a_3 = weights percent of the oxides, and b_1, b_2, and b_3 are respective numerical constants. This method is still widely used and is compatible with the concept that the constituent oxides are randomly arranged in the glass network. The approach is readily adaptable to glass manufacturing practices of expressing glass batches in weights percent of the oxides.

However there are disadvantages and limitations associated with the use of the foregoing method. These become apparent when we consider large composition areas of glass systems, and we find that property-composition relations are not linear over wide composition ranges. Depending on the fidelity of property representation desired, such linearity must be limited to smaller composition ranges. In other words, we assume linear relations within the immediate area of interest.

Reference is made to Morey and Merwin (1932), who show curved lines of equal refractive indices and dispersions on ternary weight percent plots for the Na_2O-CaO-SiO_2 glass system. These plots are typical of those for other properties where property-composition representations cover extended composition ranges. For glass formulation purposes, we can select limited composition ranges wherein departures from linearity are no greater than those to be expected from measurements of the given property. Linear equations like (4.1) can then be used within the selected composition ranges. Thus several linear equations are needed to represent all the data in a given glass system within limits imposed by experimental errors of the property measurement. The numerical constants b_1, b_2, b_3, and so on, not only vary within a given glass system, but also between different glass systems. This necessitates considerable research on preparation of glass samples and property measurements, followed by derivation of numerical constants for the constituent oxides over limited linearity ranges. The use of such procedures involves time-consuming and tedious research efforts.

Huggins and Sun (1943) made improvements in the methods of calculating specific volume, refractive indices, and dispersion. Their method involves calculating the Si/O ratio (the inverse of the O/Si ratio in Table 1.4) of silicate glasses and assuming linearity between properties and weight fractions within given ranges of this ratio. They use different numerical constants for the various oxides in the following Si:O ranges: 0.270 to 0.345, 0.345 to 0.400, 0.400 to 0.435, and 0.435 to 0.500. For example, specific volume constants for SiO_2 in these ranges are respectively 0.4063, 0.4281, 0.4409, and 0.4542. In this method the specific volumes are calculated first, and the refractive indices and dispersions are then determined as functions of specific volume. Huggins (1940) gives reasons for the breaks in linearity at Si:O ratios of 0.345, 0.400, 0.435, and 0.500. Tables 1.4 and 2.1 show that the breaks of 0.400 and 0.500 correspond, respectively, to the tetrahedral sharing of three and four oxygens, as in $Na_2O \cdot 2SiO_2$ and SiO_2. The breaks at 0.345 and 0.435 come from linearity breaks in the property data. This method recognizes departures from linearity of property-composition relations in silicate glasses, thus achieves more accurate representations of data for these properties. Blau (1951) used this method for representing thermal expansion coefficients in the binary Na_2O-SiO_2 system.

4.3 PRIMARY PHASE SUBSTRUCTURE METHOD

Use of the primary phase substructure method for quantitative representation of relations between glass compositions and properties was proposed by Babcock (1968). This work was motivated by the author's dissatisfaction with the limitations of calculation methods based on the concept that silicate glasses were simple solutions of the constituent oxides. His experience in commercial glass manufacturing further showed that research based on the random network picture of glass structures was unduly expensive in terms of time and effort. Acquaintance with the published glass literature disclosed substantial reasons for departing from the random network assumption. Some of these reasons have already been discussed, and others are cited in later chapters.

Young, Glaze, Faick, and Finn (1939) were the first to point out that specific volume-composition relations in Na_2O-SiO_2 and K_2O-SiO_2 systems change definitely at compositions approximating those at eutectics. Others, including Marinov and Modeva (1965), Winter-Klein (1959), Dietzel (1967), and Robinson (1969), have noted glass property changes at eutectics. The presence of property changes at eutectics, the reported data on structural order, and other published ideas seemed to constitute sufficient reasons for abandoning the random network postulate. However isolated experimental facts and ideas are of limited use to the glass technologist unless they are combined into a computational scheme for his practical use. The problem of developing such a scheme for silicate glasses was approached from a pragmatic standpoint: the validity of the substructure concept was to be judged in terms of its practical results. The use of such an approach is not new, one of its more notable successes being quantum mechanics, which consists essentially of rules of computation that empirically relate many physical phenomena to their postulated microstructures.

Babcock (1968, 1969, 1973) described uses of the substructure method in relating glass properties to mole fraction compositions in the following glass systems:

Na_2O-SiO_2	K_2O-SiO_2	Na_2O-CaO-SiO_2
Na_2O-Al_2O_3-SiO_2	K_2O-Al_2O_3-SiO_2	Na_2O-TiO_2-SiO_2
Na_2O-K_2O-SiO_2	CaO-Al_2O_3-SiO_2	Na_2O-SrO-Al_2O_3-SiO_2

Depending on availability of data in these glass systems, linear equations relating properties and compositions were derived for the following properties: refractive index, the various dispersions, specific volume, thermal expansion, and fluidity (the inverse of viscosity). Additional papers by the author and his colleagues covering other properties and glass systems are discussed in later chapters.

The method of property-composition correlation was made as objective as possible and essentially allowed the data to speak for themselves. The data under examination were segregated into groups according to the composition ranges and areas covered by the primary crystallization phase fields involved. The compositions were expressed in mole fractions of the oxides. The separate groups of data, one group for each primary phase field, were then subjected to least mean squares computer analysis. The computer program called for determing the role of each oxide constituent in terms of linear equations of the form

$$\text{glass property} = A\ SiO_2 + B\ CaO + C\ Na_2O + \cdots \tag{4.2}$$

where A, B, C, and so on, are numerical constants characteristic of the respective oxides, and the amounts of the oxides are expressed in mole fractions. For example refractive index n_D for glasses in the SiO_2 phase field of the $Na_2O\text{-}SiO_2$ system can be represented by the equation $n_D = 1.4581\ SiO_2 + 1.6174\ Na_2O$. The computer program indicated a standard error of 0.0002, which compares favorably with the ±0.0005 possible error in refractive index measurements. Differences between measured and calculated values could be caused by variations in chemical analysis, sample homogeneity, annealing, or measurement of refractive index.

It follows from the equations used in the substructure method that glass property values lie on straignt lines within given phase fields of binary systems. Figure 4.1 shows lines representing calculated values of n_D and specific volume in the three primary phase fields $Na_2O\cdot SiO_2$, $Na_2O\cdot 2SiO_2$, and SiO_2 in the binary $Na_2O\text{-}SiO_2$ system. The lines of equal refractive index change slope at the eutectics located at 0.63 and 0.75 mole fractions of SiO_2. Likewise property values lie on planes in ternary systems, lines of equal property values being parallel within each phase field and their distances apart being characteristic of the given phase field. Figure 4.2 gives projections of refractive index n_D data on the $CaO\text{-}SiO_2$ mole fraction composition plane. Mole fraction compositions of Na_2O are obtained by subtracting totals of $CaO + SiO_2$ from 1.00. The boundaries between adjacent phase fields are straight lines formed by the intersections of property planes. Three property planes intersect at ternary invariant points as indicated at R, Q, N, and O. Property data in ternary systems can be represented by solid models, the property data values being plotted vertically above horizontal composition planes. Data representations in quaternary systems require separate projections or solid models for each composition level of the fouth oxide. There is no analytical limit to the number of oxides that can be represented by type (4.2) equations. Equations for refractive index are derived directly from measured data and are not dependent on measurements of specific volume as required by the Huggins and Sun (1943) method. The oxides in Figures 4.1 and 4.2 have been abbreviated to save space. Designations are $S = SiO_2$, $C = CaO$, and $N = Na_2O$. Abbreviations are also used in designate glass systems (e.g., N-C-S = $Na_2O\text{-}CaO\text{-}SiO_2$). Primary phases $Na_2O\cdot 2SiO_2$ and $Na_2O\cdot 3CaO\cdot 6SiO_2$ are indicated by N2S, N3C6S, and so on. This nomenclature is used in all subsequent figures and tables. The SiO_2 phase fields are designated by S(Q), S(T), S(C), and for combined phase fields, for example, by S(*TC*). (See key accompanying Figure 4-3.)

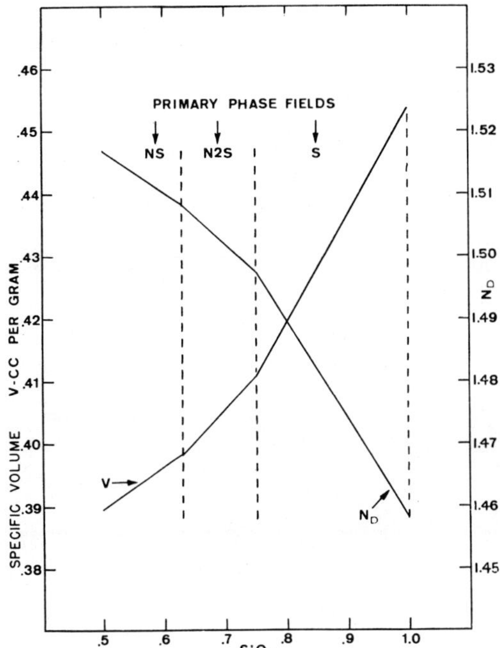

Figure 4.1

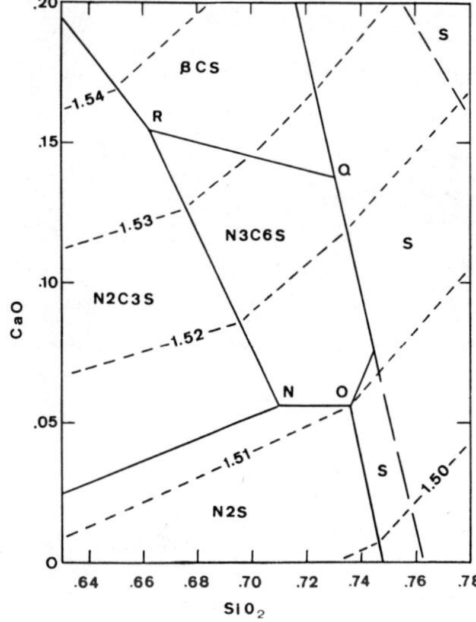

Figure 4.2

45

TABLE 4.1   Refractive Index n_D for Annealed Glasses in the Na_2O-CaO-SiO_2 System

$$n_D = A\ SiO_2 + B\ CaO + C\ Na_2O$$

Primary phase	A	B	C	Standard Error
S(TC)	1.4568	1.7850	1.6261	0.0006
N2S	1.4764	1.7513	1.5642	0.0005
N2C3S	1.4812	1.7559	1.5481	0.0009
N3C6S	1.4715	1.7638	1.5740	0.0008
α CS	1.4681	1.7813	1.5620	0.0007
β CS	1.4676	1.7518	1.6044	0.0004

Experimental data ±0.0005.

Standard error = $\left[(\Delta D)^2 / (N-1) \right]^{1/2}$, where ΔD = difference between measured and calculated data and N = number of measurements.
Data source. Morey and Merwin (1932).

Table 4.1 gives analytical representations of refractive index data on glasses in the Na_2O-CaO-SiO_2 system. Binary Na_2O-SiO_2 glasses are included in the S(TC) and N2S groups in this table because they lie in these phase fields. Therefore the oxide constants differ slightly from those shown by Babcock (1968) for ternary Na_2O-CaO-SiO_2 glasses.

Figure 4.3 presents the ternary phase diagram of the Na_2O-CaO-SiO_2 system and information on crystalline compounds and weights percent compositions of invariant points. The validity of the substructure method can be judged by relations between refractive index representations and the primary phase invariant points in Table 4.2. For example, property planes for the primary phases α CS, β CS, and S intersect at the invariant point (T). The average value of n_D for point (T) composition is obtained by using the three appropriate equations from Table 4.1 is 1.5300 ±0.0003. Mole fraction compositions are used in the calculations, but weights percent compositions are shown for easy reference to Figure 4.3. Average values of n_D for the six invariant point indicate that property planes intersect at compositions obtained by thermodynamic equilibrium methods.

Alternatively the composition of invariant point (T) can be calculated from equations of the three intersecting planes without reference to the phase diagram in Figure 4.3. Simulataneous solution of the following equations

$$n_D = 1.4681\ SiO_2 + 1.7813\ CaO + 1.5620\ Na_2O$$
$$n_D = 1.4676\ SiO_2 + 1.7518\ CaO + 1.6044\ Na_2O$$
$$n_D = 1.4568\ SiO_2 + 1.7850\ CaO + 1.6261\ Na_2O$$
$$SiO_2 + CaO + Na_2O = 1$$

shows the composition to be $SiO_2 = 73.14$, $CaO = 15.68$, and $Na_2O = 11.18$, when converted to weights percent. This composition is in excellent agreement with the one shown in Figure 4.3 which was determined by thermodynamic methods.

Refractive indices of glasses located in the six primary phase fields can be closely estimated by using appropriate equations given in Table 4.1. Compositions areas wherein the equations are valid are given in weights percent in Figure 4.3 and must be converted to mole fractions before making calculations. The substructure method is used in subsequent discussions whenever sufficient data are available for its application.

4.4 CALCULATION OF PRIMARY PHASE DIAGRAMS

The determination of primary phase diagrams in glass systems follows principles originally described by Gibbs (1906). Applications to heterogeneous equilibria in glass systems have been described by Morey (1954), Roedder (1959), and others. Results of the numerous investigations of glass systems have been compiled by Levin, Robbins, and McMurdie (1964, 1969). The devitrification of a multicomponent glass is a complex process because glass is a supercooled material. Measurement of the liquidus temperature of a glass requires the establishment of thermodynamic equilibrium between the molten glass and its crystalline primary phase. Roedder emphasizes that these measurements necessitate careful attention to sample preparation, experimental procedures, and data representation. The most commonly used method involves quenching the sample from its high temperature equilibrated condition at such a rate that further phase changes do not occur. The quenching method consists of five steps: (1) composition synthesis, (2) melt homogenization, (3) establishment of equilibrium conditions, (4) rapid quenching, and (5) crystalline identification. These data are then used to graphically determine the primary phase diagram in the form of equilibrium lines in binary systems and equilibrium surfaces in ternary systems. Gibbs (1906) and Morey (1930) have described these graphical methods.

Establishment of phase boundaries and liquidus temperatures in glass systems entails considerable time and effort. For example, Schairer and Bowen (1956)

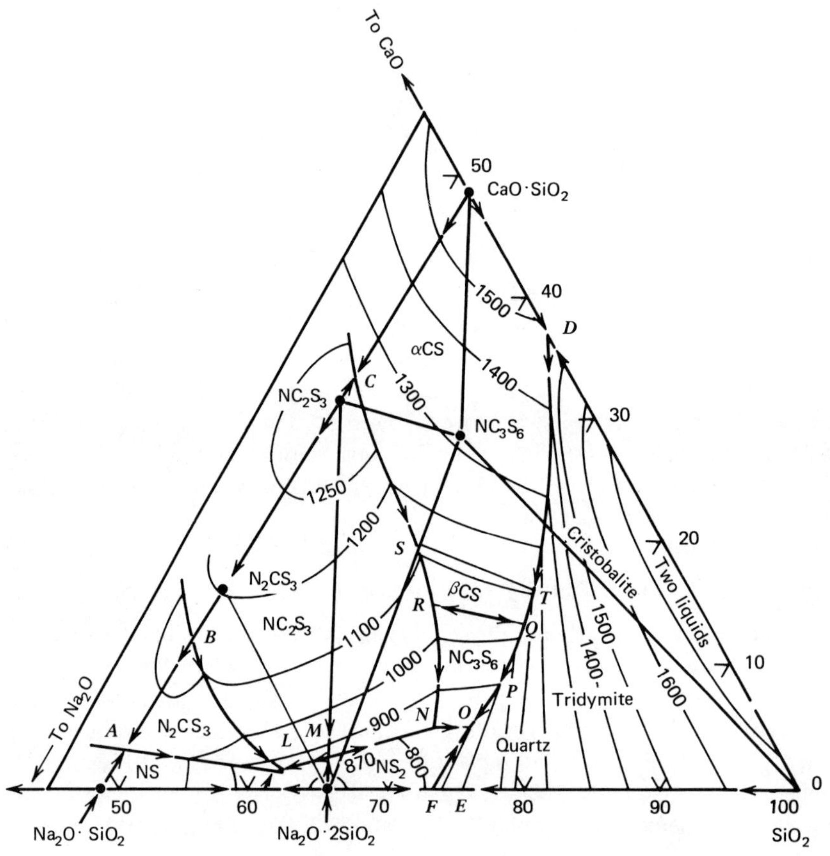

Figure 4.3 Primary phase of Na_2O-CaO-SiO_2 system, high SiO_2 corner of system. *Source.* Levin, Robbins, and McMurdie (1964). Originally from G. W. Morey and N. L. Bowen, *J. Soc. Glass Technol.,* **9** 232-233 (1925).

Compounds	CaO	Na₂O	SiO₂	Temperature (°C)	Condition*
SiO_2	—	—	100.0	1710	M
$aCaO \cdot SiO_2$	48.3	—	51.7	1540	M
$\beta CaO \cdot SiO_2$	48.3	—	51.7	1180	I
$Na_2O \cdot SiO_2$	—	50.8	49.2	1088	M
$Na_2O \cdot 2SiO_2$	—	34.1	65.9	874	M
$2Na_2O \cdot CaO \cdot 3SiO_2$	15.6	34.4	50.0	1141	D
$Na_2O \cdot 3CaO \cdot 6SiO_2$	28.5	10.5	61.0	1047	D
$Na_2O \cdot 2CaO \cdot 3SiO_2$	31.6	17.5	50.9	1284	M

*M = melting point, D = decomposition point, I = inversion point.

TABLE 4.2 Calculated Refractive Indices n_D at Invariant Points in Annealed Na_2O-CaO-SiO_2 Glasses

	Invariant point primary phases	Compositions (wt %)			n_D
		SiO_2	CaO	Na_2O	
(S)	αCS-βCS-N2C3S	62.8	19.5	17.7	1.5500 ±0.0004
(T)	αCS-βCS-S	73.0	15.6	11.4	1.5300 ±0.0003
(R)	βCS-N2C3S-N3C6S	66.5	14.5	19.0	1.5355 ±0.0006
(Q)	βCS-S-N3C6S	73.4	12.9	13.7	1.5247 ±0.0003
(N)	N2S-N2C3S-N3C6S	70.7	5.2	24.1	1.5122 ±0.0002
(O)	N2S-N3C6S-S	73.5	5.2	21.3	1.5098 ±0.0004

Source. Babcock (1968).

	Point	Crystal Phases*	CaO	Na_2O	SiO_2	Temperature (°C)
△	A	NS-N_2CS_3	3.0	—	—	1060
●	B	N_2CS_3-NC_2S_3	11.5	—	—	1141
△	C	NC_2S_3-aCS	33.0	—	—	1280
△	D	aCS-S	37.0	—	63.0	1436
☺	E	T-Quartz	—	24.3	75.7	870
△	F	Quartz-NS_2	—	26.4	73.6	790
*	K	NS_2-NS-N_2CS_3	1.8	37.5	60.7	821
○	L	N_2CS_3-NC_2S_3-NS_2	2.0	36.6	61.4	827
○	N	NS_2-NC_2S_3-NC_3S_6	5.2	24.1	70.7	740
*	O	NC_3S_6-Q-NS_2	5.2	21.3	73.5	725
☺	P	Q-NC_3S_6-T	7.0	18.7	74.3	870
○	Q	T-βCS-NC_3S_6	12.9	13.7	73.4	1035
○	R	NC_3S_6-NC_2S_3-βCS	14.5	19.0	66.5	1030
●	S	βCS-NC_2S_3-aCS	19.5	17.7	62.8	1110
●	T	aCS-S-βCS	15.6	11.4	73.0	1110
△	I	NS-NS_2		38.0		840

Key = △ Binary eutectic
  * Ternary eutectic
  ● Decomposition point
  ○ Reaction point
  ☺ Inversion point

*C = CaO, N = Na_2O, S = SiO_2, Q = quartz, T = tridymite.

prepared and measured 340 synthetic compositions in the development of a portion of the ternary Na_2O-Al_2O_3-SiO_2 system. The possibility of calculating phase diagrams has been under discussion for several years. Phase calculations are now being used increasingly in materials science research. Chayes (1968) described a method of calculating the locations of phase boundaries in binary systems by means of discriminant functions. His calculations were in excellent agreement with phase boundaries appearing in a number of published diagrams. He suggests that this procedure could be applied to multicomponent systems in which locations of phase boundaries are impossible to ascertain by mean of graphical methods. Kaufman and Bernstein (1970) describe in detail the use of computer methods in determining phase diagrams of refractory metals. They made use of lattice stability and other basic metal properties in their research. Barron (1972) used excess free energy data in the calculation of phase boundaries in silicates and provides other references on the subject.

Babcock (1968) showed that primary phase boundaries and invariant points could be closely estimated from glass property data (see Table 4.2 for such information on the Na_2O-CaO-SiO_2 system) and presented similar information on Na_2O-Al_2O_3-SiO_2, K_2O-Al_2O_3-SiO_2, Na_2O-SiO_2, and K_2O-SiO_2 glass systems. In all cases the oxide compositions of invariant points estimated from glass property data were in reasonable agreement with those in published phase diagrams. The substructure method of calculation was set forth in detail by Babcock (1973) by means of property data on the Na_2O-TiO_2-SiO_2 system published by Hamilton and Cleek (1958). The partial phase diagram of this system (Figure 4.4) is from Levin, Robbins, and McMurdie (1964). The substructure method was tested by assuming that no phase diagram existed for this system. In this case the glasses were segregated into contiguous groups wherein the linear equations represented property data within given limits. These limits, compatible with precisions of measurements, were that differences between measured and calculated values would be no greater than the following; 65×10^{-4} for specific volume, 7×10^{-4} for refractive index, 25×10^{-5} for mean dispersion n_F-n_C, and 35×10^{-5} for dispersive power. Use of a computerized least mean squares program revealed that the glasses were located within seven contiguous groups on the basis of the limits given. These groups were designated as A, B, C, D, E, F1, and F2 following the nomenclature of Hamilton and Cleek (1958).

Computerized equations for refractive index and specific volume were used to construct the ternary phase diagram of Figure 4.5. Computerized simultaneous solutions of property equations and the conditional equation $SiO_2 + TiO_2 + Na_2O = 1$ were used to determine composition locations of phase boundaries and invariant points. For example, the boundary equation between groups A and

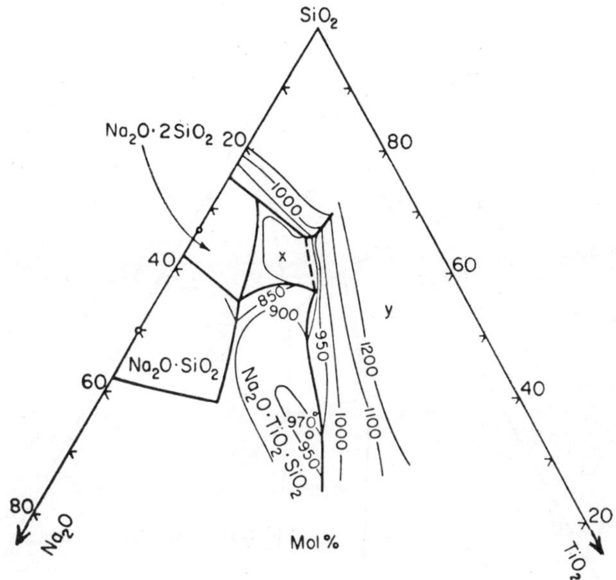

Figure 4.4 Primary phase diagram of Na_2O-TiO_2-SiO_2 System. From Levin, Robbins, and McMurdie (1964).

B was obtained from simultaneous solution of the equations

(A) n_D = 1.45058 SiO_2 + 2.24879 TiO_2 + 1.93299 Na_2O (4.3)

(B) n_D = 1.49204 SiO_2 + 2.18124 TiO_2 + 1.51324 Na_2O (4.4)

      SiO_2 + TiO_2 + Na_2O = 1   (4.5)

The boundary equation (A-B) from refractive index data was

$$SiO_2 = 0.7428 - 0.3238 \, TiO_2 \tag{4.6}$$

Fourteen such boundary equations were determined from refractive index data, and a like number were found from specific volume data for glasses located in the seven primary phase fields. It was learned that the property planes for glasses in phase fields *A, B,* and *D* intersected in a triple point. The composition of this point was obtained by computerized solution of six boundary equations, that is (A-B), (A-D), and (B-D) for both n_D and specific volume. The composition of the triple point A-D-F2 was determined in the same manner, using boundary equations (A-D), (A-F2), and (D-F2) for refractive index and specific volume.

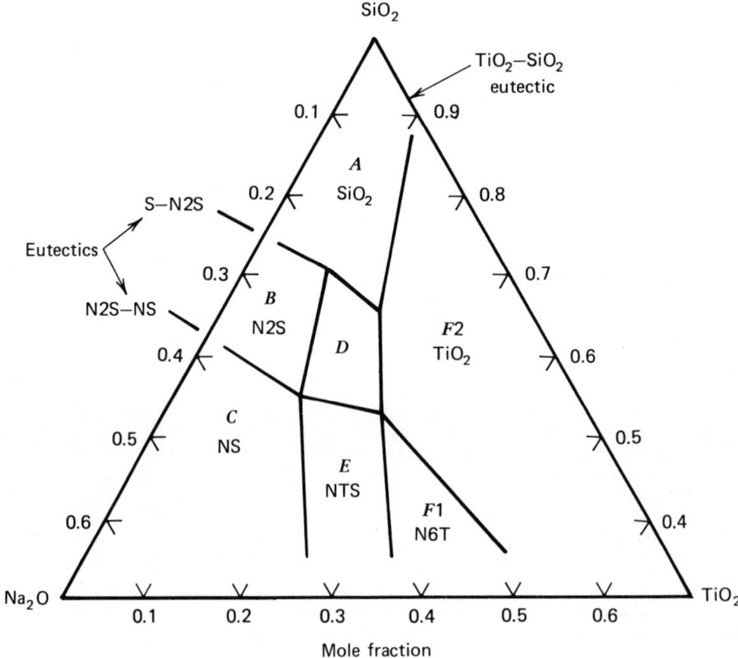

Figure 4.5 Computerized Na_2O-TiO_2-SiO_2 primary phase diagram: $S=SiO_2$, $T = TiO_2$, $N = Na_2O$. From Babcock (1973).

Table 4.3 lists computerized compositions of the triple points A-B-D and A-D-F2 and average property values with average deviations. In the same manner, the compositions of quadruple points B-C-D-E and D-E-F1-F2 were determined to be SiO_2, 0.5443; TiO_2, 0.1492; Na_2O, 0.3064; and SiO_2, 0.5296; TiO_2, 0.2459; Na_2O, 0.2245, respectively. Alternatively the compositions of invariant points can be calcuated by simultaneous solutions of the property equations for the given primary phase fields as demonstrated in Section 4.3.

The substructure method of approximating compositions of invariant points and primary phase boundaries is not meant to replace the classical thermodynamic method of Gibbs (1906). However such calculated diagrams (1) provide information on glass systems where few or no primary phase data are available, and (2) can serve as a preliminary time-saving step toward obtaining more exact data by the classical method.

4.5 SUBSTRUCTURE CLASSIFICATION OF GLASSES

Glasses are presently classified in terms of their major oxide constituents. This method has evolved naturally from common usage in the glass industry. This

TABLE 4.3 Calculated Compositions and Property Data at Invariant Points

Mole fraction compositions and properties*	Invariant points	
	A-B-D	A-D-F2
SiO_2	0.7028	0.6552
TiO_2	0.1009	0.1802
Na_2O	0.1963	0.1646
n_D	1.5651 ±0.0017	1.6254 ±0.0009
Density (g/cc)	2.5714 ±0.0008	2.6696 ±0.0026
Mean dispersion	0.01289 ±0.00027	0.01779 ±0.00007
Dispersive power	0.02900 ±0.00016	0.02871 ±0.00020

*Property values are averages of those calculated for the separate phase fields.

classification will undoubtedly continue to be used by glassmakers and users of glass products. However from the substructure concept it is not satisfactory for the use of glass scientists and technologists. The term "soda-lime-silica glasses" refers only to major amounts of Na_2O, CaO, and SiO_2 contained by such glasses. Their compositions may be located anywhere within the composition area in Figure 4.3. Furthermore such glasses may contain minor amounts of the oxides as MgO, Al_2O_3, Fe_2O_3, and so on, which have substantial effects on their properties. If these glasses contain substantial amounts of Al_2O_3, they may be designated as "soda-lime-alumina-silica glasses." This classification is equally indefinite.

Babcock (1969) recommended that silicate glasses be classified in terms of their composition locations in primary phase fields. On this basis the Na_2O-CaO-SiO_2 system in Figure 4.3 consists of 10 different kinds of glass located in the following primary phase fields:

S(C) cristobalite	NS sodium metasilicate
S(T) tridymite	α CS pseudowollastonite
S(Q) quartz	β CS wollastonite
N2S sodium disilicate	N2C3S
2NC3S	N3C6S devitrite

Previously discussed data show that glass properties are linearly related to mole fraction compositions within given primary phase fields of single glass systems. The information given in Tables 4.4 and 4.5 demonstrates that these relations are valid within similar phase fields in different glass systems. The specific

TABLE 4.4   Specific Volume of Oxides in Annealed Glasses*

Primary phase field	Volumes of oxides (cc/g) in five glass systems					
	N-S	N-C-S	K-S	K-N-S	N-A-S	Average
Volume of SiO_2						
S(T)	0.45405	0.45338	0.45158	0.45256		0.45289 ±0.18%
S(Q)	0.44453		0.44395	0.44742	0.44922	0.44628 ±0.46%
N2S	0.43630	0.43758		0.43744	0.43546	0.43670 ±0.19%
NS	0.42538				0.42510	0.42524 ±0.03%
Volume of Na_2O						
S(T)	0.27885	0.27332		0.28204		0.27807 ±1.14%
S(Q)	0.30814			0.29930	0.29611	0.30118 ±1.54%
N2S	0.33268	0.33318		0.32853	0.33452	0.33223 ±0.56%
NS	0.35159				0.35171	0.35165 ±0.02%

*Average value for N-S system; $K = K_2O$; $A = Al_2O_3$.

Source. Babcock (1968-1969).

TABLE 4.5 Refractive Index n_D of Oxides in Annealed Glasses

Primary phase field	Glass systems					Averages
	N-S	N-S	N-C-S	N-A-S	N-L-S	
Refractive index of SiO_2						
N2S	1.4763	1.4770	1.4739	1.4764	1.4745	1.4756 ±0.08%
NS	1.4837	1.4818		1.4843		1.4833 ±0.07%
Refractive index of Na_2O						
N2S	1.5625	1.5638	1.5658	1.5642	1.5657	1.5644 ±0.07%
NS	1.5500	1.5561		1.5410		1.5490 ±0.35%

Sources. Data for N-L-S glasses (L = Li_2O) for Kracek (1939), other data from Babcock (1968).

volumes of oxides in Table 4.4 are identified with the constants A, B, C, and so on, in equation (4.2). These constants were derived by least squares methods from data obtained by six different investigators. Thus the specific volume for glasses in the S(T) primary phase field can be represented by the equation $V = 0.45289\ SiO_2 + 0.27807\ Na_2O$. The constants for refractive index in Table 4.5 were derived in the same way from data obtained by five different investigators. Refractive index data in N2S primary phase fields for these systems can be represented by the equation $n_D = 1.4756\ SiO_2 + 1.5644\ Na_2O$. This information indicates that the spatial arrangements of SiO_2 and Na_2O are unique for each primary phase glass substructure.

This type of classification tells at once the general nature of a given glass composition. Glasses located in the NS, N2S, 2NC3S, and N2C3S phase fields in Figure 4.3 are known to be unsuitable for glass products because of their physical and chemical properties. Glasses in the S(T) field are known for their high chemical resistance and low thermal expansion. It thus seems reasonable to use substructure nomenclature and to designate glasses as S(T) or tridymite glasses, N2S or sodium disilicate glasses, and so on.

4.6 PROPERTIES AND STRUCTURES OF VITREOUS SILICA

Vitreous silica, containing close to 100% SiO_2, deserves special consideration because its properties are uniquely affected by changes in temperature and pressure. The effects of temperature and pressure on this glass have received much attention in the literature and continue to be active subjects of research. According to representative published data, the properties of vitreous silica deviate from those expected from a material containing only one homogeneous phase. Thus properties of vitreous silica as they relate to changes in temperature and pressure are referred to as "anomalous." This glass has been called quartz glass, fused quartz, or fused silica, and the ambiguous nomenclature has led to confusion in both commercial and scientific uses of the material. Following Sosman (1965), the term "vitreous silica" seems to be most descriptive of this type of glass.

Vitreous silica can be made by flame or electrical fusion of naturally occurring quartz crystals. Sosman (1965) points out that natural silica holds surprisingly constant at 46.75 ±0.05% silicon and 53.25 ±0.05% oxygen. These figures correspond closely to a mass number of 28.09 for silicon and 16 for oxygen, totaling 60.09 for the atomic weight of SiO_2. Our data reviews on vitreous silica refer to this material unless noted otherwise. The nine possible kinds of SiO_2 formed by three isotopes each of silicon and oxygen are outside our present interests.

Vitreous silica made from quartz usually contains small amounts of impurities such as Fe, Cr, Al, and Ca, ranging from parts per million up to a few tenths percent. The water—or, more precisely, the OH anion content—can be reduced to low levels if electrical fusion manufacturing methods are used. Vitreous silica made by synthetic hydrolized $SiCl_4$ processes can have controlled low levels of heavy metal impurities, but the OH content can be as high as 0.10%. However methods are available for reducing the OH content to low levels for special uses of vitreous silica.

In the author's view the anomalous behavior of vitreous silica with changes in temperature and pressure is related to stable and metastable phases in crystalline SiO_2 as reflected in strongly supercooled vitreous SiO_2. These relations appear reasonable in terms of the complex nature of crystalline SiO_2. Each of the silica minerals quartz, tridymite, and cristobalite has its SiO_4 tetrahedra linked according to different crystal systems, and each has both high and low temperature forms. Quartz is a common mineral and next to the feldspars is the most abundant mineral in the earth's crust. Tridymite and cristobalite are less abundant but are widely distrubuted in volcanic rocks. Frondel (1962) and Sosman (1965) give comprehensive descriptions of the various modifications of crystalline SiO_2. Sosman also covers vitreous SiO_2 in considerable detail. Sosman lists nine modifications of tridymite with their respective temperature ranges, in addition to high and low temperature forms of quartz and cristobalite. Relative to data on vitreous silica it is sufficient here to speak simply of the temperature or temp-ranges where high-low modifications occur.

1. High and low temperature quartzes are hexagonal and rhombohedral, respectively. Quartz is stable in the temperature range from -273 to 870°C and is capable of metastable existence at higher temperatures. The change from high to low temperature forms is at 573°C.

2. High and low temperature forms of tridymite are hexagonal and ortho-rhombic, respectively. Tridymite is stable from 870 to 1470°C and is capable of metastable existence below 870°C and above 1470°C. Changes from high to low temperature forms occur at 64, 117, and 163°C.

3. High and low temperature cristobalites are cubic and tetragonal, respectively. Cristobalite is stable from 1470°C to its melting point at 1723°C. It is capable of metastable existence at any temperature below 1470°C. Changes from high to low forms occur from about 175 to 272°C.

All the foregoing temperatures for the various modifications of crystalline SiO_2 refer to a pressure of 1 atm. Sosman (1965) discusses the effects of pressure on the modifications of crystalline SiO_2.

Changes from one polymorphic form of crystalline SiO_2 to another, as from tridymite to quartz at temperatures below 870°C and to cristobalite at temperatures above 1470°C, involve the breaking of very strong Si-O bonds. High energies are required for these changes, which are termed "reconstructive." Changes from one polymorphic form to another require long periods for completion and are subject to supercooling. Levin, Robbins, and McMurdie (1964) have demonstrated that glasses with high SiO_2 contents have SiO_2 primary phase fields; in most systems the phase field embraces separate areas or regions of quartz, tridymite, and cristobalite. The primary phase of vitreous silica is cristobalite. The properties of silicate glasses approach more and more closely those of vitreous silica as the SiO_2 contents increase. Inversions from high to low temperature forms are quite different from those involved in polymorphic changes. High-low inversions require small energy exchanges characteristic of changes in the Si-O-Si angles between adjoining SiO_4 tetrahedra. These rapid and reversible changes are termed "displacive." With regard to the high-low inversion of quartz, Frondel (1962) states,

> As the inversion point is approached on rising temperature, the properties of low-quartz begin to change relatively rapidly some $50^\circ C$ or more below the inversion point, in contract to the behavior on cooling high-quartz down toward the inversion point. There is an abrupt change in nearly all the physical properties at the inversion point. The changes are reversible and none of the properties are permanently altered after cycling back through the inversion.

As outlined in our previous discussions, the Babcock primary phase substructure postulate deals specifically with relations between glass properties, compositions, and structures for silicate glasses under temperature equilibrated conditions. Clarification of property-structure relations in vitreous silica as they are affected by changes in temperature and pressure presents quite different problems. However in the author's view these phenomena can be reasonably related to substructures associated with the high-low temperature inversions in crystalline SiO_2. It is believed that they enhance the validity of the substructure postulate and extend the area of its usefulness. Results of principal publications that reinforce this opinion are outlined as follows.

Rinne (1914) measured refractive indices of vitreous silica for four visible frequencies of helium in the temperature range from -160 to $1000^\circ C$. Sosman (1927) recalculated Rinne's values to absolute indices and observes that "In the three curves which are carried to $1000^\circ C$, there is an easily observed irregularity in the vicinity of $600^\circ C$, which is reminiscent of the less marked irregularity in the dilatation of vitreous silica in this same region, and which is suggestively near the inversion temperature of quartz." As noted earlier, this high-low inversion for quartz is $573^\circ C$. Lebedev (1921) reported that glasses high in SiO_2 exhibited breaks in the refractive index curves between 520 and $590^\circ C$.

Babcock, Barber, and Fajans (1954) reviewed data on density, thermal expansion, of elasticity moduli, and compressibility of vitreous silica. Douglas and Isard (1951) observed measurable differences in density at room temperature caused by quenching the glass from temperatures ranging between 1000 and

1500°C. The equilibrium state was established at each temperature before the quench. The density of vitreous silica quenched from 1500°C was greater than densities of glasses quenched from lower equilibrium temperatures. They found that the time necessary to reach these equilibrium states increased exponentially as the equilibrium temperatures decreased. Equilibrium times were a few seconds at 1500°C, approximately one month at 1000°C, and were estimated to be several thousand years at room temperature. Density data obtained at room temperature were considered to be characteristic of the frozen-in states, and the relatively fast density measurements left the states uncharged. This behavior cannot be understood in terms of a single continuous network. Coefficients of thermal expansion of vitreous silica reported by Sosman (1927) are considerably lower than those for quartz over a wide temperature range. Coefficients of expansion of vitreous silica become negative below -73°C and remain at low positive values up to about 1200°C. This behavior is also unique among most substances. Young's modulus and the rigidity modulus increase with temperature from room temperature to 700°C; both reach maximum values between 700 and 950°C. This may be related to the minimum in coefficient of thermal expansion between 625 and 825°C. Calculated compressibilities of vitreous silica from data published by Birch and Dow (1936) showed unusual effects of temperature and pressure. Compressibility decreased with increasing temperature to 400°C at five different pressures. Compressibility increased at all temperatures as pressures were increased to high values. All the above-mentioned properties are related to volume, and their unique behaviors are mutually consistent. Babcock et al. suggested that the rapid and reversible structural changes during property measurements were analogous to similar changes that take place in high-low temperature inversions of crystalline SiO_2. For example, they noted that the high-low inversion of cristobalite occurs in seconds at 270°C. These anomalies may be rationalized in terms of the presence of two or more different ionic arrangements in vitreous silica, assuming that each arrangement behaves "normally" the relative amounts of each change continuously with temperature and pressure.

Owen and Douglas (1959) reported a sharp maximum in the dielectric constant of vitreous silica at 570°C, which is near the high-low temperature inversion of quartz at 573°C. Doborzynski (1938) found a minimum in the dielectric constant of vitreous silica at -73°C, which is just at the temperature at which the coefficient of thermal expansion becomes negative. Deeg (1957) found discontinuities in internal friction data on vitreous silica in temperature regions where the crystalline forms of SiO_2 undergo phase changes. Westbrook (1960) observed an inflection in the microhardness-temperature curve at 573°C in vitreous silica made by fusion of quartz. Vitreous silica made from cristobalite showed an irregularity in microhardness around 150°C, which is near the temperature at which cristobalite undergoes a high-low temperature inversion.

Vitreous silica made by the synthetic $SiCl_4$ process showed an inflection near $573°C$, even though fusion of quartz was not involved.

Bruckner (1970, 1971) gave a most comprehensive and detailed coverage of the anomalous properties of vitreous silica and their structural implications. He condensed the extensive research of the past few decades into 93 pages and furnished 243 references. His review includes the following properties and phenomena: electrical conduction and dielectric loss, refractive index and optical absorption, elastic and internal friction behavior, volume and pressure effects, viscosity, thermal expansion, heat capacity, heat conduction, relaxation behavior, and effects of irradiation on properties. Bruckner sums up his review by observing that

> Extensive agreement is found between different physical properties . . . which show anomalies occurring near the transformation points of the crystalline modifications and which also indicate, to a certain degree. the structural heterogeneity of this simple one-component glass. This heterogeneity is related to a variety of structural arrangements of silicon tetrahedra for which a certain probability appears to predominate, that a preference of those arrangements exists which are related to some extent to the crystalline modifications.

He further observes that unanswered questions remain on the properties and structures of vitreous silica.

Vukcevich (1972) gives results of his careful study of various proposed relations between properties and structures of vitreous silica. He observes that every theory suggested up to now is based on one of the following two premises.

1. Bending of the Si-O-Si angles in the SiO_4 tetrahedra is not only possible but is responsible for the anomalous behavior. Examples include the acoustic loss mechanism described by Anderson and Bommel (1955); compressibility treated by Smyth, Londeree, and Lorey (1953); and thermal properties covered by Smyth, Skogen, and Harsell (1953), Smyth (1955), and Gaskell (1966).

2. The structure of vitreous silica is metastable and may exist in two or more equivalent states, at least locally. Examples are the equivalent positions of the Si-O-Si angle discussed by Anderson and Bommel (1955); two oxygen positions in elongated bonds investigated by Strakna (1961); and coexisting subgroups examined by Babcock, Barber, and Fajans (1954).

Vukcevich concludes that these two premises may well become important parts of the ultimate theory of vitreous silica. On this basis, he proposes a new model to explain the anomalous properties of vitreous silica. Principal features of his proposal can be outlined as follows. We have used the easily understandable terms of low and high temperature modifications of quartz, tridymite, and cristobalite following the practice of Sosman (1965). Vukcevich uses α and β in referring to low and high temperature forms. He notes that low temperature α forms of quartz, tridymite, and cristobalite have higher densities than the high temperature β forms. Experimental evidence indicates that density changes between α and β forms are caused by changes in the Si-O-Si angle between neighboring tetrahedra, referred to simply as the oxygen angle. The oxygen angle gives a measure of the relative orientation of two neighboring tetrahedra. Present x-ray diffraction data show that the Si-O distances do not change significantly during the reversible α-β transitions. The fact that α and β forms have two equilibrium densities at different temperatures indicates two minima in the potential energy curve. This property can be visualized in a plot of potential energy versus oxygen angle with minima for α and β forms. This curve is characteristic of second-order solid state transformations and can be compared to Figure 1.1, which depicts first-order transformations. Reversible transitions between α and β forms would be expected to be gradual, not sharp at one definite temperature. This accepted interpretation is in agreement with this transition in quartz and with the gradual transition in cristobalite, which extends over a temperature region of about 40°C. The author points out that these transitions are characteristic of those in vitreous silica noted earlier. Vukcevich proposes that an energy barrier exists between the α and β states. Upon heating the α phase, enough thermal energy must be available to jump the barrier to the β state. Following the proposed angular potential, the structure of vitreous silica can be interpreted as an essentially random network of SiO_4 tetrahedra between which the oxygen angles have a definite preference for either of the two characteristic oxygen angles ϕ_α and ϕ_β. Vukcevich points out that the oxygen angles in vitreous silica are characterized by a broad distribution of angles from $120°$ to $180°$, with the most probable values being near $145°$ as shown by Mozzi and Warren (1969). He postulates a corresponding distribution or spectrum of energy states and discusses them in some detail. Vukcevich describes his model in comprehensive mathematical terms and shows that it reasonably accounts for data on volume, compressibility, thermal expansion, and specific heat. He concludes that although his work offers good guidelines, it leaves room for further improvements. He recommends further experimental and theoretical research and outlines some directions in which it should be pursued.

Bruckner and Vukcevich conclude from their separate reviews that a variety of structural arrangements exist in vitreous silica which are related to their crystalline modifications. This conclusion is somewhat similar to relations between substructures in silicate glasses and substructures in their respective primarly crystalline phases as proposed by Babcock (1968). It is of interest to compare the principal features of the author's proposal with the one advanced by Vukcevich. Earlier discussions in this chapter demonstrate that a number of property-composition relations in silicate glasses are uniquely related to the primary crystalline phase fields in which they are located. These categorical differences led to the assumption that different ionic substructures existed in the different phase fields. For example, consider glasses located in the contiguous SiO_2 (quartz) and $Na_2O \cdot 2SiO_2$ (sodium disilicate) primary phase fields illustrated in Figure 4.3. Substructures of glasses and their crystalline primary phases in the two fields or areas are qualitatively different. Going from the quartz field to the sodium disilicate field requires changes in the number and rearrangement of strong primary Si-O bonds. These changes call for relatively large amounts of energy and are termed first-order reactions. Thus the substructure method involves categorical differences between primary phase fields in terms of first-order reactions of Si-O bonds and their spatial arrangements. Though the substructure method is very useful for glass technology purposes, it gives us no information on details of the different substructures. Vukcevich gives details of quartz, tridymite, cristobalite, and vitreous silica structures in terms of oxygen angles and their associated energy states. Changes in structures between α and β modifications require relatively small amounts of energy and are termed second-order reactions.

Taken together, the works of Vukcevich and the author may be considered as useful guidelines for further research in silicate glasses. Both approaches leave many unanswered questions. Many crystalline silicates that can be melted into glasses have both stable and metastable forms that are dependent on both temperature and pressure. Sodium disilicate easily comes to mind as an example. Could relations between the structures of crystalline $Na_2O \cdot 2SiO_2$ and a glass of this composition be clarified by using the method of Vukcevich? Table 4.4 demonstrates that specific volume data can be used to distinguish between substructures in the tridymite and quartz primary phase fields. Structures of glasses in the two phase fields can be reproduced by annealing procedures, but one may ask how α and β modifications may influence oxide constants in the descriptive equations. Property constants in substructure equations reflect particularly the contributions of oxides in stable phases. Influences of second-order metastable phases on oxide contributions to properties cannot be evaluated by the substructure method. Although they are thought to be relatively small, they are nevertheless essential to an understanding of silicate glass structures. Answers to these and other questions, which must come from materials scientists are pertinent for future advancements in glass technology.

II

Electromagnetic Properties

I have a paper afloat, with an electromagnetic theory of light, which, till I am convinced to the contrary, I hold to be great guns.

James Clark Maxwell (1831-1879)

Electromagnetic properties of materials are integrated into one comprehensive subject in terms of Maxwell's electromagnetic theory, which distinguishes two principal types of material; namely, dielectrics and conductors. Glasses are highly ionized materials, unique in their ability to act simultaneously as dielectrics and conductors, depending on temperature and the imposed electrical fields. They are dielectrics at room temperature but become more and more conductive as the temperature is raised. These characteristics derive from their electronic and ionic microstructures, which contribute to their behavior as supercooled liquids.

Silicate glasses are intrinsically polarized materials; the bond or charge polarization is caused by interactions of electrons and ions as noted in Part I. The imposition of external electromagnetic fields results in additional polarization of glasses. It is important to distinguish between the intrinsic charge polarization and the wave polarization due to external force fields. Imposition of either electrical or mechanical force fields can result in wave polarization. Isotropic glasses become birefringent when subjected to uniaxial mechanical pressure as in the stress-optical effect. Likewise the imposition of electromagnetic fields causes rotation of the plane of polarization as in the Faraday and Kerr effects.

The effects of imposed force fields on glasses may be either temporary or permanent, depending on the stress, the time of duration, the temperature of the glass, and the microstructure of the glass under consideration. Interactions between force fields and glasses may be conservative, leaving the sample unchanged; they may be dissipative with energy absorption; or the sample may be destroyed in the test. In the latter case dielectric breakdown and mechanical fracture are cataclysmic phenomena and cannot be described analytically. Dielectric and mechanical strengths depend not only on material constants but also on the size and shape of the sample or product, the rate of stress application, and other test boundary conditions. This complex subject is outside the scope of the book. Only studies on the strength of materials can help to determine what part of the strength depends on the material and what parts on test circumstances.

5

Dielectric Properties

Glasses are rather ideal dielectric materials and are widely used in modern technology. Their dielectric properties are dependent on glass composition and on frequency, temperature, and glass microstructures. Published property data are not sufficient to comprehensively describe these relations. Isolated sets of data can be used to indicate relations between properties, compositions, frequencies, and temperature. Further research on systematically arranged glass compositions is needed to clarify information on dielectric properties. Relations between macroscopic dielectric properties and glass microstructures continue to be investigated.

5.1 EFFECTS OF STATIC ELECTRICAL FIELDS

Further polarization of glasses occurs if they are subjected to the action of externally applied electrical fields. Oxide glasses (e.g., the silicates) are particularily susceptible to such polarization because of the high polarizability of oxygen anions. The polarizabilities of the cations vary, being least for silicon, intermediate for dibasic cations like calcium, and greatest for such monovalent cations as sodium. The total polarization can be indicated by the general equation

$$P = P_e + P_i + P_o \tag{5.1}$$

where P_e relates to displacement of electrons, P_i to that of ions, and P_o to orientation of ionic groups in the lattice framework or substructure. The latter is usually referred to as dipole orientation. Equation (5.1) oversimplifies the process; the separate effects of the electrical field on the glass cannot be distinctly isolated.

The static electrical field and its polarization effects have the same directions in isotropic materials such as glasses. The integrated polarization is proportional to the magnitude of the field and can be expressed by

$$P = \epsilon E \tag{5.2}$$

where ϵ the electric susceptibility is practically constant for steady fields. Displacements are related to the field by

$$D = (1 + 4\pi\epsilon)E \qquad (5.3)$$

If we put $k = 1 + 4\pi\epsilon$, the relation between displacement D and the field E simplifies to

$$D = kE \qquad (5.4)$$

where k is the specific inductive capacity or dielectric constant. Since D, E, and P have the same physical dimensions, ϵ and k are pure numbers. The dielectric constant of a vacuum is 1.

The effects of a static field on a glass depend on glass composition and temperature. Glass behaves as a solid dielectric at room temperature and as an electrolytic conductor in the molten condition. When glass acts as the dielectric in a condenser, the following sequence of events take place when a static field is imposed. The field produces a surge of electric charge from one plate to the other, and the charging current reduces exponentially to a steady value, usually in a matter of seconds. If the charging process is then interrupted and the condenser short-circuited, a discharge current of reverse sign is observed. The discharge current resembles the charge current both in time and in magnitude. The charge and discharge current characteristics depend on the composition in terms of its electronic and ionic structure. The time delays are results of the polarization of the glass.

If a static field is impressed on molten glass, electrolytic conduction occurs; the ions acting as electrical carriers. In this case the dielectric properties are negligible. At intermediate temperatures glass acts simultaneously as a dielectric and a conductor.

5.2   EFFECTS OF CYCLIC ELECTRICAL FIELDS

The majority of dielectric applications involve cyclic or alternating fields in which the reversibility of polarization is time dependent. Reversibility depends on whether there is sufficient time available for electronic, ionic, and ionic group displacements to return to their original positions after the field is removed. The cyclic field can be expressed by

$$E = E_m \sin \omega t \qquad (5.5)$$

where E_m is the maximum field, t is time, and ω is the angular velocity of field

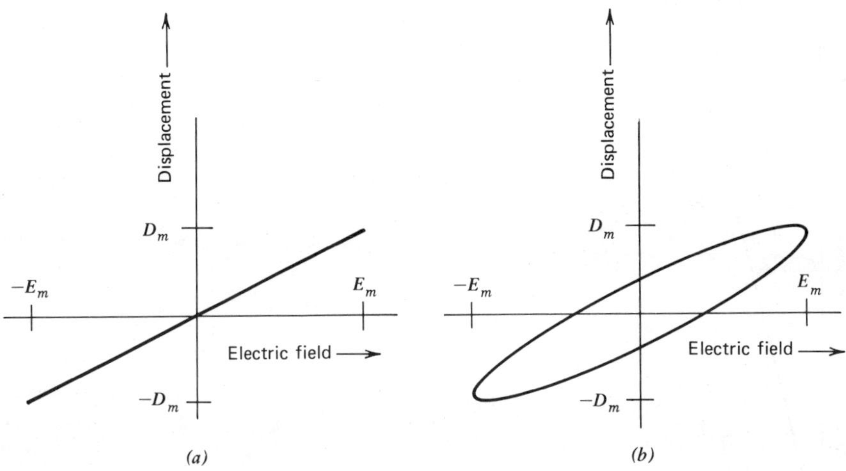

Figure 5.1 E and D cycle. (*a*) Loss free. (*b*) High loss.

reversal. If there were no time delays, the electrical displacement would be

$$D = D_m \sin \omega t \tag{5.6}$$

In this case D and E are in phase and the loss-free condition is illustrated in Figure 5.1 *a*. If there is a time lag in polarization, the displacement becomes

$$D = D_m \sin(\omega t - \delta) \tag{5.7}$$

where δ is called the loss angle. The area of the hystersis loop in Figure 5.1 *b* represents the accompanying energy loss per cycle. Energy losses occur when the period of cycling approximates the various time delays of electrons, ions, and ionic groups in the glass structure. In high frequency applications where δ (radians) has low values, sin δ may be replaced by tan δ. In these terms tan δ is called the loss tangent and k tan δ the loss factor. The loss tangent is also called the power factor.

The dielectric constant k can be separated into the in-phase component k' and the out-of-phase component k'' in terms of tan δ as illustrated in Figure 5.2, where tan $\delta = k''/k'$. Van Vlack (1970) uses the out-of-phase component to analyze the power loss as a function of frequency by starting with the relation

$$\alpha = \alpha_0 / 1 + i\omega\tau \tag{5.8}$$

where α_0 is the static polarizability, α is the complex polarizability for cyclic

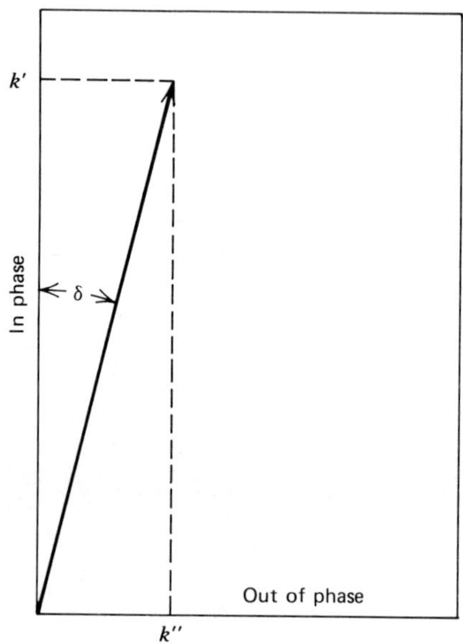

Figure 5.2 Loss tangent $\delta = k''/k'$, where k' and k'' are the real and imaginary parts of the complex dielectric constant.

fields, ω is the angular velocity, and τ is the dielectric relaxation time. The dielectric relaxation time τ is equal to $1/\omega_0$ at the natural or resonant frequency ω_0. Remembering that i represents $\sqrt{-1}$. it can be shown that

$$k = k' - ik'' \qquad (5.9)$$

where k' and k'' are, respectively, the real and imaginary components of the dielectric constant.

Figure 5.3 shows the effects of frequency on the dielectric constant k and the loss factor $k'' = k \tan \delta$. Three special conditions can be distinguished.

1. At low frequencies where ω is much smaller than ω_0; $\omega\tau$ and k'' approach zero and $k' = k$. There are negligible losses in this case, since the out-of-phase component $\tan \delta$ and the loss factor $k \tan \delta$ drop to zero.

2. At high frequencies where ω is much larger than ω_0; k' approaches 1 and k'' approaches 0. Likewise k''/k' and the loss factor $k \tan \delta$ become very small, and little energy is absorbed.

3. Near resonant frequencies ω is close to ω_0 and $\omega\tau$ approaches 1. Power losses are at a maximum when k'' is maximum and $\omega\tau = 1$.

As discussed in Section 3.5, the measured macroscopic properties of glasses reflect the integrated effects of the microscopic constituents of the glass structure.

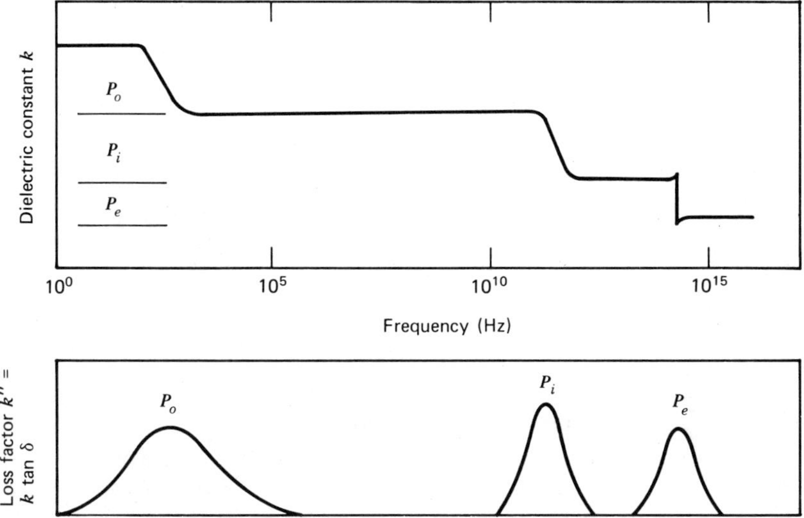

Figure 5.3 Effect of frequency on dielectric constant and loss factor.

Thus it is evident that Figure 5.3 gives an idealized and categorical description of the effects of a cyclical electrical field on the dielectric properties of glasses. However some clarification of property-structure relations is achieved by relating certain glass properties to the separate effects of electrons, ions, and ionic groups. Such relations are qualitative, and the reader will appreciate that this approach is oversimplified and leaves much to be desired.

5.3 DIELECTRIC LOSSES AND MICROSTRUCTURES

Dielectric theory describes the effects of cyclic or periodic electrical fields on materials and encompasses the complete electromagnetic spectrum. Common usage has led to discussion of dielectric and optical properties as two separate subjects. This procedure is followed here, but it must be kept in mind that such a separation is one of convenience. Dielectric losses in glasses can be qualitatively related to the presence of ions and ionic groups in the frequency range where glasses are used as insulators. Optical properties relate generally to the presence of electrons and are discussed in later chapters.

Equation (5.9) shows the dielectric constant to be a complex quantity. The real and imaginary components k' and k'' relate to the dielectric constant and dielectric absorption, respectively. All the wave aspects of dielectrics, including optics, involve complex quantities and are described by various types of wave equations. The real component represents the real velocity that is propagated with constant amplitude. The imaginary component represents evanescent or transitory wave propagation and is associated with absorption. When glasses are used as insulating materials these absorptions result in dielectric losses, usually in the form of heat. When glasses are used for transmission optics these absorptions result in luminescence, color, and so on. Both types of absorption result in relaxation phenomena involving τ; the relaxation time of the process.

Stevels (1957) described dielectric absorption in glasses in terms of four postulated constituents of their microstructures. Following the suggestion of Owen (1963), the second type is called "ionic relaxation losses" because it is more descriptive than "dipole losses." Stevels considers dielectric losses in glasses to consist of an integrated spectrum of losses caused by the following categorized mechansims.

1. Conduction losses. These losses, which are related to the movement of ions, are low at room temperature and account for only a part of the total loss. Conduction losses are important at low frequencies and increase rapidly with temperature. The conduction current is in phase with the applied voltage, and as a rule the dielectric loss is independent of other absorption losses.

2. Ionic relaxation losses. These losses relate the bonding forces of ions in the structure to the frequency of the applied field. Losses shift to lower frequencies as the temperature is decreased. Generally these losses in glasses occur at frequencies less than 10^6 Hz. These losses depend not only on the mobile ions, which contribute to conduction losses, but also to relatively immobile ions in the glass substructure.

3. Deformation losses. Stevels (1957) proposed that these losses arise from small deformations of the glass network or, as the author would say, from deformation of the glass substructure. The precise nature of this deformation is not known. Deformation losses in vitreous silica are very low because of the high strengths of its silicon and oxygen bonds. The addition of other cations to SiO_2 results in increased losses because of reduction of the number of Si-O bonds. Maximum deformation losses are believed to occur near 10^{13} Hz, although at this frequency they are submerged in the vibration losses.

4. Vibration losses. It is expected that all components of the glass structure will vibrate under the action of a cyclic or periodic electrical field. Resonances may occur in the frequency range from about 10^8 to 10^{13} Hz. It is believed that lattice or substructure vibrations in this range are related to this type of dielectric loss, the infrared absorption edge, and specific heat as visualized by Debye. Ions of different masses and restoring forces will vibrate with characteristic frequencies depending on their locations and bonding in the glass microstructure.

Stevels (1957) showed a graphic generalized spectrum of losses in glasses in the frequency range from one to 10^{14} Hz and at temperatures of 50° and 300° Kelvin. This graphic plot resulted from consideration of some experimental data, their interpolations, and reasonable inferences. It is useful in sorting out categorical aspects of the problem of dielectric loss.

5.4 PROPERTIES OF COMMERCIAL GLASSES

Glasses are unique among solid dielectrics because of their range of more than 4:1 in dielectric constant and their wide range in dielectric losses over 14 decades of frequency. These ranges are of particular importance when it becomes necessary to match the dielectric properties of other materials in dielectric devices. The adaptability of glasses to a variety of manufacturing processes makes possible an extensive variety of useful dielectric products. Glass manufacturers thus provide the design engineer with products that enable him to solve most of the problems encountered in the storage and control of electrical energy. Space does not permit discussion of available commercial glasses. However data on representative glasses illustrate the extent of possible ranges in glass composition, frequency, and temperature.

Table 5.1 presents the effects of frequency on dielectric constant and loss tangent for selected glasses made by the Corning Glass Works. Vitreous silica glass 7940 is made by the synthetic aqueous $SiCl_4$ process and contains appreciable water in the form of OH ions. Vitreous silica has the lowest dielectric constant and loss tangent of all the silicate glasses, although these property values vary appreciably with the manufacturing process. Its property changes with frequency are also lowest of the silicate glasses. The data in Table 5.1 cover a small frequency range, but commercial glasses are available in the frequency range from zero to 10^{20} Hz and higher. It is usual for the dielectric constants of glasses to decrease with frequency, but we cannot generalize about the effect of frequency on dielectric loss. The loss tangent of glass 0080 in Table 5.1 shows a minimum value at a frequency of 10^7 Hz, whereas the values for the other glasses increase with frequency. Morey (1954) reviews other data giving minima in loss tangent. Our inability to explain the effect of frequency on dielectric absorption becomes understandable in terms of our discussion in Section 5.3.

Table 5.2 shows the effects of temperature on dielectric properties of selected Corning glasses. There are exceptions to the generalization that both dielectric constant and loss tangent increase with temperature. The dielectric constants of glasses 7940 and 7911 are unchanged over the indicated temperature range, but those of the other glasses increase with temperature. The loss factor of 7940 decreases, then increases with temperature, while the others steadily increase with temperature. The published literature shows examples of both minima and maxima in the dielectric properties as the temperature is changed. Section 4.6 noted that Owen and Douglas (1959) reported a sharp maximum in the dielectric constant of vitreous silica at 570°C, which is near the high-low quartz inversion

TABLE 5.1   Effect of Frequency on Dielectric Properties of Selected
Commercial Glasses

	Composition (wt %) of four glasses			
Oxide	7940	0080	7570	8363
SiO_2	99.9	73.0	3.0	5.0
H_2O	0.1			
Al_2O_3		1.0	11.0	3.0
B_2O_3			11.0	10.0
MgO		4.0		
CaO		5.0		
Na_2O		17.0		
PbO			75.0	82.0

Frequency (Hz)
at which values
of k and
tan δ obtained

		7940	0080	7570	8363
10^6					
k		3.78	6.90	14.52	19.0
tan δ	$\times 10^4$	0.1	100	19	27
10^7					
k		3.78	6.82	14.50	19.0
tan δ	$\times 10^4$	0.1	85	24	37
10^8					
k		3.78	6.75	14.42	19.0
tan δ	$\times 10^4$	0.3	90	33	57
3×10^9					
k			6.71		17.8
tan δ	$\times 10^4$		126		124
10^{10}					
k		3.78	6.71	14.20	
tan δ	$\times 10^4$	1.7	170	98	

Measurements at $25°$C.

sources. Hutchins and Harrington (1966); *Corning Glass Works Bulletin.*

at $573°$C. Doborzynski (1938) reported a minimum in the dielectric constant
of vitreous silica at $200°$K ($-73°$C), which is just at the temperature where the
coefficient of thermal expansion becomes negative. Stevels (1957) shows in his
review that maxima and minima are quite prevalent in loss tangent-temperature
curves in glasses. Here again the complexity of these phenomena as shown in
Section 5.3 rules against assigning reasonable causes for the measured data.

TABLE 5.2　Effect of Temperature on Dielectric Properties of Selected Commercial Glasses

Oxide	Composition (wt %) of four glasses			
	7940	7911	7070	1990
SiO_2	99.9	96.0	71.0	41.0
H_2O	0.1			
Al_2O_3		0.5	1.0	
B_2O_3		3.5	26.0	
Li_2O			0.5	2.0
Na_2O			0.5	5.0
K_2O			1.0	12.0
PbO				40.0
Temperature ($^\circ$C) at which values of k and $\tan \delta$ obtained				
100				
k	3.76	3.56	4.0	7.95
$\tan \delta$　$\times 10^4$	1.3	7	20	42
200				
k	3.76	3.56	4.0	7.95
$\tan \delta$　$\times 10^4$	0.9	8	23	42
300				
k	3.76	3.56	4.0	7.95
$\tan \delta$　$\times 10^4$	0.7	10	29	43
400				
k	3.76	3.555	4.0	8.40
$\tan \delta$　$\times 10^4$	0.7	13	40	300
500				
k	3.76	3.55	4.2	9.7
$\tan \delta$　$\times 10^4$	0.9	17	70	1000

Measurements at 10^{10} Hz.

sources.　Hutchins and Harrington (1966); *Corning Glass Works Bulletin.*

Principal efforts have been directed toward developing low loss glasses for use in products, and reference is made to Volf (1961) on this subject. The dielectric properties are dependent on glass composition, but the existing data are not sufficient to permit development of quantitative effects of composition. It is possible, however, to assign directional effects of the oxides and by experimental research to develop glasses for special dielectric uses.

5.5 PROPERTIES OF EXPERIMENTAL GLASSES

Publications on measurement methods and experimental data on dielectric properties include those of Littleton and Morey (1933), Morey (1954), Von Hipple (1954a, 1954b), Sharbaugh and Roberts (1959), Owen (1963), and Eitel (1965). Further reference is made to publications listed in the Appendix. These and other references available to the reader provide sufficient information on measurement methods. However experimental property data are meager and are not arranged for easy interpretation by materials scientists and technologists. Subsequent discussions have outlined the basic theory of dielectric materials, emphasizing the complexity of the various factors involved in the dielectric properties of glasses. Existing data can be interpreted in terms of more than one assumed mechanism for these processes. This is especially true for dielectric absorption and the related relaxation phenomena. Further research is needed to establish meaningful relations between macroscopic property data and details of glass microstructures. Such research requires property measurements on glass compositions arranged so that the separate effects of each glass constituent can be quantitatively evaluated. It is suggested that compositions be expressed in either mole fractions or mole percentages for ease of interpretation by different investigators.

Published data on the effects of frequency, temperature, and glass composition on dielectric constant and dielectric loss are not extensive enough to account for these relations. This section reviews sets of experimental data in the effort to clarify some of these relations. The data are evaluated in terms of the substructure postulate, which relates properties and compositions in terms of linear equations of the form (4.2). Although this method gives relations with compositions located in single primary phase fields, it is extended here to groups of glasses located in more than one phase field for purposes of comparison. This extended use amounts to representing the data in terms of the random network theory following equations of the form (4.1), except that compositions used here are in mole fractions rather than weights percent. This extended use gives values of the given glass oxides which are averaged over the different primary phase fields. Glass compositions in the literature are arranged without regard to their locations in primary phase fields. Thus in many cases we must treat the glasses as simple solutions of oxides, being satisfied with average contributions of the oxides as indicated by their respective constants in the equations. Appen (1956) gives oxide constants for dielectric constant and other glass properties on this basis. Such oxide constants are useful in estimating properties of glasses, but experience shows that they do not represent the data as well as those obtained when all the glasses in the group are located in one phase field. This is particularly evident when the glasses cover extended composition areas over several primary phase fields.

TABLE 5.3   Dielectric Constant k and tan δ for Na_2O-TiO_2-SiO_2 Glasses in $Na_2O \cdot 6TiO_2$ Primary Phase Field

Oxide	Values for five glasses				
	7	11	12	16	17
SiO_2	0.468	0.395	0.362	0.337	0.300
TiO_2	0.302	0.314	0.358	0.341	0.400
Na_2O	0.230	0.291	0.280	0.322	0.300
k_{meas}	11.30	11.80	12.60	12.50	13.10
k_{calc}	11.32	11.87	12.45	12.46	13.19

$$k_{calc} = 5.470\, SiO_2 + 20.128\, TiO_2 + 11.662\, Na_2O$$

	7	11	12	16	17
tan δ_{meas}	0.0160	0.0200	0.0180	0.0200	0.0200
tan δ_{calc}	0.0158	0.0200	0.0187	0.0219	0.0196

$$\tan \delta_{calc} = 0.00281\, SiO_2 - 0.00874\, TiO_2 + 0.07431\, Na_2O$$

Measurements at 0.96×10^{10} Hz and $29°C$.

Data source.   Hirayama and Berg (1961).

Na_2O-TiO_2-SiO_2 Glasses.

Hirayama and Berg (1961) supply data on the effects of composition, frequency, and temperature on dielectric constant and loss tangent for glasses in this system. Their glasses lie in the seven phase fields described by Hamilton and Cleek (1958) and Babcock (1973). Table 5.3 gives data on five of the glasses located in the $Na_2O \cdot 6TiO_2$ phase field with computerized equations representing the data. Average differences between measured and calculated values for k and tan δ of 0.075 and 0.0003, respectively, indicate satisfactory representation of the data. Data on nine other glasses located in the other six phase fields can be represented by the following equations:

$$K = 4.610\, SiO_2 + 22.515\, TiO_2 + 10.951\, Na_2O \tag{5.10}$$
$$\tan \delta = 0.01040\, SiO_2 + 0.00146\, TiO_2 + 0.05388\, Na_2O \tag{5.11}$$

Average differences between measured and calculated values for these properties are 0.213 and 0.0015, respectively: higher than those for glasses in the $Na_2O \cdot 6TiO_2$ phase field.

Figure 5.4 indicates that dielectric constant and loss tangent for glass number 7 in Table 5.3 decrease rapidly with increases in frequency. Data on the effect

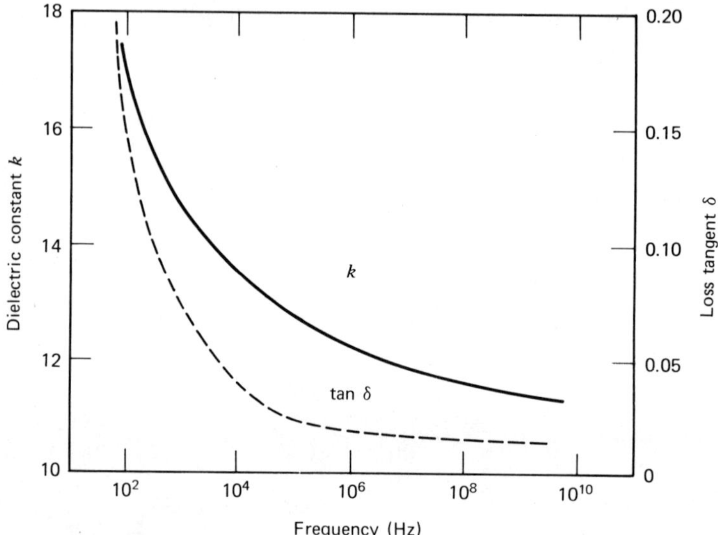

Figure 5.4 Effect of frequency on dielectric constant and loss tangent on glass 7 in Table 5.3 at 29° C. Tabular data from Hirayama and Berg (1961).

of frequency on dielectric properties for glasses in the $Na_2O \cdot 6SiO_2$ phase field can be represented by linear equations (Table 5.4). The relatively high values of constants for Na_2O, 226.298 and 7.75664, at $10^{1.78}$ Hz or 60 cycles reflect the high conduction losses caused by ionic movement as discussed in Section 5.3. It is well known that conduction losses in glasses decrease with increases in frequency. Conduction losses, as noted in Section 5.3, are caused by ionic movement, particularly by monovalent cations such as Na_2O. Changes in values of the constants for Na_2O with increases in frequency are in good agreement with this generalization. It is noted that constants for SiO_2 increase regularly as the frequency is increased.

Hirayama and Berg (1961) published data on electrical resistivity on four glasses in the $Na_2O \cdot 6SiO_2$ phase field at 29°C. Their data, expressed in terms of $\rho \times 10^{-9}$ ohm-cm, can be represented by the equation

$$\log_{10}(\rho \times 10^{-9}) = 6.87\, SiO_2 + 5.96\, TiO_2 - 14.09\, Na_2O \qquad (5.12)$$

Figure 5.5 plots calculated values for electrical resistivity and dielectric constant at frequencies of $10^{1.78}$ and 0.96×10^{10} Hz for the three indicated glass compositions. These data reveal that electrical resistance (the inverse of condutivity) has controlling influence on the dielectric constant at the low frequency and little

TABLE 5.4   Effect of Frequency on Dielectric Properties of Glasses
in the $Na_2O \cdot 6TiO_2$ Primary Phase Field

Glass property = A SiO_2 + B TiO_2 + C Na_2O

Frequency (Hz)	A	B	C	Average difference from measured value	Difference (%)
Dielectric constant k					
$10^{1.78}$	-59.530	-22.605	226.298	0.21	0.8
10^3	-5.050	18.021	48.698	0.03	0.2
10^4	1.774	22.175	25.145	0.02	0.1
10^5	4.064	23.854	16.106	0.02	0.1
0.96×10^{10}	5.470	20.128	11.662	0.08	0.1
Loss tangent δ					
$10^{1.78}$	-2.19210	-1.86120	7.75664	0.0060	1.3
10^3	-0.49120	-0.45457	1.92108	0.0020	1.4
10^4	-0.15523	-0.14455	0.65930	0.0010	1.7
10^5	-0.04669	-0.06701	0.27380	0.0002	0.6
0.96×10^{10}	0.00281	-0.00874	0.07431	0.0003	1.5

Measurements at $29°C$.

Data source. Hirayama and Berg (1961).

influence at the high frequency. This is caused by the movement of Na cations
under the action of a cyclic electrical field, the ease of movement being greater
at the low frequency. It is noted that electrical resistivity and dielectric constant
are inversely proportional to each other at both frequencies.

The equation in Table 5.5 represents dielectric property data at five different
temperatures for glasses in the $Na_2O \cdot 6SiO_2$ phase field. Both properties increase
in value as the temperatures increase. Figure 5.6 presents the effect of temperature on these properties for glass number 7 in Table 5.3 at a frequency of 10^5
Hz. The increases in value of the constants for Na_2O reflect the increased ease
of movement of Na cations as temperatures increase. These glasses act as dielectric insulators at $29°C$ but take on more and more the character of electrolytic conductors as the temperatures are raised.

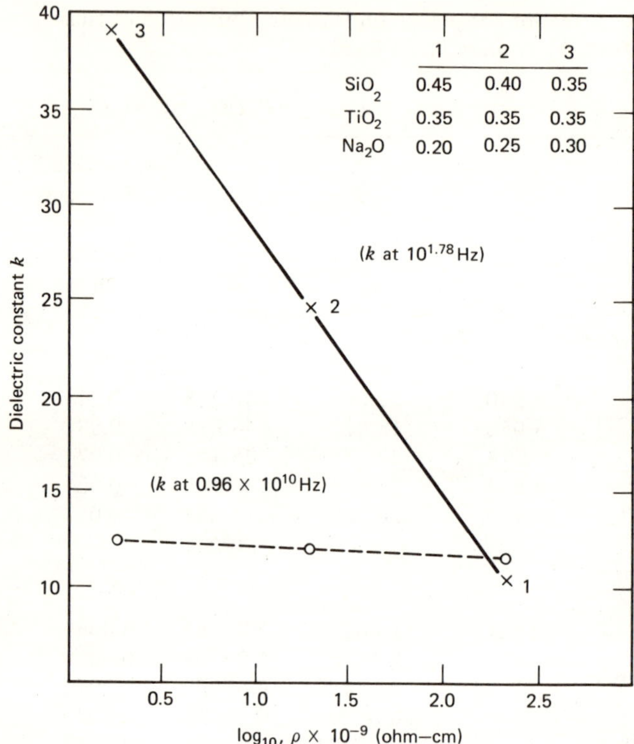

Figure 5.5 Calculated values of dielectric constant and electrical resistivity.

Na$_2$O-K$_2$O-PbO-SiO$_2$ Glasses.

Navias and Green (1946) published data on dielectric constant and loss tangent at room termerature and at frequencies of 10^{10} and 3×10^{10} Hz. Their 105 glasses incorporated 11 oxides combined in different ways. The glasses contained up to six oxides expressed in weights percent and are not arranged for easy interpretation. The equations in Table 5.6 represent their data for six glasses in the SiO$_2$ phase field of the above-mentioned system. It is noted that the constant for SiO$_2$ in the equation representing dielectric constant data has the same value as the measured dielectric constant of glass number 1 (vitreous silica).

Li$_2$O-Na$_2$O-K$_2$O-PbO-SiO$_2$ Glasses.

Data on three glasses containing Li$_2$O reported by Navias and Green were added to the group of six glasses in Table 5.6. Data on this group of nine glasses can be represented by the equations

TABLE 5.5   Effect of Temperature on Dielectric Properties of Glasses in the $Na_2O \cdot 6TiO_2$ Primary Phase Field

Glass property $= A \ SiO_2 + B \ TiO_2 + C \ Na_2O$

Temperature ($^\circ$C)	A	B	C	Average difference from measured value	Difference (%)
Dielectric constant k					
29	4.064	23.854	16.106	0.02	0.1
61	2.422	21.706	24.834	0.00	0.0
89	-1.678	20.825	37.443	0.02	0.2
125	-14.052	16.824	74.056	0.25	1.3
147	-45.459	-8.089	178.261	0.33	1.3
Loss tangent δ					
29	-0.04669	-0.06701	0.27380	0.0002	0.6
61	-0.14844	-0.15634	0.68156	0.0014	2.5
89	-0.37491	-0.30359	1.47174	0.0025	2.0
125	-1.09419	-0.81378	3.94648	0.0093	3.2
147	-1.98260	-1.64239	7.28985	0.0109	2.1

Measurements at 10^5 Hz.

Data source. Hirayama and Berg (1961).

$$k = 3.887 \ SiO_2 + 19.905 \ PbO + 16.755 \ Na_2O + 12.367 \ K_2O + 5.331 \ Li_2O \tag{5.13}$$

$$\tan \delta = 0.00158 \ SiO_2 + 0.02584 \ PbO + 0.07797 \ Na_2O - 0.01655 \ K_2O - 0.00626 \ Li_2O \tag{5.14}$$

Average differences between measured and calculated values of k and $\tan \delta$ are 0.098 and 0.0012, respectively. Although primary phase locations of the glasses are not definitely known, the values of oxide constants suggest that they are located in the SiO_2 primary phase field.

BaO-PbO-Al_2O_3-B_2O_3-SiO_2 Glasses.

Data on 17 glasses measured by Navias and Green at room temperature and 10^{10} Hz can be represented by the equations

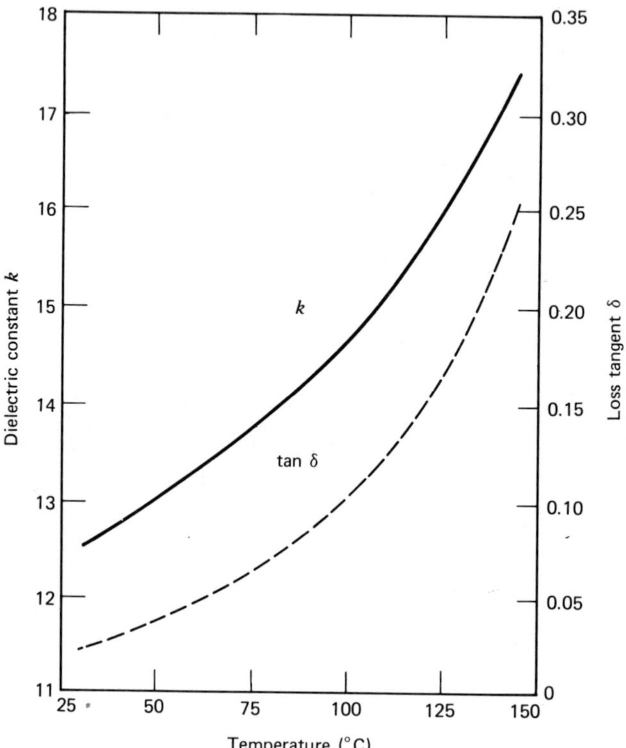

Figure 5.6 Effect of temperature on dielectric constant and loss tangent on glass 7 in Table 5.3 at 10^5 Hz. Tabular data from Hirayama and Berg (1961).

$$k = 3.816\ SiO_2 + 2.949\ B_2O_3 + 8.015\ Al_2O_3 + 20.046\ PbO \qquad (5.15)$$
$$+\ 16.773\ BaO$$

$$\tan\delta = 0.00083\ SiO_2 + 0.00080\ B_2O_3 + 0.00201\ Al_2O_3$$
$$+\ 0.02668\ PbO + 0.01768\ BaO \qquad (5.16)$$

Average differences between measured and calculated values of k and $\tan\delta$ are 0.119 and 0.0007, respectively. Values of oxide constants suggest that these glasses are in the SiO_2 phase field; confirming data are not available.

Na_2O-CaO-SiO_2 Glasses.

Moore and de Silva (1952) measured dielectric constant on a number of glasses at 10^3 Hz and at temperatures of 25, 50, and 100°C. Data on seven glasses in the $Na_2O \cdot 3CaO \cdot 6SiO_2$ phase field appear in Table 5.7. The average difference between measured and calculated values of k is 0.06.

TABLE 5.6   Dielectric Constant k and tan δ for Na_2O-K_2O-PbO-SiO_2 Glasses in SiO_2 Primary Phase Field

Oxide	Glass number					
	1	78	79	90	92	93
SiO_2	1.0000	0.7758	0.8006	0.6667	0.6722	0.7065
PbO		0.0482	0.0308	0.3333	0.2714	
Na_2O		0.0694	0.0665			0.2935
k_2O		0.1066	0.1001		0.0564	
k_{meas}	3.81	6.65	6.17	9.14	8.74	7.67
k_{calc}	3.81	6.61	6.21	9.14	8.74	7.69

$$k_{calc} = 3.806\ SiO_2 + 19.814\ PbO + 16.972\ Na_2O + 14.277\ K_2O$$

	1	78	79	90	92	93
tan δ_{meas}	0.0005	0.0069	0.0077	0.0112	0.0075	0.0240
tan δ_{calc}	0.0007	0.0074	0.0067	0.0104	0.0085	0.0241

$$\tan \delta_{calc} = 0.00074\ SiO_2 + 0.02983\ PbO + 0.08042\ Na_2O - 0.00210\ k_2O$$

Measurements at 10^{10} Hz and room temperature.

Data source. Navias and Green (1946)

TABLE 5.7   Dielectric Constant k for Na_2O-CaO-SiO_2 Glasses in $Na_2O \cdot 3CaO \cdot 6SiO_2$ Primary Phase Field

				Glass number			
Oxide	3	4	5	7	8	9	10
SiO_2	0.7188	0.7091	0.6993	0.7391	0.7398	0.7406	0.7286
CaO	0.1070	0.1070	0.1070	0.0963	0.0857	0.0751	0.1069
Na_2O	0.1742	0.1839	0.1937	0.1646	0.1745	0.1843	0.1645
k_{meas}	8.42	8.40	8.55	8.01	8.00	8.05	8.35
k_{calc}	8.35	8.45	8.55	8.10	8.05	7.99	8.26

$$k_{calc} = 5.022\ SiO_2 + 19.943\ CaO + 15.029\ Na_2O$$

Measurement at 10^3 Hz and $25°C$.

Data source. Moore and De Silva (1952).

Na_2O-Al_2O_3-SiO_2 Glasses.

Topping and Isard (1971) report data on six glasses in this system at frequencies of 9.1×10^9 (X) band) and 34×10^9 Hz (Q band) and at temperatures in the range 25 to 300°C. The compositions are arranged on the basis of Na_2O-xAl_2O_3 -$(5\text{-}2x)SiO_2$ with x values of 0, 0.25, 0.50, 0.75, 1.0, and 1.1. As x increases, SiO_2 is replaced by Al_2O_3 + Na_2O. This is an unsatisfactory way of composition representation, yet some qualitative information can be gained from the data, which are plotted graphically. At 25°C the data of Topping and Isard show that dielectric constant increases as x increases and the data curves at both frequencies have changes in slope when $x = 1.0$. Their loss tangent results at both frequencies indicate sharp maxima at the same composition. The weight percent composition of this glass is close to 52 SiO_2, 30 Al_2O_3, and 18 Na_2O, and this composition is near the boundary between the primary phases of $Na_2O \cdot Al_2O_3 \cdot 2SiO_2$ (carnegieite) and Al_2O_3 (corundum) (Levin, Robbins, and McMundie, 1964, Fig. 501). In the author's view, the changes in properties at the composition where $x = 1.0$ are caused by changes in glass substructures when compositions move from one primary phase field to another. The mole fraction composition of this glass is as follows: 0.60 SiO_2, 0.20 Al_2O_3, and 0.20 Na_2O.

Information on glasses in the six glass systems just cited illustrates the problems involved in determining quantitative relations between dielectric properties, glass composition, frequency, and temperature. It must be assumed that the quoted property measurements were made on homogeneous, strain-free samples. It is well known that dielectric properties are sensitive to heat treatment procedures used in sample preparation. All the compositions were theoretical; that is, they were calculated from the glass batches and were not chemically analyzed. This is not to question the precision of the data, but rather to emphasize requirements for glass samples. These data are representative of those found in the literature, which is lacking in dielectric property information. Demonstrated methods of evaluating data on glasses in the six systems indicate quantitative relations between glass compositions and the effects of frequency and temperature. There are indications that the separate contributions of oxides, in terms of constants in the linear equations, depend on the primary phase field in which the given glass is located. Examples given here are not sufficient to confirm the presence of primary phase substructures in silicate glasses as postualted by the author. This question must be subjected to further research on dielectric properties, and these efforts should be planned to ensure that groups of glasses located within given primary phase fields are compared with groups in adjacent fields, to test the validity of the substructure concept for dielectric properties.

5.6 DIELECTRIC RELAXATION IN GLASSES

The review of Owen (1963) on electric conduction and dielectric relaxation in glasses covers the results of principal investigators. The accompanying list of references treats developments up to 1963. Owen observes that we are still some distance removed from a quantitative understanding of even the simpler electrical phenomena in glasses. The uncertainty about the ultimate structures of glasses is a major obstacle, and data on dielectric relaxation can be made to fit almost any postulated mechanism. Though Owen generally follows the random network theory in his discussion, he states, "But while the random network theory may still perhaps be regarded as a good first approximation, there is now some reason to believe that there exist in glass rather larger ordered regions than had previously been supposed." This statement agrees with the author's view, expressed throughout this book, that these ordered regions are qualitatively related to more highly ordered regions in crystalline primary phases of given glasses.

More recent research on dielectric relaxation has placed greater emphasis on systematic experimental planning and more careful analysis of results. The research of Hakim and Uhlmann (1973) is a good example for investigators to follow in this respect. They measured dc conductivity and dielectric properties on six binary Cs_2O-SiO_2 and Na_2O-SiO_2 glasses; the compositions were expressed in mole fractions. The authors describe the preparation of samples, the experimental procedures, and their method of data analysis. They assumed that the total dielectric loss was equal to the sum of dc and ac losses. The ac loss was separately evaluated in terms of frequency. Their loss values are expressed in terms of k'', where $k'' = k' \tan \delta$. This follows the nomenclature used in Section 5.2.

Table 5.8 presents data on the effects of composition, frequency, and temperature on dielectric losses in Cs_2O-SiO_2 glasses. It is noted that values of k'' reach maxima that are dependent on frequency and temperature. The maxima increase as the temperatures increase and at increasing frequencies for both glasses. Comparisons of data for the two compositions suggest that losses in the high SiO_2 glass are less than those of the low SiO_2 glass, although direct comparisons cannot be made. The loss curves of these and other glasses measured are broader than those expected for single relaxation time behavior. The curves show a distinct asymmetry toward the high frequency side of the loss maxima. Table 5.9 gives relaxation times in seconds, as determined by Hakim and Uhlmann from their data. The ranges and maxima of relaxation times are approximated from their curves, which exhibit definite asymmetry toward the low τ values. There interpretation of the data indicated the existence of two distinct relaxation times τ_1 and τ_2. The computer program used was subjected to the condition that the maximum in the calculated dielectric loss coincided with the experi-

TABLE 5.8 Effects of Composition, Frequency, and Temperature on Dielectric Losses: $k'' = k'\tan\delta$

SiO$_2$ 0.85 and Cs$_2$O 0.15 glass

	Frequency (Hz)	k''		Frequency (Hz)	k''		Frequency (Hz)	k''
80°C	$10^{1.4}$	2.0	*125°C*	10^2	1.1	*175°C*	$10^{2.8}$	1.6
	$10^{1.6}$	*2.1*		$10^{2.5}$	2.7		10^3	2.3
	10^2	2.0		*10^3*	*3.4*		*$10^{3.7}$*	*4.3*
	10^3	1.4		10^4	2.2		10^4	4.1
	10^4	0.8		10^5	1.0		10^5	2.7

SiO$_2$ 0.95 and Cs$_2$O 0.05 glass

	Frequency (Hz)	k''		Frequency (Hz)	k''		Frequency (Hz)	k''
253°C	10^2	1.0	*301°C*	$10^{2.5}$	0.6	*411°C*	10^4	0.5
	$10^{2.5}$	1.6		10^3	1.7		$10^{4.5}$	1.0
	$10^{2.7}$	*1.7*		*$10^{3.5}$*	*2.0*		10^5	2.1
	10^3	1.6		10^4	1.7		*$10^{5.5}$*	*2.3*
	10^4	1.0		10^5	0.8		10^6	1.7
	10^5	0.5						

Values are estimated from graphical plots; maximum values are given in italics.

Data source. Hakim and Ulhmann (1973).

TABLE 5.9   Distribution of Relaxation Times for the High Si Glass
(SiO_2, 0.85, Cs_2O, 0.15)

Temperature ($^{\circ}$C)	Relaxation time (sec)		Calculated	
	Range	Maximum	τ_1	τ_1
80	10^{-4} - 10^{-2}	$10^{-2.5}$	1.5×10^{-2}	6.9×10^{-4}
125	10^{-5} - 10^{-3}	10^{-4}	7.4×10^{-3}	3.4×10^{-5}
175	10^{-6} - 10^{-4}	$10^{-4.5}$	1.3×10^{-4}	7.8×10^{-6}

Source of data and calculated values. Hakim and Uhlmann (1973).

mental maximum. Their calculated values of τ_1 and τ_2 appear in Table 5.9. It is noted that the relaxation times decrease rapidly as the temperatures increase and that they are all in fractional parts of a second. These data and those on seven other glasses studied by Hakim and Uhlmann show broad distributions of relaxation times. They note that other investigators have found similar broad distribution curves for dielectric losses in glasses and state: "The nature of such processes remains to be elucidated satisfactorily."

Although data on dielectric processes in glasses cannot be reasonably related to microstructure constituents, they clearly indicate the presence of a spectrum of loss mechanisms responding to cyclic electrical fields.

6

Refractive Index
and Dispersion

Data on refractive indices and dispersions form the bases of selecting commercial optical glasses for use in transmission optical systems. Refractive index controls reflection losses at interfaces between glasses and other materials. This chapter briefly covers the research and commercial manufacturing aspects of the development of optical glasses, which has led to international adoption of definitions and nomenclature for these properties. Representative information on available commercial glasses is related to some aspects of optical design. Extensive data on experimental glasses are discussed in some detail and related to the presence of primary phase substructures in glasses. Some examples of its use in formulating optical glasses indicate the merits of the substructure method. Attention is given to effects of hydrostatic pressure and temperature on refractive index in terms of precision optical systems.

6.1 DEVELOPMENT OF OPTICAL GLASSES

The first makers of microscopes and telescopes used glasses then being manufactured for ordinary glass products such as windows, bottles, tableware, and art glass. The types of glass available consisted of so-called crown glasses, characterized by low refractive indices and low dispersions, and flint glasses (containing PbO), which had high refractive indices and high dispersions. Optical designers had only these two general types until about the last quarter of the nineteenth century. Limited research efforts over the years by Newton, Harcourt, Fraunhofer, Stokes, Scheele, Lavoisier, and Faraday had indicated the feasibility of making special glasses for optical purposes. Some aspects of this historical development have been outlined by Hovestadt (1902), Wright (1921), Glaze (1953), Morey (1954), and Maloney (1968).

Hovestadt describes the research of Ernst Abbe and Otto Schott, which resulted in the manufacture of new and improved optical glasses of the required quality and sizes. This development was aided materially by Carl and Roderich Zeiss. Knowledge of the optical properties of oxides in glasses was limited to

SiO_2, CaO, Na_2O, K_2O, and PbO when Abbe and Schott began their research. Their studies on these and a number of additional oxides led directly to establishment of the Jena Glass Works, which published its first price list of optical glasses in 1886. This pioneering work established guidelines for glass research that have been followed by subsequent investigators. For example, Abbe and Schott used platinum melting crucibles and stirrers; a practice that has continued to the present time. During the same general period Chance Brothers in England and Parra-Mantois in France became active in optical glass manufacturing. However the work of Abbe and Schott was of greatest influence on development of glasses for optical and other technical uses because it was made public in the book by Hovestadt.

Early optical designers were handicapped by having to use either low index-low dispersion glasses or those having high index-high dispersion properties. Abbe and Schott, followed by subsequent investigators, made improvements in this situation. Morey (1939) and others introduced the use of rarer elements, particularly lanthanum and thorium, in glasses to achieve high index-low dispersion characteristics. Other elements, particularly fluorine, have been found to contribute important optical properties. Research efforts continue on refractive index-dispersion relations in glasses. Laboratories of optical glass manufacturers are especially active in this field.

Optical glass differs considerably from ordinary glass. These differences are spelled out in terms of tight specifications, which relate to different levels of perfection needed for the various intended uses. The precise manufacturing controls required include the ability to reproduce specified property values in subsequent meltings of given standard optical glasses. Optical glasses should have the following characteristics and properties.

1. They should be free of chemical inhomogeneities, which give rise to variable refractive index in terms of light scattering. These physical imperfections include those from incompletely melted materials from the glass batch or the melting container. Such flaws cause regions of variable refractive index and are referred to as stria, ream, or cord. Gas bubbles are objectionable especially if they lie in the plane of the real image.

2. They should be isotropic. Their preparation requires precise annealing procedures that minimize anisotropy in terms of low birefringence levels.

3. Their optical properties (e.g.) refractive index, dispersion ratios, and extinction coefficient should be specified.

4. They should be chemically stable in the environment in which they are to be used. This requirement includes resistance to attack by water vapor, carbon dioxide, and bacterial and plant organisms.

Wright (1921) describes optical glass manufacturing that was carried out in large refractory pots for a number of years. The melting, refining, and homogenizing of the glass was done in closed pots, which were indirectly heated. The pot with its contents was then cooled very slowly to room temperature, and the broken pieces of glass were inspected and segregated in terms of optical properties. Pieces close to the walls and bottom of the pot were discarded because of contamination by solution of materials in the pot. This wasteful and expensive process has generally been replaced by that using platinum melting crucibles or pots and by continuous melting in platinum or platinum-lined electrical furnaces. Another reason for the change in melting methods resulted from new optical glasses developed by Morey (1939) and others. These glasses, containing the oxides of lanthanum, thorium, and tantalum, required higher melting temperatures that dissolved the most refractory ceramic pots available. Eden (1961) has described the new melting methods as used by Schott and Genossen in Mainz, Germany. Corning, Pittsburgh Plate, Bausch & Lomb, and other manufacturers have developed their own continuous melting methods.

Advancements in optical glass making demonstrate the interdependency of research, technology, and manufacturing in making glasses available to the optical designer. Mutual appreciation by manufacturers and designers of their respective problems is necessary for continuing advances in this area.

6.2 DEFINITIONS AND NOMENCLATURE

When electromagnetic radiation meets the boundary between one material and another, some of it is reflected; then passes into the second material, where some is absorbed and the remainder is transmitted. The ratio of the velocity of light in a vacuum v_0 to that in the given material v is defined as the refractive index of the material.

$$n = v_0/v \tag{6.1}$$

Figure 5.3 categorizes the effects of cyclic electromagnet fields on microstructure constituents in the material. In the case of glasses these constituents are electrons, ions, and ionic groups. The refractive index derives principally from the polarization of electrons in glasses, which occurs in the range from about 10^{14} to 10^{15} Hz. It follows from Maxwell's theory that for the same frequencies in this range the dielectric constant k is equal to n^2. Corresponding to equation (5.9), the refractive index is a complex quantity

$$n = n' - in'' \tag{6.2}$$

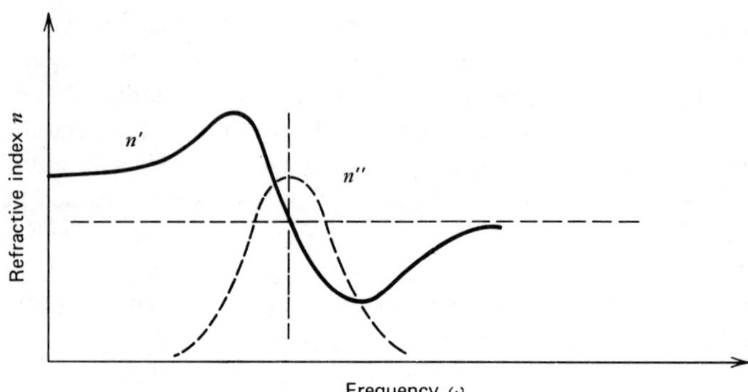

Figure 6.1   Effect of frequency on n' and n''.

where n' and n'' are the real and imaginary components, respectively. The real part is the refractive index as defined previously, and the imaginary part relates to absorption in terms of the extinction coefficient. Refractive index values of glasses decrease smoothly with decreases in frequency. Such normal dispersion curves provide needed data for image formation in optical systems. Thus we are presently interested in cases of glasses that react passively to the incoming radiation, where little absorption occurs, and where refractive index is expressed by a single number at a given frequency. In other words we are speaking of glasses that act as linear dielectrics under imposed test conditions. We know from nonlinear optics that the refractive index depends also on the intensity of the light. In such cases the refractive index is represented by a series of terms, with all but the linear term being very small. Pursuit of these and other related phenomena would lead into the interesting and useful subject of nonlinear optics, an extensive subject that is beyond the scope of this book. Those wishing an introduction to nonlinear optics can consult Baldwin (1969).

Figure 6.1 plots relations between n' and n'' as functions of frequency. Two important features can be observed. (1) The refractive index becomes large at frequencies just below resonance and drops rapidly to a value less than 1 just above the resonant frequency. (2) This anomalous dispersion region, where $dn'/d\omega$ becomes negative, is accompanied by absorption and may correspond to that of an emission line in the spectra of gases and vapors. Whether liquid or solid glasses are in condensed states and give rise to bands or regions of absorption. Absorption phenomena in glasses are covered in a later chapter.

Values of refractive index are designated by the letter n, with a subscript indicating the wavelength at which it was measured. Values are measured at room temperature and atmospheric pressure unless noted otherwise. Optical glass man-

TABLE 6.1   Refractive Index Nomenclature

Spectral line	Element	Color	Wavelength (Å*)
C	H	Red	6563
C'	Cd	Red	6439
D	Na	Yellow	5893
d	He	Yellow	5876
e	Hg	Green	5461
F	H	Blue	4861
F'	Cd	Blue	4800
g	Hg	Blue	4358
h	Hg	Violet	4047

*1 $\text{Å} = 10^{-1}$ millimicron (or 10^{-1} nm) $= 10^{-4}$ micron $= 10^{-8}$ cm.

ufacturers give refractive indices at several wavelengths, but it is sufficient here to indicate representative ones, as in Table 6.1. Following Abbe, most of the published data give values for the C, D, and F lines. The differences $n_F - n_D$, $n_D - n_C$, and $n_F - n_C$ are called partial dispersions, the later being designated as the mean dispersion. The derived values $(n_F - n_D)/(n_F - n_C)$ and $(n_D - n_C)/(n_F - n_C)$, termed partial dispersion ratios, are measures of the relative lengths of the partial spectra in given glasses. Differences between partial dispersion ratios of two glasses indicate the degree of similarity of their partial spectra. The Abbe or ν value is the ratio of $n_D - 1$ (the effective or excess refractivity) to the mean dispersion

$$\nu_D = n_D - 1/n_F - n_C \tag{6.3}$$

Its reciprocal, called the dispersive power, expresses the mean dispersion in terms of the effective refractivity. Refractive index data for the C, D, and F spectral lines are not sufficient for designers of precise optical systems. Catalogs of optical glass manufacturers list data for many spectral lines and usually show six or more partial dispersion ratios. An internationally accepted six-figure numbering system is used to show refractive index and Abbe value. Thus a glass with $n_d = 1.51835$ and $\nu_d = 60.28$ is numbered 518603. Manufacturers may also list values for n_e and ν_e. This method of describing optical glasses was suggested by Zschokke (1918).

6.3 PROPERTIES OF COMMERCIAL GLASSES

At the time Abbe and Schott began their research, crown optical glasses contained major amounts of SiO_2 and no PbO. Flint types contained major amounts of both SiO_2 and PbO and can be designated as lead silicate glasses. Abbe and Schott developed new glass types, including those containing major amounts of B_2O_3 and P_2O_5, but no SiO_2. Glasses representative of early production of the Jena Glass Works are listed in Table 6.2. The first three numbers of the indicated types give values of n_D, and the last three show values of ν_D. The indicated compositions in weights percent relate to glass type and are not necessarily those of the sample measured. Refractive index is significantly dependent on heat treatment; differences up to 0.001 are possible for some glass types. Presently available data on relations between optical properties and compositions of commercial glasses do not permit the derivation of quantitative effects of the various oxides. However qualitative contributions of oxides to optical properties can be noted in Table 6.2. Manufacturers now make use of some 15 to 20 different oxides in meeting requirements of optical designers. Most of the glasses are of the oxide type, but they also contain fluorine and other nonoxide constituents. Table 6.3 compares property data on typical crown and flint glasses manufactured by Chance-Pilkington with vitreous silica. It has been noted that the use of rare earth oxides results in glasses having high refractive indices and relatively low dispersions. Schott and Genossen offer more than 40 glasses of this type. It is of interest to compare properties of their glass LaK N19, type 755531, which contains 21% ThO_2 with type 748278 (Table 6.3). A wide variety of glasses with refractive indices in the approximate range from 1.45 to 2.0 and Abbe values from about 20 to 85 are now manufactured by Corning in the United States., Chance-Pilkington in Great Britain, Schott in West Germany, Hoya in Japan, and others. The Schott catalog lists more than 200 glass types.

6.4 PROPERTIES OF EXPERIMENTAL GLASSES

Published literature provides ample coverage of data on refractive index and density, the inverse of specific volume, in silicate glass systems. These data are more extensive than those for any other glass properties because of their important uses in the optical glass industry. Many publications furnish data on dispersion in terms of mean dispersion and Abbe number. These data can be found in Morey (1954) and publications listed in the Appendix. Properties treated in this section are those measured at room temperature and at a pressure of 1 atmosphere. The following discussion covers representative information on these properties in terms of the author's substructure method.

$Na_2O-CaO-SiO_2$ Glasses

Morey and Merwin (1932) give data on refractive index n_D, mean dispersion $n_F - n_C$, and density for these glasses. Table 6.4 gives analytical representations

TABLE 6.2 Representative Optical Glasses Manufactured by Jena Glass Works

Oxide	Composition (wt %) of eight glass types							
	496644*	513573*	540598[†]	610574[†]	627391[†]	778265[†]	510600[†]	558670*
SiO_2	71	71	60	35	42.8	27.3		
B_2O_3	14		3	10			63.8	3.0
P_2O_5								59.5
Na_2O	10	17	3		0.7		8.0	
K_2O			10		7.5	1.5	3.5	
BaO			19	42	10.8		3.5	28.0
ZnO		12	5	8	5.1			
PbO					32.6	71.0	3.0	
Al_2O_3	5				5		18.0	8.0

*Data source. A. Winkelmann and O. Schott (1894), *Ann. Phys. Chem.*, **51**, 697;
[†]*Data source.* E. Zschimmer (1912), in C. Doelter, *Handbuch der Mineralchemie,* Vol. 1, p. 869, T. Sleinkopff, Dresden and Leipzig.

TABLE 6.3 Optical Properties of Selected Commercial Glasses

	Vitreous SiO_2, 458678	BSC, 510644	DEDF, 748278
Property			
n_d	1.45846	1.50970	1.74842
$n_f - n_C$	0.00676	0.00791	0.02687
γ_d	67.83	64.44	27.85

Wavelength (Å)	Frequency $\times 10^{-14}$ Hz	Refractive indices		
		Vitreous SiO_2	BSC	DEDF
3,650	8.22	1.47539	1.52882	
4,047 (h)	7.41	1.46962	1.52290	1.79811
4,358 (g)	6.88	1.46669	1.51941	1.78373
4,800 (F')	6.25		1.51562	1.76917
4,861 (F)	6.17	1.46313	1.51518	1.76753
5,461 (e)	5.49	1.46008	1.51159	1.75477
5,876 (d)	5.11	1.45846	1.50970	1.74842
6,438 (C')	4.66	1.45670	1.50766	1.74186
6,563 (C)	4.57	1.45637	1.50727	1.74066
7,800	3.85		1.50419	1.73168
10,140	2.96	1.45024	1.50041	1.72253
18,131	1.65	1.44070	1.49064	1.70882
24,374	1.23	1.43095		
25,628	1.17		1.47873	1.69791

Data sources. SiO_2 from Malitson (1965); others from Chance-Pilkington catalog.

of optical and specific volume data derived from the measurements of these investigators. Their glasses were located in the six indicated primary phase fields. Respective values of standard error, defined in Table 4.1, are in good agreement with accuracies of the experimental data. Computerized information in Table 6.4 was obtained by using data on 94 glasses distributed as follows in the primary phase groups: S(TC), 23; N2S, 9; N2C3S, 25; N3C6S, 8; αCS, 20; βSC, 9. Oxide constants from Table 6.4 can be used to closely estimate properties of glasses whose compositions are located in the primary phase fields indicated in Figure 4.3. All compositions can be formed into ordinary single liquid glasses with the exception of binary CaO-SiO_2 glasses located in the cristobalite primary phase field (Figure 4.3). These compositons form milky-looking immiscible materials containing two liquids. Weights percent compositions must be converted to mole fractions before property calculations are made.

TABLE 6.4   Properties of Annealed Na_2O-CaO-SiO_2 Glasses

Primary phase	Glass property = A SiO_2 + B CaO + C Na_2O			
	A	B	C	Standard error
Refractive index, n_D				
S(TC)	1.4568	1.7850	1.6261	0.0006
N2S	1.4764	1.7513	1.5642	0.0005
N2C3S	1.4812	1.7559	1.5481	0.0009
N3C6S	1.4715	1.7638	1.5740	0.0008
αCS	1.4681	1.7813	1.5620	0.0007
βCS	1.4676	1.7518	1.6044	0.0004
Specific volume, V (cc/g)				
S(TC)	0.45403	0.21247	0.27503	0.0006
N2S	0.43559	0.24535	0.33390	0.0003
N2C3S	0.42778	0.26693	0.34948	0.0005
N3C6S	0.43726	0.25282	0.32597	0.0005
αCS	0.43352	0.25569	0.33671	0.0007
βCS	0.43731	0.26095	0.31533	0.0004
Mean dispersion, $n_F - n_C$				
S(TC)	0.00680	0.01459	0.01455	0.00006
N2S	0.00692	0.01371	0.01412	0.00005
N2C3S	0.00677	0.01534	0.01441	0.00007
N3C6S	0.00697	0.01452	0.01395	0.00007
αCS	0.00675	0.01623	0.01300	0.00008
βCS	0.00633	0.01636	0.01502	0.00007
Abbe number, ν				
S(TC)	66.80	46.21	28.37	0.33
N2S	65.91	53.85	31.26	0.32
N2C3S	66.01	46.48	31.14	0.32
N3C6S	66.75	48.45	28.16	0.39
αCS	66.31	43.21	34.99	0.31
βCS	67.99	41.05	29.61	0.37
Dispersive power, $1/\nu$				
S(TC)	0.01491	0.02021	0.02507	0.0001
N2S	0.01471	0.01852	0.02577	0.0001
N2C3S	0.01439	0.02103	0.02651	0.0001
N3C6S	0.01465	0.02005	0.02609	0.0001
αCS	0.01484	0.02183	0.02358	0.0001
βCS	0.01429	0.02223	0.02570	0.0001

Experimental data: n_D ±0.0005; density ±0.0010.

Substructure representations: Babcock (1968).

Data source. Morey and Merwin (1932); Figure 4.3, primary phase diagram.

TABLE 6.5   Glass Composition Formulations (wt %)

Oxide	Glass				
	A	1	2	3	4
SiO_2	0.8200	0.7895	0.7609	0.7798	0.8005
CaO	0.0800	0.1105	0.0800	0.1001	0.1228
Na_2O	0.1000	0.1000	0.1591	0.1201	0.0767
Property					
n_D	1.5000	1.5100	1.5100	1.5100	1.5100
v	61.32	60.69	59.04	60.13	61.32

Representation of properties in terms of linear equations makes the formulation of glasses having desired property values simple and direct. Table 6.5 presents examples of such composition formulations for glasses in the S(TC) primary phase field using equations indicated in Table 6.4. It is desired to raise the n_D value of glass A from 1.5000 to 1.5100. This can be done in a number of ways by substitution of one oxide for another. In glass 1 CaO for SiO_2 = 0.0100/(1.7850 − 1.4568) = 0.0305. In glass 2 Na_2O for SiO_2 = 0.0100/(1.6261 − 1.4568)= 0.0591. In glass 3 equal amounts of CaO and Na_2O for SiO_2 = 0.0100/(1.7056 − 1.4568) = 0.0402. These substitutions are 0.0201 CaO for SiO_2 and 0.0201 Na_2O for SiO_2. Calculated values of Abbe number v emphasize that all glass properties change with composition. Let us assume that the values of v for glasses 1 to 3 are too low for our purpose. Say we want a glass having n_D = 1.5100 and v = 61.32. This glass composition can be determined by solving the following equations simultaneously:

$$1.5100 = 1.4568\ SiO_2 + 1.7850\ CaO + 1.6261\ (1\text{-}SiO_2\text{-}CaO)$$
$$61.32 = 66.80\ SiO_2 + 46.21\ CaO + (1\text{-}SiO_2\text{-}CaO)$$

Thus glass 4 has the required properties. The Na_2O-CaO-SiO_2 glasses have low refractive indices and low dispersions. Property contributions of the oxides appear in Table 6.4. Glasses in this system formed the basis of early ordinary crown compositions. Following the research of Abbe and Schott, other oxides have been incorporated in crown glasses to achieve needed variations in refractive indices and dispersion.

Na_2O-TiO_2-SiO_2 Glasses

Computerized information on properties in seven primary phase fields of this system are given in Table 6.6. The high refractive indices and high dispersions of

TABLE 6.6 Properties of Annealed Na_2O-TiO_2-SiO_2 Glasses

Glass property = A SiO_2 + B TiO_2 + C Na_2O

Primary phase	A	B	C	Average difference between measured and calculated values
Refractive index, n_D				
A	1.45058	2.24879	1.63299	4×10^{-4}
B	1.49204	2.18124	1.51324	2×10^{-4}
C	1.49520	2.16666	1.51040	2×10^{-4}
D	1.48252	2.28742	1.47607	2×10^{-4}
E	1.51456	2.26183	1.42118	3×10^{-4}
F1	1.52091	2.34668	1.31844	2×10^{-4}
F2	1.47235	2.30784	1.48561	3×10^{-4}
Specific volume V, (cc/g)				
A	0.45004	0.18665	0.28062	26×10^{-4}
B	0.42753	0.17663	0.35994	13×10^{-4}
C	0.41120	0.25480	0.36533	21×10^{-4}
D	0.42452	0.21683	0.35044	22×10^{-4}
E	0.39277	0.24491	0.39322	20×10^{-4}
F1	0.39771	0.24830	0.38140	28×10^{-4}
F2	0.42568	0.24826	0.31211	22×10^{-4}
Mean dispersion, $n_F - n_C$				
A	0.00604	0.06507	0.01231	11×10^{-5}
B	0.00787	0.05431	0.00991	1×10^{-5}
C	0.00542	0.05875	0.01273	7×10^{-5}
D	0.00627	0.07329	0.00345	3×10^{-5}
E	0.00655	0.07499	0.00050	5×10^{-5}
F1	0.00692	0.09023	-0.1616	5×10^{-5}
F2	0.00520	0.08801	-0.00921	1×10^{-5}
Dispersive power, $1/\nu$				
A	0.01588	0.08562	0.01700	14×10^{-5}
B	0.01656	0.07408	0.01928	1×10^{-5}
C	0.01371	0.07386	0.02435	2×10^{-5}
D	0.01624	0.08804	0.01208	4×10^{-5}
E	0.01777	0.08354	0.01143	6×10^{-5}
F1	0.02012	0.08979	0.00137	10×10^{-5}
F2	0.01745	0.09072	0.00741	6×10^{-5}

Identity of primary phases: A, SiO_2 (cristobalite); C, $Na_2O \cdot SiO_2$; E, $Na_2O \cdot TiO_2 \cdot SiO_2$; F2, TiO_2 (rutile); B, $Na_2O \cdot 2SiO_2$; D, unknown; F1, $Na_2O \cdot 6TiO_2$. Substructure representations: Babcock (1973).

Data source. Hamilton and Cleek (1958); Figures 4.4 and 4.5, primary phase diagrams.

TABLE 6.7   Properties of Annealed Na_2O-PbO-SiO_2 Glasses in SiO_2 Primary Phase Field

Glass property $= A$ SiO_2 $+ B$ PbO $+ C$ Na_2O

Property	A	B	C	Average difference between measured and calculated values $\times 10^{-4}$
n_C	1.45187	2.23030	1.63100	2
n_D	1.45365	2.24714	1.63571	2
n_F	1.45818	2.29037	1.64562	2
$n_F - n_C$	0.00631	0.06007	0.01462	2
Dispersive power, $1/\nu$	0.01543	0.08227	0.02183	1
Specific volume V, (cc/g)	0.41522	-0.37430	0.35440	23

Average difference between measured and calculated densities = 0.02.

Data sources. Peddle (1920); Levin, Robbins, and McMurdie (1964), primary phase diagram.

these glasses are caused by the presence of TiO_2. Hovestadt (1902) notes that TiO_2 was one of the 28 oxides used in the research of Abbe and Schott. The Schott optical glass catalog lists relatively few glasses containing TiO_2. The absorption of light in the visible region by TiO_2, especially in the presence of some oxides like those of iron, is one reason for its limited use. An oxide that does not contribute visible color, PBO, is widely used in glasses having high refractive indices and dispersions.

Na_2O-PbO-SiO_2 Glasses

Table 6.7 contains information on glasses located in the SiO_2 primary phase field. The use of linear equations permits direct representation of mean dispersion n_F - n_C simply by subtracting values of constants for n_F and n_C. Equations for partial dispersions n_F - n_D *and* n_D - n_C can be obtained in the same manner. It is of interest to compare values of oxide constants of these PbO glasses with those of TiO_2 glasses in the SiO_2 primary phase field. The two sets of constants for n_D, mean dispersion, and dispersive power are nearly the same, especially those of PbO and TiO_2. Densities of the PbO glasses are much higher, as indicated by respective specific volumne constants. Property comparisons of TiO_2 and PbO glasses in the SiO_2 primary phase field appear in Table 6.8.

TABLE 6.8   Property Comparisons of TiO_2 and PbO Glasses in SiO_2 Primary Phase Field

Oxide	Composition (mole fraction)		
SiO_2	0.75	0.75	0.75
X	0.05	0.10	0.15
Na_2O	0.20	0.15	0.10
Property			
Refractive index, n_D			
$X = PbO$	1.5297	1.5603	1.5909
$X = TiO_2$	1.5269	1.5577	1.5885
Dispersive power, $1/v_D$			
$X = PbO$	0.02005	0.02307	0.02609
$X = TiO_2$	0.01959	0.02302	0.02645
Mean dispersion, $n_F - n_C$			
$X = PbO$	0.01065	0.01293	0.01520
$X = TiO_2$	0.01024	0.01289	0.01552
Specific volume (cc/g)			
$X = PbO$	0.36358	0.32715	0.29071
$X = TiO_2$	0.40298	0.39829	0.39359
Density (g/cc)			
$X = PbO$	2.750	3.057	3.440
$X = TiO_2$	2.482	2.511	2.541

Ternary Glasses Containing BaO and SiO_2

Cleek and Babcock (1973) published property data on six systems containing TiO_2, La_2O_3, ZnO, Nb_2O_5, and Al_2O_3 as the third oxide. Data on refractive indices, dispersion, and specific volume were reviewed in accordance with the substructure method. The glasses were evaluated in terms of computer groups, with limitations on values of standard error, because primary phase information was not available at the time of publication. Information on glasses containing the rarer oxides La_2O_3, Ta_2O_5, and Nb_2O_5 are of particular interest.

BaO-La_2O_3-SiO_2 Glasses

Figure 6.2 shows compositions of glasses used in the two groups of this series. Data representations are given in Table 6.9 The lanthanum oxide La_2O_3 contributes high refractive index, high dispersion, and high density to those glasses. Property constants can be used to closely estimate properties within the composition areas of Figure 6.2.

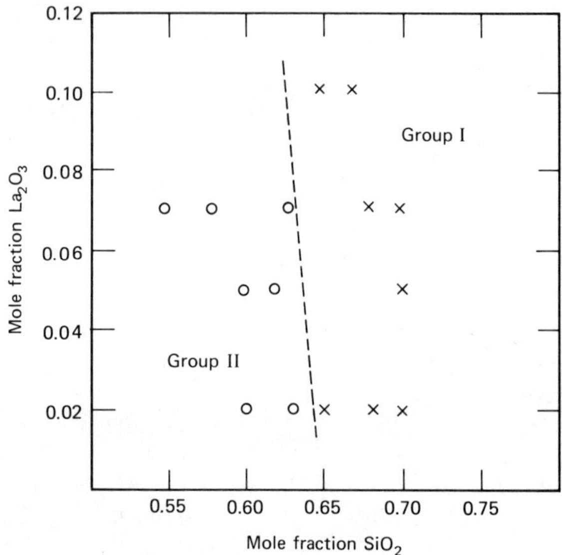

Figure 6.2 Composition locations of BaO-La_2O_3-SiO_2 glasses. Adapted from Cleek and Babcock (1973).

BaO-Ta_2O_5-SiO_2 Glasses

Figure 6.3 indicates composition locations of glasses in the three groups in this series, and computer equations representing the reviewed data can be found from Table 6.10. The oxide Ta_2O_5 contributes high values of refractive index, dispersion, and density to these glasses.

BaO-Nb_2O_5-SiO_2 Glasses

Composition locations of glasses in the two groups reviewed in this series are given in Figure 6.4. Equation constants for optical properties appear in Table 6.11. Glasses in this system owe their high values of refractive index and dispersion to NB_2O_5.

Glasses in SiO_2 Primary Phase Fields

Previous discussions have shown certain glass properties to be linearly related to mole fraction compositions within given primary phase fields. Tables 4.5 and 4.6 demonstrate that these relations are also valid within similar primary phase fields in different glass systems, indicating that property constants of oxides are characteristic of the primary phase field in which the composition is located. For example, from Table 4.6 refractive index n_D for binary N-S glasses in N2S primary phase fields of N-S, N-C-S, N-A-S, and N-L-S systems can be

TABLE 6.9   Properties of Annealed BaO-La_2O_3-SiO_2 Glasses

Glass property = A SiO_2 + B La_2O_3 + C BaO			

Property	A	B	C	Standard error
Group I glasses *				
n_C	1.48825	2.51281	1.83983	0.0007
n_D	1.49036	2.52380	1.84518	0.0007
n_F	1.49538	2.55158	1.85810	0.0007
$n_F - n_C$	0.00713	0.03877	0.01827	
Abbe number, ν_D	63.00	6.90	41.88	0.091
Specific volume (cc/g)	0.35428	−0.17875	0.10807	0.0008
Group II glasses[†]				
n_C	1.51660	2.46619	1.79980	0.0004
n_D	1.51872	2.47735	1.80510	0.0004
n_F	1.52372	2.50559	1.81819	0.0004
$n_F - n_C$	0.00712	0.03940	0.01839	
Abbe number, ν_D	63.54	5.35	40.78	0.097
Specific volume (cc/g)	0.32645	−0.10865	0.14791	0.0003

*Average of differences between measured and calculated densities = 0.0009.
[†] Average differences between measured and calculated densities = 0.0005.

Data source and substructure representations. Cleek and Babcock (1973).

represented by the equation n_D = 1.4756 SiO_2 + 1.5644 Na_2O. The data validate this equation, even though refractive index is a function of heat treatment; the different investigators may have used dissimilar annealing procedures. Differences in heat treatment may cause refractive index differences of the order of 0.001. Such equations may be safely used to estimate refractive index values to within this order of magnitude. Facts given earlier in this section show that refractive index differences from data of single investigators are in the fourth decimal place. Two ways of using such information for formulation of optical glasses can be outlined.

Tables 6.4, 6.6, and 6.7 reveal that SiO_2 constants for n_D in SiO_2 primary phase fields are of closely equal value; the same is true for Na_2O constants for n_D. Assume that we wish to increase the n_D value of a given N-C-S glass by substituting TiO_2 for one of the three oxides. This can be done by using appropriate constants for SiO_2, CaO, Na_2O, and TiO_2 as outlined previously.

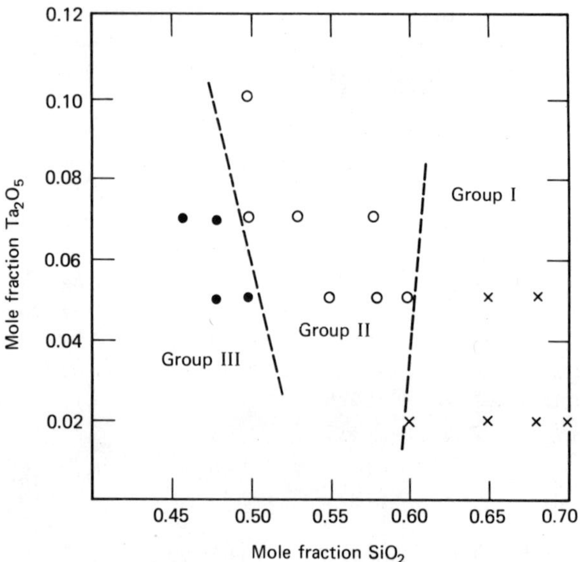

Figure 6.3 Composition locations of BaO-Ta_2O_5-SiO_2 glasses. Adapted from Cleek and Babcock (1973).

Such property calculations will be valid if the substitutions made do not move the composition into another primary phase field. Changes in n_D and other properties can be effected in N-C-S glasses by substitutions of PbO or by combinations of TiO_2 and PbO. Such property estimations minimize the amount of experimental research needed otherwise in glass composition formulation.

A second property estimation method is illustrated in Table 6.12, using data on glasses located in SiO_2 phase fields. Oxide constants for the 48 annealed glasses were obtained by combining glasses in the N-S, N-C-S, N-A-S, and K-C-S systems into one computer group. Constants for the 45 quenched glasses resulted from combining glasses in the N-L-S, K-M-S, and N-A-S systems. Note that the SiO_2 constant for quenched glasses is higher than that for annealed glasses and is close to the n_D value of 1.4584 for vitreous silica published by Malitson (1965). We pointed out in Section 4.6 that Douglas and Isard (1951) found the density of vitreous silica to decrease as the temperature from which it was quenched was decreased. From general relations between refractive index and density, it follows that the n_D value for quenched vitreous silica is reduced when this material is annealed. Thus the n_D constants for SiO_2 in annealed silicate glasses in SiO_2 phase fields are lower than those for quenched glasses.

TABLE 6.10   Properties of Annealed BaO-Ta_2O_5-SiO_2 Glasses

	Glass property = A SiO_2 + B Ta_2O_5 + C BaO			
Property	A	B	C	Standard error
Group I glasses *				
n_C	1.43819	2.68156	1.95482	0.0003
n_D	1.43974	2.70166	1.96134	0.0003
n_F	1.44331	2.75161	1.97761	0.0003
$n_F - n_C$	0.00512	0.07005	0.02279	
Abbe number, ν_D	65.77	−96.27	34.13	0.221
Specific volume (cc/g)	0.37394	−0.34930	0.06008	0.0008
Group II glasses [†]				
n_C	1.50395	2.48739	1.85520	0.0005
n_D	1.50627	2.50586	1.86048	0.0005
n_F	1.51030	2.57296	1.87262	0.0005
$n_F - n_C$	0.00635	0.08557	0.01742	
Abbe number, ν_D	59.09	−85.41	47.28	0.425
Specific volume (cc/g)	0.31595	−0.16059	0.14165	0.0007
Group III glasses [‡]				
n_C	1.65915	2.31775	1.69455	0.00003
n_D	1.66596	2.33911	1.69456	0.00000
n_F	1.67132	2.36772	1.70992	0.00004
$n_F - n_C$	0.01217	0.04997	0.01537	
Abbe number, ν_D	50.87	−12.64	46.87	0.201
Specific volume (cc/g)	0.24761	0.02726	0.19721	0.00009

*Average of differences between measured and calculated densities = 0.007.
[†] Average differences between measured and calculated densities = 0.010.
[‡] Average differences between measured and calculated densities = 0.002.
Data source and substructure representation. Cleek and Babcock (1973).

The n_D values of most silicate glasses are increased by the annealing process, as reflected by constants for Al_2O_3, K_2O, and Na_2O for annealed and quenched glasses. Such differences are usual for oxides other than SiO_2. Constants for all seven oxides were then determined by combining annealed and quenched glasses into one group. Constants in Table 6.12 can be used to estimate n_D values of glasses within 0.001. Calculations will be valid as long as the glasses remain in the SiO_2 primary phase field.

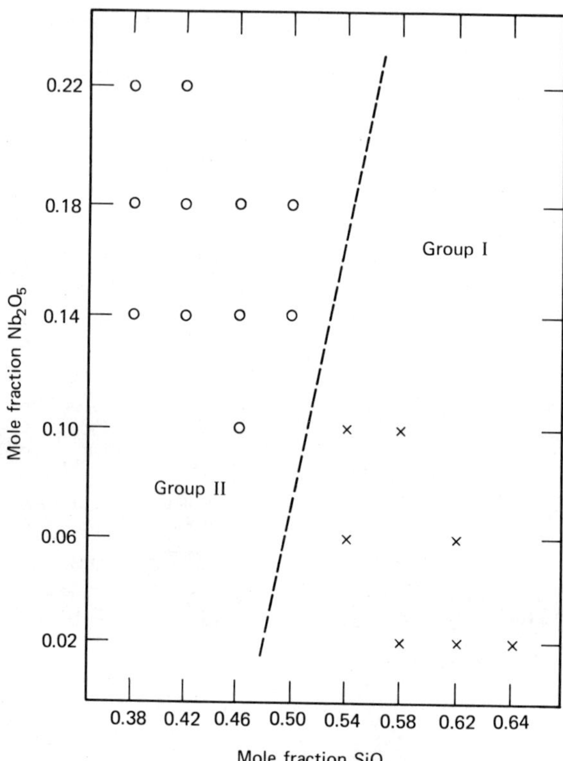

Figure 6.4 Composition locations of BaO-Nb$_2$O$_5$-SiO$_2$ glasses. Adapted from Cleek and Babcock (1973).

Other Methods of Calculating Refractive Indices

Many methods for the calculation of refractive indices have been published. Baak (1969) comprehensively reviews these earlier approaches. The reader may be interested in comparing them in terms of data representation and mathematical complexity with the substructure method preferred by the author.

6.5 REFRACTIVE INDEX AND DENSITY RELATIONS

Refractive index values for glasses derive from interactions of their integrated electronic and ionic structures with electromagnetic radiation. Development of empirical relations between refractive index and density or its inverse specific volume have been extensively covered in the literature. Morey (1954), Baak (1969), and others have published detailed reviews of this research. Relations

TABLE 6.11   Properties of Annealed BaO-Nb_2O_5-SiO_2 Glasses

Glass property = A SiO_2 + B Nb_2O_5 + C BaO			
Property	A	B	C	Standard error
Group I glasses				
n_C	1.49209	2.78035	1.84569	0.0005
n_D	1.49408	2.80691	1.85111	0.0005
n_F	1.49870	2.87595	1.86462	0.0006
$n_F - n_C$	0.00661	0.09560	0.01893	
Abbe number, ν_D	59.38	-103.22	43.03	0.371
Group II glasses				
n_C	1.56263	2.66145	1.78253	0.0011
n_D	1.56558	2.69145	1.78584	0.0011
n_F	1.57170	2.77114	1.79431	0.0011
$n_F - n_C$	0.00907	0.10969	0.01178	
Abbe number, ν_D	43.36	-35.25	47.86	0.300

Data source and substructure representations. Cleek and Babcock (1973).

such as those proposed by Gladstone and Dale (1863), Lorenz (1880), and Lorentz (1880) have received major attention.

Gladstone and Dale:

$$r = (n - 1)/d \qquad (6.4)$$

$$R = M(n - 1)/d \qquad (6.5)$$

Lorenz-Lorentz:

$$r = [(n_2 - 1)/(n_2 + 2)]/d \qquad (6.6)$$

$$R = M/d [(n_2 - 1)/(n_2 + 2)] \qquad (6.7)$$

Where r = specific refraction, R = molar refraction, n = refractive index, d = density, and M = molecular weight. Thus $r = R/M$ from these equations. The equations were first described to relate refractive index and density in gases and vapors. Their extension to solids and materials in the condensed state such as glasses has not been verified. Although such relations are useful for glass technology purposes, they do not lead to clarification of the roles of electronic and ionic structures in glasses. The refractive index of glasses usually increases with density and decreases with specific volume. To say that refractive index is simultaneously influenced by changes in desnity and electronic systems is to oversimplify the problem. Density is related to how tightly the ions and ionic groups are packed together in the substructure and to their influence on attendant electronic systems. Density and specific volume values which are a

TABLE 6.12   Refractive Index of Glasses in SiO_2 Primary Phase Fields

Refractive index = A SiO_2 + B Al_2O_3 + C CaO + D MgO + . . .

	Oxide constants for 93 glasses		
Oxide constants	Annealed (48)	Quenched (45)	Combined (93)
SiO_2	1.4562	1.4579	1.4567
Al_2O_3	1.5538	1.5475	1.5526
CaO	1.7844		1.7837
MgO		1.6305	1.6344
K_2O	1.6399	1.6243	1.6313
Na_2O	1.6288	1.6177	1.6256
Li_2O		1.6945	1.6963
Average difference between measured and calculated values	0.0005	0.0007	0.0006

Babcock (1968); Na_2O-SiO_2, Na_2O-CaO-SiO_2, Na_2O-Al_2O_3-SiO_2.
Morey (1954); K_2O-CaO-SiO_2.
Murthy and Hummel (1955); Li_2O-MgO-SiO_2.
Roedder (1951); K_2O-MgO-SiO_2.
Data sources. Levin, Robbins, and McMurdie (1964); primary phase diagrams.

measure of the integrated spatial arrangements of all constituents of the structure, do not provide information on structural details.

Empirical substructure equations relating refractive index and specific volume to composition express these relations in more direct forms. Constants in refractive index equations show refractive index per mole for each of the oxides involved. Likewise constants in specific volume equations show directly the specific volume per mole of each oxide. Equations for refraction over that in a vacuum $(n_D - 1)$ can be obtained directly by subtracting 1 from appropriate n_D oxide constants. For example, in the SiO_2 phase field of the Na_2O-CaO-SiO_2 system, we have

$$n_D - 1 = 0.4568\ SiO_2 + 0.7850\ CaO + 0.6261\ Na_2O$$

Relations between refractive index and density, in terms of specific volume, are given in Table 6.13 for a number of experimental glasses. Constants in the equations were determined by a least mean squares computer program. Note that oxide constants for Na_2O-CaO-SiO_2 glasses are closely the same for all primary phase groups. In fact the equation obtained by combining all glasses into one group represents the data quite as well. Oxide constants in the Na_2O-

TABLE 6.13   Relations Between Refractive Index and Density

Phase field	A	B	C	Average difference between measured and calculated values
		$(n_D - 1)/d = A\ SiO_2 + B\ CaO + C\ Na_2O$		
SiO_2	0.20840	0.22381	0.19232	0.0001
N2S	0.20856	0.22214	0.19225	0.0001
N2C3S	0.20800	0.22514	0.19323	0.0002
N3C6S	0.20835	0.22817	0.19000	0.0001
αCS	0.20811	0.22509	0.19152	0.0002
βCS	0.20767	0.22334	0.19646	0.0001
Combined group of 95 glasses	0.20817	0.22455	0.19305	0.0002
		$(n_D - 1)/d = A\ SiO_2 + B\ TiO_2 + C\ Na_2O$		
A, SiO_2	0.21031	0.37463	0.17830	0.0003
B, N2S	0.21260	0.34460	0.18260	0.0001
C, NS	0.20630	0.36071	0.18865	0.0002
D, unknown	0.21196	0.38201	0.16425	0.0003
E, NTS	0.21133	0.37222	0.17098	0.0002
F1, N6T	0.21681	0.38949	0.14421	0.0003
F2, TiO_2	0.21353	0.39576	0.14341	0.0001
Combined group of 51 glasses	0.21252	0.37491	0.17036	0.0008
		$(n_D - 1)/d = A\ SiO_2 + B\ PbO + C\ Na_2O$		
SiO_2	0.19557	0.02691	0.22176	0.0007

TiO_2-SiO_2 glasses vary somewhat for different phase fields, and their differences are distinctive. Here again, the equation for the combined group represents the data reasonably well. Constants for SiO_2 in SiO_2 phase fields in the three glass systems are distinctly different; the same is true for constants for Na_2O in this phase field. This suggests that constants for $(n_D - 1)/d$ are characteristic of given glass systems, but not for glasses located in given primary phase fields.

6.6   EFFECTS OF PRESSURE AND TEMPERATURE

The dependence of refractive index on hydrostatic pressure and temperature is important in design of precision optical systems. These phenomena relate to athermalization of optical elements, self-focusing of laser beams, and molecular scattering in optical transmission lines. The application of hydrostatic pressure increases the refractive index and decreases specific volume. The effects of temperature on refractive index, which simultaneously cause changes in specific volume, are more complex. Comparative effects of temperature on refractive index, specific volume, and temperature coefficient of refractive index are illustrated in Figure 6.5. These curves are typical of optical glasses. Refractive index reaches minimal values, usually below 0°C, at temperatures characteristic of glass compositions. The temperature region in which refractive index decreases rapidly is also unique for a given glass composition. Rapid decreases in $\Delta n/\Delta T$ occur in the same temperature zone and may become negative. Changes in $\Delta V/V$ are always positive and attain high values in the same characteristic temperature zones. The development of an athermalized glass element involves selecting a glass composition that has minimal changes in refractive index and specific volume in the temperature range in which is to be used. It must also be pointed out that dn/dT varies with the wavelength of light and reaches minimal values at wavelengths characteristic of specific optical glasses. Table 6.14 shows changes of dn/dT in the range 15 to 25°C for glasses selected from the Chance-Pilkington

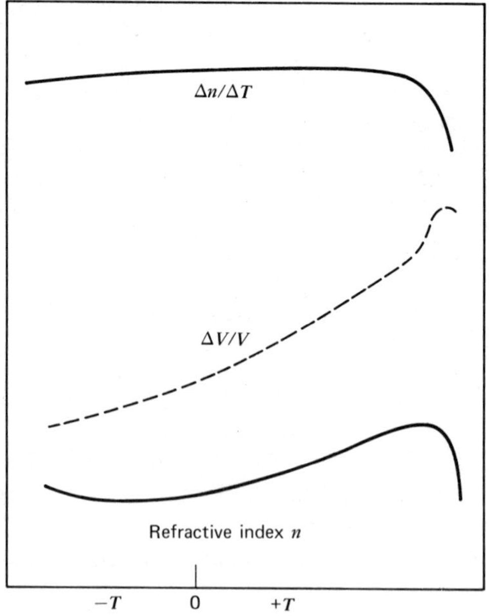

Figure 6.5 Comparative effects of temperature T on glass properties. Adapted from Waxler and Cleek (1973).

TABLE 6.14   Effect of Wavelength on dn/dT per °C $\times$ 10^5 for Selected Optical Glasses

Glass: Type:	BSC 510644	HC 519604	ZC 508612	DEDF 748278	ELF 548456
Wavelength (Å)					
3,650	0.49	0.43	0.83		0.64
4,047	0.37	0.32	0.74	1.65	0.57
4,800	0.26	0.23	0.69	1.04	0.48
5,461	*0.21*	0.19	0.67	0.87	0.44
6,438	0.22	*0.17*	*0.66*	0.81	0.40
7,800	0.25	0.19	*0.66*	*0.76*	0.37
10,140	0.33	0.25	0.67	0.78	*0.35*
18,131	0.52	0.49	0.71	1.05	0.41
25,628	0.66	0.68	0.77	1.36	0.51

Minimum values of dn/dT are italicized.

Source. Chance-Pilkington catalog.

catalog. The relevance of these data to precision optical design is apparent.

Morey (1954) reviewed early research on the effects of hydrostatic pressure and temperature. Waxler and Weir (1965) measured changes in refractive index caused by hydrostatic pressure on vitreous silica and two commercial glasses. Vedam, Schmidt, and Roy (1966) reported measurements on vitreous silica. Baak (1969) reviewed pressure and temperature effects on refractive index. The refractive index of vitreous silica at low temperatures was reported by Waxler and Cleek (1971). Waxler, Cleek, Malitson, Dodge, and Hahn (1971) published data on optical and mechanical properties of laser glasses, together with mathematical background material.

The definitive paper by Waxler and Cleek (1973) related pressure and temperature effects on refractive index in terms of measured data and their analytical expressions for vitreous silica and four experimental glasses. This treatment of the subject furnishes a method for use in further research. The investigators data and calculated values of parameters for vitreous silica and four experimental glasses appear in Table 6.15.

The quantity $d(\partial n/\partial d)$ was calculated from the equation

$$d\partial n/\partial d = 1/\beta \, (dn/dp) \tag{6.8}$$

where d = density, n = refractive index, β = compressibility, and P = pressure.

TABLE 6.15   Changes in Refractive Index and Polarizability for Glasses at Wavelength 5876 Å

	Composition (wt %) of five glasses				
Oxide	Vitreous silica	Calcium aluminate F75	Phosphate F1329	Barium Borate E1583	Germanate F998
SiO_2	100.0	5.0			
GeO_2					35.9
B_2O_3				59.0	
P_2O_5			77.7		
Al_2O_3		41.5	12.0		
MgO		5.0	10.3		
CaO		48.5			
BaO				41.0	
BaF_2					5.3
La_2O_3					17.4
Ta_2O_5					3.4
TiO_2					6.1
ZnO					6.2
ZrO_2					4.7
Property					
Refractive index n_{0D}:	1.45846	1.67155	1.48905	1.5822	1.8550
$dn/dT \times 10^5$:	0.87	0.75	0.46	-0.11	0.86
$\gamma \times 10^5$:	0.13	2.15	1.48	1.85	2.18
$d(\partial n/\partial d)$:	0.32	0.19	0.33	0.40	0.39
$(\partial n/\partial T)_d \times 10^5$:	0.91	1.16	0.95	0.63	1.71
Λ_0:	0.17	0.65	0.19	0.16	0.41
$P \times 10^5$:	-0.05	-1.20	-0.61	-0.88	-1.40
$Q \times 10^5$:	0.01	0.74	0.12	0.14	0.58
$R \times 10^5$:	0.91	1.20	0.95	0.63	1.70

Source. Waxler and Cleek (1973).

Assuming polarizability to remain constant, they differentiated the Drude equation

$$(n^2 - 1)/4\pi N = \alpha \qquad (6.9)$$

to determine the dependence of refractive index on density

$$d(\partial n/\partial d) = (n^2 - 1)\,/2n \tag{6.10}$$

where N = number of oscillators per unit volume and α = polarizability. Values of the strain polarizability constant

$$\Lambda_0 = (d/\alpha)\,(\partial\alpha\,/\,\partial d) \tag{6.11}$$

introduced by Mueller (1935) and Ramachandran (1958) were calculated and are shown in Table 6.15. Decreases in polarizability given by Λ_0 increase with density and are positive for all glasses. This behavior is typical for cubic crystals and glasses.

Waxler and Cleek attribute changes in refractive index caused by hydrostatic pressure to two opposing effects: (1) the increase in N, the number of scattering centers per unit volume, which always results in an increase, and (2) contraction of the electronic cloud, which causes a decrease in atomic polarizability. They note that for MgO, diamond, and ZnS, the second effect outweighs the first and $d(\partial n/\partial d)$ values are negative for these crystals.

Waxler and Cleek used the following relation to calculate values of the change of refractive index with temperature

$$dn/dT = (\partial n/\partial T)_d - (d\partial\,n/\partial d)_T^2 \tag{6.12}$$

where γ is the coefficient of volume expansion. The first term in this equation (at constant density d) is usually much smaller than the second (at constant temperature T) for most substances, and when this is true, dn/dT is closely equivalent to the second term. Table 6.15 lists positive values of dn/dT for all glasses except the barium borate composition. The minimum value of the index for this glass occurs at about 150°C, and it is understandable that dn/dT is negative near room temperature. Most optical glasses have minimum indices below 0°C. Relative to developing glasses with low values of dn/dT for optical use, each term in equation (6.12) is uniquely dependent on glass composition. Data are not now available for calculating such values in terms of specific glass oxides. Contributions of constituent oxides to these parameters and their relations with glass structures must be determined from further research. The very low value of γ, the coefficient of volume expansion for vitreous silica, indicates that a high value of γ cannot be caused by volume changes. The author suggests that this behavior of vitreous silica is related to its unique structural changes with temperature as discussed in Section 4.6. Changes in refractive index and volume in terms of the structure of vitreous silica are covered in detail by Vukcevich (1971) and by Bruckner (1970, 1971).

Waxler and Cleek (1973) outline the work of Ramachandran (1958), who suggested that dn/dT consists of three contributions designated as P, Q, and R. The P is dependent only on changes in density in terms of the numbering of scattering centers; Q denotes the change in polarizability caused by change in density, and R represents the change in polarizability caused by temperature change. The correspinding equations are

$$P = -\gamma(n^2 - 1)\big/2n \tag{6.13}$$

$$Q = -\gamma[d(\partial n/\partial d) - (n^2 - 1)/2n] \tag{6.14}$$

$$R = dn/dT + \delta(d\partial n/\partial d) \tag{6.15}$$

Values of P, Q, and R for the five glasses are given in Table 6.15. The P is always positive and reflects the decrease in scattering centers with increase in temperature. Values of P and Q for vitreous silica are very small because of the low coefficient of expansion of this material. Values of R are positive and considerably larger than Q for all glasses, indicating that the change in polarizability is largely dependent on temperature. Waxler and Cleek (1973) outline implications of the various parameters to problems of athermalization in optical elements, self-focusing and electrostriction effects in lasers, and molecular scattering in optical transmission lines.

Waxler et al. (1971) determined data on physical and chemical properties of five commercial neodymiumdoped laser glasses. Their measurements included thermooptic properties, photoelasticity, refractive index, optical homogenity, transmittance, thermal conductivity, hardness, density, and chemical analyses. Thermal changes in refractive index at constant volume or density were related to selffocusing of laser light. The investigators used the following equasion to relate pertinent parameters to the use of glass lasers:

$$dn/dT = [d(\partial n/\partial d)_T \ 1/d(\partial d/\partial T) + (\partial n/\partial T)_d] \tag{6.16}$$

where d = density. The third term on the right is equal to the negative of the volume coefficient of expansion, which is equal to three times the linear coefficient of expansion. The terms dn/dT and $(\partial n/\partial T)d$ are important in calculating the thermal self-focusing of laser beams. The first part of the laser pulse heats the glass and causes changes in refactive index and thermal expansion. When there is sufficient time for thermal expansion to occur, dn/dT becomes important. If a Q-switched laser pulse pulse passes before acoustic relaxation occurs, $(\partial n/\partial T)d$ becomes important. Data on $d(\partial n/\partial d)T$ are used to calculate self-focusing induced by electrostriction. Table 6.16 lists chemical analyses of the five commercial glasses, together with calculated values of the pertinent parameters.

In summary we can compare data on hydrostatic pressure and temperature effects on refractive index from the tabular information. Table 6.16 shows that dn/dT for commercial optical glasses varies with wavelength and attains minimal values at wavelengths characteristic of specific glasses. Thus we infer that minima are related to glass compositions, which are not known in this case. Data in Tables 6.15 and 6.16 relate to wavelengths of 5876 and 6438 Å, respectively. Waxler et al. assumed that dn/dT at 6438 Å was reasonably close to that at 10,600 Å. With increasing wavelength, the sign of dn/dT changes from negative to positive at a given wavelength. With increasing temperature, the sign of

TABLE 6.16   Optical Characteristics of Commercial Laser Glasses at Wavelength 6438 and 25°C

Oxide	Composition (wt %) of commercial laser glasses				
	A	B	C	D	E
SiO_2	66.6	61.3	67.8	66.3	66.1
Li_2O	1.0	1.0			14.5
Na_2O	6.7	6.2	7.9	3.4	
K_2O	10.0	17.2	13.8	18.2	
CaO					10.1
BaO	5.5	3.0	3.3	3.5	
PbO			1.1	1.8	
ZnO	1.6	1.8			
Al_2O_3	1.8	1.9			4.4
Nd_2O_3	5.4	5.8	5.5	3.5	3.4
Sb_2O_3	0.8	0.8	0.9	3.2	
CeO_2					0.5
Fe_2O_3	0.011	0.010	0.0070	0.0090	0.0055
TiO_2			0.19	0.39	

Property					
Refractive index n_d:	1.51663	1.52116	1.51437	1.51781	1.56480
$dn/dT \times 10^5$:	−0.18	−0.16	−0.03	−0.40	−0.28
$\gamma \times 10^5$:	2.94	3.01	2.65	2.90	2.52
$d(\partial n/\partial d)$:	0.34	0.34	0.34	0.36	0.30
$(\partial n/\partial T)_d \times 10^5$:	0.82	0.86	0.87	0.64	1.03

Source. Waxler, Cleek, Malitson, Dodge, and Hahn (1971).

dn/dT changes from negative to positive at a given temperature. Data are not available for calculating specific effects of individual oxides on these two variables. Tabular data indicate that dn/dT varies both in magnitude and sign for the different glasses. Values of volume coefficient of expansion ℓ are of the same order of magnitude excepting vitreous silica. Values of $d(\partial n/\partial d)$ and $(\partial n/\partial T)_d$ can also be compared for glasses in the two groups.

The calcium aluminate, phosphate, barium borate, and germanate glasses in Table 6.15 are representative of nonsilicate glass-forming systems. Thus the five glasses have compositions located in five different primary phase fields, and their property-composition relations cannot be determined using the substructure method. Specific oxide contributions to the properties would be only approximate and of limited use, whatever method was used to determine them. Glasses in Table 6.16 are all silicates, but primary phase locations of the composition are not definitely known. Equation constants of the oxides could not be quantitatively determined by the substructure method. However subjecting the group of glasses to least squares analysis would provide useful information for estimation purposes.

Glass manufacturers have their own proprietary methods for formulating commercial glasses. It is evident from Table 6.16 that manufacturers of glasses A to E have used somewhat different approaches to the problem of making laser glasses. Manufacturers continue to provide the optical community with suitable glasses for use in optical systems of ever-improving precision. Technology always runs ahead of basic understanding of optical parameters in meeting demands of optical designers. It is evident that understanding of optical parameters and their relations to glass compositions and structures is an important part of this technology. Clarification of these dimensions lies in the area of materials science research in glasses, taking both experimental and theoretical approaches.

7

Stress-Optical Effects
in Glasses

Effects of mechanical pressure on the refractive index of glasses depend on the magnitude of the externally applied pressure and the manner in which it is applied. Two types of application can be distinguished. (1) Glass under non-uniform pressure, as in tension or compression, becomes birefringent; and the resulting difference in refractive index is proportional to the magnitude of the stress. Such measurements can be used to study the material constants of the glass or to evaluate internal strain in glass as a guide to annealing procedures. (2) Uniform hydrostatic pressure results in increases in refractive index, but the glass does not become birefringent.

Other methods of producing birefringence in glass are briefly outlined. Interest in these methods lies particularly in material constants of glasses and their contributions to glass microstructures.

7.1 PHOTOELASTIC PROPERTIES

When not under stress other than atmospheric pressure, annealed glass is isotropic. If it is subjected to a uniaxial pressure P as in Figure 7.1, it becomes doubly refracting or birefringent. When viewed in polarized light between crossed Niclos prisms it shows interference colors. Glass under such a stress behaves as an anisotropic crystal, and the birefringence is proportional to the intensity of the stress. This phenomenon is the basis of the method in common use for the detection and estimation of internal strain in glass. Internal strain caused by too rapid cooling can be removed by annealing; strain caused by composition inhomogeneity cannot.

Relations between elastic deformation and the propogation of light in glass can be described in terms of the Neumann (1841) equations

$$
\begin{aligned}
v_x &= v + qX_x + pY_y + pZ_z \\
v_y &= v + pX_x + qY_y + pZ_z \\
v_z &= v + pX_x + pY_y + qZ_z
\end{aligned}
\tag{7.1}
$$

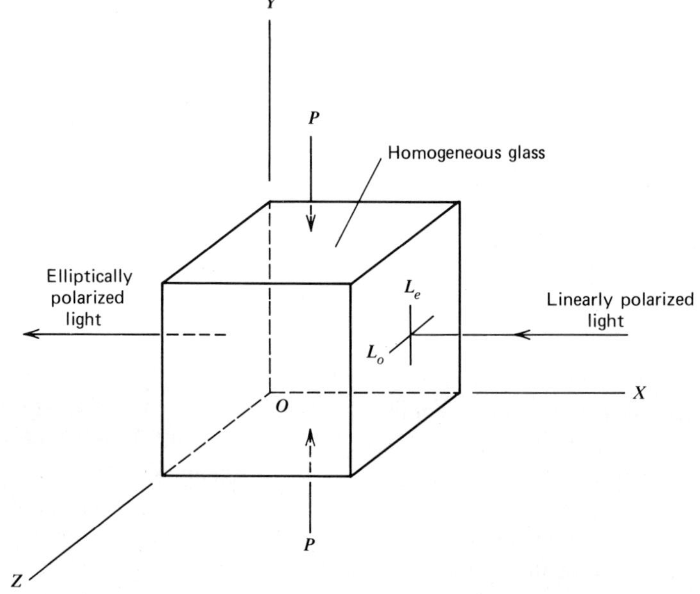

Figure 7.1 Effect of uniaxial compression on light propagation in homogeneous glass.

where X_x, Y_y, and Z_z are dilations in the three directions parallel to the three principle axes, v is the velocity of light in the unstressed glass, v_x, v_y, and v_z are the velocities of light in the stressed glass whose vibrations are parallel to the three axes, and p and q are material coefficients of the glass under consideration and can be determined experimentally.

A compressive stress p acting on the glass in Figure 7.1 paralleled to the Y axis will produce dilations

$$\begin{aligned}
X_x &= \sigma P/E \\
Y_y &= -P/E \\
Z_z &= \sigma P/E
\end{aligned} \tag{7.2}$$

in which E is Young's modulus and σ is Poisson's ratio. If n is the refractive index of the unstressed glass and n_x, n_y, and n_z are the refractive indices of the stressed glass for light vibrating in the three principal directions, then

$$nv = n_x v_x = n_y v_y = n_z v_z \tag{7.3}$$

By combination of equations (7.1) to (7.3), it follows that

$$(ny - nz)/n = P/E \ (1+\sigma) \ (q/v - p/v) \qquad (7.4)$$
$$(ny - nz)/n = P/2S \ (q/v - p/v) \qquad (7.5)$$

where S is the shear modulus.

The change of n_x is the same as that of n_z, and no birefringence is observed for light passing through the glass in the direction of the optic axis OY. Light vibrating at right angles to the optic axis is called the ordinary ray L_O; that vibrating through the optic axis is called the extraordinary ray L_e. The birefringence is by definition equal to $n_y - n_z$. It is evident from the relations just given that the birefringence is proportional to the difference of the cofficients p and q. However, p and q cannot be determined by measurements of birefringence alone. It is necessary to measure the absolute retardation of a ray of light vibrating, say, in the direction OY, to uniquely determine p and q. This measurement plus measurements of refractive indices, elastic moduli E or S, pressure P, and Poisson's ratio σ provides complete photoelastic information on a given glass. Two stress-optical coefficients can be determined from such measurements and are defined by the equations

$$n_y - n = C_1 Py \qquad (7.6)$$
$$n_z - n = C_2 Py \qquad (7.7)$$
$$\therefore \quad C = C_2 - C_1 \qquad (7.8)$$

where C is the relative stress-optical coefficient used by glass manufacturers in evaluating annealing procedures and by customers of the products.

Photoelastic research on glass following Neumann's equations provides information on coefficients p and q and their relations with elastic moduli. In addition the technologically important relative stress-optical coefficient C can be derived from such studies. Although relatively simple methods are available for direct measurements of the relative coefficient C, they cannot furnish data on material constants p and q. Studies of both types, principally on commercial glass compositions, have been published. Morey (1954) and Baak (1969) outline descriptions of work in this area. Schaefer and Nassenstein (1953) used a static interferometric method to make a complete photoelastic study of 154 optical glasses manufactured by Schott. Borrelli and Miller (1968) chose a dynamic method employing a quartz resonator for a similar study of three glasses manufactured by Corning. From a materials science standpoint such studies need to be made on systematically arranged experimental compositions.

7.2 RELATIVE STRESS-OPTICAL COEFFICIENT

Considerable data on the coefficient C have been published because of its technological importance. It is in order to examine in detail the effect of a uniaxial compression P on isotropic glass as illustrated in Figure 7.1. The thrust P compresses the glass in the direction OY and extends it in equal amounts in the directions OX and OZ. The velocity of light is altered when the light enters the glass. The wave front is split into two components, which are polarized in directions at right angles to each other. The compressed glass acts like an anisotropic crystal and causes the light to become elliptically polarized. The ray of elliptically polarized light can be treated as the resultant of two rays vibrating along OY and OZ and differing in phase. A thrust parallel to OZ will produce a similar effect. However a thrust parallel to OX will produce no phase difference in the rays vibrating along OY and OZ because the glass is extended equally in these two directions. Under conditions of Figure 7.1 after the ordinary ray L_o and the extraotdinary ray L_e emerge from the glass, they will be out of phase, exhibiting a relative retardation of $(n_z - n_y)n$ centimeters per centimeter. The unit commonly used for the stress-optical coefficient is the brewster, which represents a fractional difference in retardation of one part in 10^{13} per dyne per square centimeter in a distance of 1 cm. This coefficient also can be expressed in terms of relative retardation in microns per centimeter for stress in kilograms per square centimeter.

$$Py - Pz = [(nz - ny)/n] \ 10^{13}/C \tag{7.9}$$
$$= [(nz - ny)/n] \ 10^{7}/C' \tag{7.10}$$

where $(n_z - n_y)/n$ = fractional difference of retardation (cm/cm)

$\qquad C$ = stress-optical coefficient (brewsters)

$\qquad C'$ =stress-optical coefficient [(microns x cm)/kg]

For example, a stress difference $P_y - P_z$ of 75 kg/cm^2 when applied to a glass with a refractive index of 1.55 and a stress-optical coefficient C' of 2.5 (microns x cm)/kg will result in the following birefringence:

$$n_z - n_y = (P_y - P_z)\, C'\, n \times 10^{-7}$$
$$= 75 \times 2.5 \times 1.55 \times 10^{-7}$$
$$= 0.000029 \text{ or } 29 \times 10^{-6}$$

It is evident from the foregoing that the coefficient $C' = 0.980\ C$. Thus for practical purposes the two coefficients are numerically equivalent.

The first studies on the stress-optical effect were reported by Pockels (1902, 1903). His research on flint lead silicate glasses showed that coefficient C decreased rapidly as the PbO content increased. Later work by Adams and Williamson (1919) and Waxler and Napolitano (1957) confirmed Pockels' data. Combined results of these authors closely fit the curve in Figure 7.2. Although most glasses under stress behave as uniaxial negative crystals, the data cited showed a reversal of this effect for glasses containing large amounts of

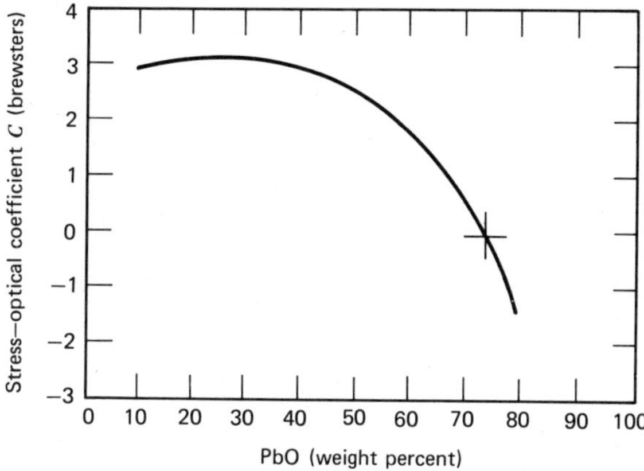

Figure 7.2 Variation of stress-optical coefficient C with PbO
content. Adapted from Waxler and Napolitano (1957).

PbO. It is noted in Figure 7.2 that glasses containing about 75% PbO have C
values of zero; those with higher PbO contents have negative values of C. No
birefringence is observed when C is zero, and internal strain cannot be evaluated
by means of polarized light. Relations between these experimental data on the
two different types of glass and their ionic and electronic microstructures are
not well understood. It is reasonable to assume that a uniaxial compression
would decrease ionic distance in the direction of the pressure and increase such
distances in directions normal to it. Simultaneous compressions and extensions
of electronic systems about the ions would occur particularly for those about
the oxygen anion: SiO^2 has very low polarizability, whereas PbO is rather
easily polarized because of its structutural makeup, as noted by Fajans and
Kreidl (1948). Following the discussion in Section 4.6, it is suggested that
birefringence in vitreous silica is caused by changes in Si-O-Si angles and distortion of electronic systems.

Van Zee and Noritake (1958) described measurements of the coefficient C on
commercial container and plate glasses in the range 23 to 650°C. Data on three
glasses showed that C increased with temperature and reached definite maxima
in the range 570 to 590°C, where the glasses lost the ability to support stress.
Their research indicated that values of C under compressive stresses were measurably larger than those obtained under tensile stresses throughout the temperature range. Maximum values of C were obtained at the same temperature, and
values under compression were markedly larger than those under tension. Figure
7.3 shows the effect of temperature on the coefficient C for plate glass under
compression. Data in Table 7.1 together with those in Figure 7.2 indicate that
PbO and BaO lower the coefficient C whereas SiO_2 and B_2O_3 increase it. Data

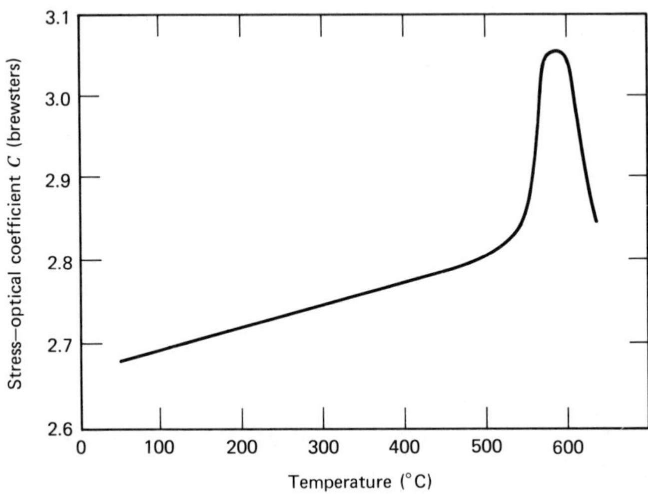

Figure 7.3 Effect of temperature on stress-optical coefficient C for plate glass. Data from Van Zee and Noritake (1958).

on commercial glasses are not suitable sources for evaluating composition effects on this property. Published data on systematically arranged experimental compositions are not available for the most commonly used oxides. The paper by Nissle and Babcock (1973) on two ternary glass systems outlines methods of obtaining quantitative relations between the coefficient C and oxide composition.

Na_2O-TiO_2-SiO_2 Glasses

Nissle and Babcock measured the coefficient C for glasses located in two primary phase fields of this system. Data were segregated into two groups D (primary phase unknown) and E ($Na_2O \cdot TiO_2 \cdot SiO_2$). Twenty sets of data in each group, five for each glass, were then subjected to least squares computer analysis. Table 7.2 presents measured and calculated data, as well as glass compositions. The composition location of the boundary between phases D and E was calculated to be

$$1.91967 = 2.86874\ SiO_2 + 1.81937\ TiO_2$$

from simultaneous solution of the equations for C given in Table 7.2 and $1 = SiO_2 + TiO_2 + Na_2O$. Satisfactory solution of these equations indicates that relations between the coefficient C and composition are measurably different for glasses in groups D and E. Figure 7.4 plots measured values of C with calculated lines of equal values of C. The broken line indicates the calculated boundary between primary phases D and E. The author suggests that the primary phase of group D glasses is probably Na_2O-TiO_2-$2SiO_2$.

TABLE 7.1   Relative Stress-Optional Coefficient C for Commercial Glasses of Various Compositions

| | Composition (wt %) of seven glasses | | | | | | |
| | 1 | 2 | 3 | 4 | 5 | 6 | 7 |
Oxide	Vitreous silica	522595	563509	620603	755276	Plate	Chemical
SiO_2	100.0	73.7	51.6	30.8	28.4	70.9	74.2
B_2O_3		5.0		17.9			9.74
Al_2O_3		1.0		1.4		0.19	5.59
Na_2O			1.5	0.3		12.11	6.57
K_2O		11.1	9.5		2.5	2.16	0.54
MgO						1.64	
CaO						12.00	0.95
BaO			14.0	48.7		0.01	2.19
ZnO			12.0				
PbO		6.0	11.0		69.0		
SO_3						0.44	
$As_2O_3 + Sb_2O_3$		0.3		0.9		0.31	
C (brewsters):	3.47	3.05	1.87	1.10	0.94	2.68	3.18

Glasses	Source	Wavelength
1	Shand (1958)	
2 to 5	Schaefer and Nassenstein (1953); Baak (1969)	5535 Å
6	Van Zee and Noritake (1958)	White light
7	McGraw and Babcock (1959)	White light

TABLE 7.2 Relative Stress-Optical Coefficient C in Na_2O-TiO_2-SiO_2 Glasses

| | Glass | | | |
Oxide	D1	D2	D3	D4
SiO_2	0.600	0.600	0.600	0.670
TiO_2	0.150	0.150	0.200	0.120
Na_2O	0.200	0.250	0.200	0.210
C (brewsters)				
Measured (ave.)	2.8388	2.6600	2.9335	2.7372
	±0.0071	±0.0084	±0.0147	±0.0055
Calculated*	2.8351	2.6595	2.9351	2.7399

Oxide	E1	E2	E3	E4
SiO_2	0.500	0.450	0.400	0.550
TiO_2	0.200	0.250	0.300	0.200
Na_2O	0.300	0.300	0.300	0.250
C (brewsters)				
Measured (ave.)	2.6848	2.8993	2.9898	2.7376
	±0.0132	±0.0125	±0.0106	±0.0082
Calculated[†]	2.7054	2.8580	3.0105	2.7376

$*C = 3.23707\ SiO_2 + 5.23850\ TiO_2 - 0.27407\ Na_2O$

$†C = 2.28800\ SiO_2 + 5.33880\ TiO_2 - 1.64560\ Na_2O$

Source. Nissle and Babcock (1973); wavelength = 5461 Å.

Na_2O-Al_2O_3-SiO_2 Glasses

Nissle and Babcock measured the coefficient C for glasses in the nepheline and corundum primary phase fields of this system. Data were analyzed in terms of least squares equations. Figure 7.5 gives the effect of substituting Al_2O_3 for Na_2O in glasses containing 0.620 SiO_2. Values of indicated measured values are averages of five separate measurements. This substitution increases the coefficient C in both phase fields. The increases are linear with composition and unique for each phase field. Straight lines representing the data change slope at the composition $SiO_2 = 0.62$, $Al_2O_3 = 0.19$, and $Na_2O = 0.19$. Indeed, compositions with $Na_2O = Al_2O_3$ lie on a line that is almost coincident with the phase boundary between nepheline and corundum (Levin, Robbins, and Mc-Murdie, 1964, Fig. 501). For glass technology purposes it is reasonable to assume that the compositions are coincident. Deer, Howie, and Zussman (1966) describe the microstructure of crystalline nepheline and state that linked SiO_4 and AlO_4 tetrahedra are important features. Bragg (1937) gives details on crystalline

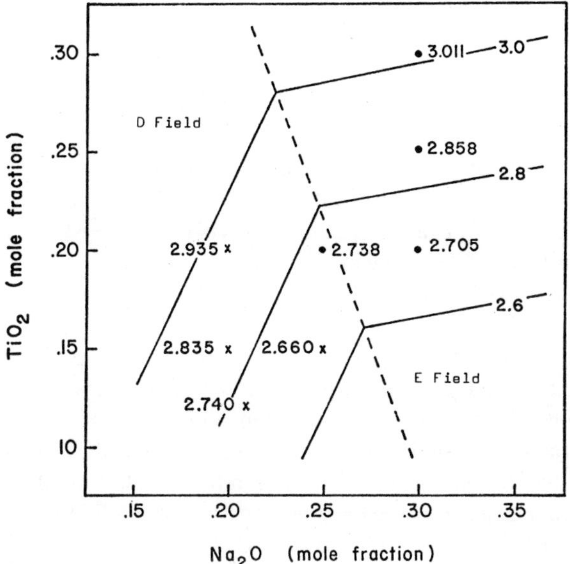

Figure 7.4

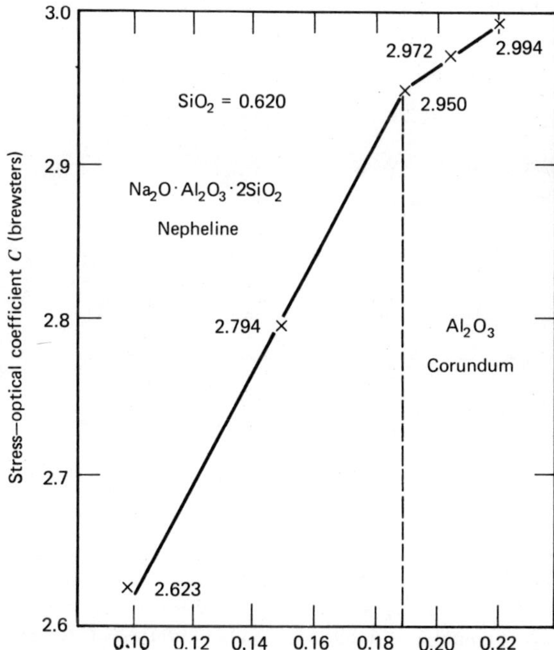

Figure 7.5 Effect of substituting Al_2O_3 for Na_2O on coefficient C. From Nissle and Babcock (1973).

123

corundum and notes that the coordination number of aluminum is 6.

Schairer and Bowen (1956) found that lines of equal refractive index in these glasses change slope abruptly at compositions where $Na_2O = Al_2O_3$. Babcock (1968) published computer calculations of their data indicating abrupt changes in slope at primary phase boundaries nepheline-corundum, albite-mullite and silica-mullite. Previous explanations of these data have been given in terms of a change in coordination of Al from 4 to 6 in a randomly arranged glass structure. This discussion indicates that these data can be understood in terms of different glass substructures existing in the adjacent primary phase fields.

Table 7.3 gives measured and calculated data on C for glasses in the nepheline phase field. The least squares equation gives good representation of the measured data. An equation representing data in the corundum phase field was derived by assuming that glasses in which $Na_2O = Al_2O_3$ are located on the nepheline-corundum phase boundary and can be considered to be located in both phase fields. Table 7.4 lists information on three measured glasses and three calculated glasses that lie on the nepheline-corundum phase boundary. Calculated values were obtained from the nepheline equation. The corundum equation represents data on the six glasses and shows good agreement with glasses in the group. The corundum equation can be used for property estimation purposes, but it must be understood that further experimental verification is needed. These calculations outline advantages of the substructure method in deriving property-composition relations for glasses in contiguous phase fields.

TABLE 7.3   Relative Stress-Optical Coefficient C of Glasses in the Nepheline Primary Phase Field

Oxide	Glass				
(mole fractions)	A1	A2	A3	A6	A7
SiO_2	0.620	0.620	0.620	0.650	0.590
Al_2O_3	0.100	0.150	0.190	0.125	0.125
Na_2O	0.280	0.230	0.190	0.225	0.285
C (brewsters)					
Measured (ave.)	2.6233	2.7944	2.9495	2.8057	2.6510
	±0.0144	±0.0067	±0.0131	±0.0091	±0.0114
Calculated*	2.6311	2.8070	2.9476	2.7964	2.6417

$*C = 3.25921\ SiO_2 + 4.19831\ Al_2O_3 + 0.68054\ Na_2O$

Source. Nissle and Babcock (1973).

Waxler and Napolitano (1957) measured relative stress-optical coefficients on a number of optical glasses. Table 7.5 contains compositions and measured values of C for seven glasses located in the SiO_2 primary phase field. The linear equation representing these data was determined by the previously described substructure method. Note that the equation gives satisfactory representation of the measured data. Comparisons of constants for SiO_2 and Na_2O in Tables 7.3 to 7.5 reveal their variations in magnitude depending on phase fields in which the glasses are located.

7.3 OTHER TYPES OF BIREFRINGENCE IN GLASSES

Birefringence in isotropic glass can be produced in additional ways, including applications of electric or magnetic fields, ionic migration, and structural or orientation birefringence. The sign convention used here is that uniaxial compression, as illustrated in Figure 7.1, results in positive birefringence in the direction of the optic axis, whereas uniaxial tension causes negative birefringence. Refractive index and density are increased in the direction of compression and decreased in the direction of tension. The respective signs are reversed for optical flint glasses containing more than about 75 weight percent PbO.

Effects of Electric Fields

Subjection of glass to the action of an electric field directed normal to a light beam causes the glass to become birefringent. The resulting birefringence can be expressed simply as

$$n_e - n_o = B \lambda E^2 \tag{7.11}$$

where B is Kerr's constant, λ is the wavelength, and E is electric field intensity. Relatively few data on glasses have been published. Morey (1954) outlines early work. Borrelli (1971) measured the birefringence induced in a number of experimental glasses by an applied electric field. Birefringence values for lead silicate glasses were low, but those for glasses with high concentrations of T^+, Nb^{5+}, and Ta^{5+} cations were much higher. Morey discussed underlying theory of the phenomena and pointed out that observed birefringence values were dominated by the electrooptic effect, the electrostrictive contributions being relatively small. Borrelli used a polarized light beam from a He-Ne gas laser (6328 Å) and noted that the very high birefringence of As_2S_3 glass could be explained in part by dispersion. The onset of intrinsic absorption is in the red region, that for oxide glasses is in the ultraviolet.

Effect of Magnetic Fields

Faraday (1830) discovered that the plane of polarization was rotated when polarized light was passed through glass along the direction of lines of force in

TABLE 7.4 Relative Stress-Optical Coefficient C of Glasses in the Corundum Primary Phase Field

Oxide (mole fractions)	Glass					
	A3	A4	A5	C1	C2	C3
SiO_2	0.620	0.620	0.620	0.640	0.660	0.680
Al_2O_3	0.190	0.205	0.220	0.180	0.170	0.160
Na_2O	0.190	0.175	0.160	0.180	0.170	0.160
C (brewsters)						
Measured (ave.)	2.9492	2.9732	2.9936			
	±0.0131	±0.0110	±0.0088			
Calculated* (nepheline)				2.9641	2.9805	2.9969
Calculated[†] (corundum)	2.9493	2.9718	2.9943	2.9650	2.9807	2.9965

*C = 3.25921 SiO_2 + 4.19831 Al_2O_3 + 0.68054 Na_2O
[†]C = 3.24825 SiO_2 + 3.21246 Al_2O_3 + 1.71035 Na_2O

Data source. Nissle and Babcock (1973).

TABLE 7.5 Relative Stress-Optical Coefficient C for Glasses in the SiO_2 Phase Field

Oxide (mole fractions)	Glass						
	4	5	6	7	8	9	10
SiO_2	0.7195	0.6976	0.7136	0.6989	0.6819	0.6676	0.6461
PbO	0.1647	0.1760	0.1896	0.2305	0.2522	0.2800	0.3155
Na_2O	0.0319	0.0675	0.1406	0.0011	—	—	—
K_2O	0.0839	0.0589	0.0562	0.0695	0.0695	0.0524	0.0383
C (brewsters)							
Measured	2.98	2.77	2.89	2.65	2.45	2.18	1.77
Standard error	0.016	0.014	0.031	0.015	0.026	0.020	0.010
Calculated*	3.04	2.74	2.89	2.64	2.40	2.16	1.82

*C = 6.527 SiO_2 − 7.102 PbO − 4.813 Na_2O − 3.999 K_2O

Data source. Waxler and Napolitano (1957).

a magnetic field. The resultant birefringence can be expressed as

$$n_e - n_O = CM_\lambda H_2 \tag{7.12}$$

where CM is the Cotton-Mouton constant, λ is the wavelength, and H is magnetic field intensity. Morey (1954) discusses early research on the subject. Relatively few data are available on glasses, and specific effects of oxides cannot be evaluated.

Effects of Ionic Migration

Glasses at elevated temperatures undergo electrolytic conduction when under the action of a dc electric field. This effect is most pronounced in glasses containing monovalent cations such as Na. Prod'homme (1964) demonstrated that an ordinary crown glass containing 15% Na_2O became permanently birefringent under such conditions. The migration of Na^{1+} cations caused changes in the glass structure which could not be removed by annealing. He observed similar effects in other glasses containing Na_2O.

Structural or Orientation Birefringence

Filon and Harris (1923) found that optical glasses of unstated composition became birefringent if they were heated to about 400°C, then cooled to room temperature under a compressive load. Data on two glasses showed birefringences of 10^{-6}. These authors proposed a diphase structure for the two glasses to account for their observations. Okuda (1927) made a detailed study of the same nature using an Na_2O-CaO-SiO_2 glass. He observed a birefringence of 5×10^{-6}, advancing assumptions to account for the uniform birefringence and suggesting that orientation of structural elements was most plausible. Raman (1950) studied this effect in glasses and preferred to describe it with the term "structural birefringence". Published data indicate that this type of birefringence is of the order of 10^{-6} for vitreous silica and silicate glasses. Okuda reported that orientation birefringence in the glass he studied could be removed by annealing. Van Zee and Noritake (1958) observed this effect in studying commercial glasses and attributed the birefringence to "frozen stress patterns." Structural design engineers use models made of organic polymers to evaluate three-dimensional strain. Stress-optical constants of these materials are many times larger than those for inorganic oxide glasses. After the models have been cooled to room temperature while under stress, they can be cut and polished without disturbing the frozen stress patterns. The resulting birefringence is usually said to be due to orientation of polymer chains and other features of their microstructures. Stress-optical coefficients of glasses are too low to take advantage of this method of internal strain analysis. However it seems certain that this phenomenon in glasses involves directional changes in their microstructures. The effect of directional microstrains on glass structures and the effects of oxide composition on these phenomena are currently under research.

Different points of view have been put forward to relate the effects of directional microstrains on glass structures. Indenbom (1953) suggested that the glass structure consists of two phases or frameworks having different melting points and that the birefringence is related to interactions between the two phases. Bogdyk'yants (1955) postulated the presence of chemically heterogeneous regions in the glass to explain the effect. Most explanations of this effect, incidentally, depart from the assumptions of the random network picture of glass structures. Botvinkin and Ananich (1958) studied a series of Na_2O-B_2O_3-SiO_2 glasses and observed that the numerical value of the birefringence under compression was larger than that under tension. They found interesting results on 100% B_2O_3 glass and a binary containing 35% B_2O_3 and 65% ZnO. The authors established compressive strains in the glasses then heated the glasses to given temperatures. Samples were then cooled to room temperature. Restoration of microstrains (amount of) depended on the temperature to which they had been heated. On heating these glasses, after compressive microstrains had been established, the birefringence gradually disappeared. On further cooling to room temperature, however, the birefringence was either wholly or partially resored, depending on the temperature to which the glass had been heated. Botvinkin and Ananich obtained similar results on certain borosilicate glasses. They suggested that such birefringence in inorganic glasses could be associated with orientation of structural aggregates or frameworks.

8

Magnetic Properties

The electromagnetic theory describes macroscopic effects of electrical and magnetic fields on materials. The interaction of electrical fields with glasses acting as insulators was outlined in Chapter 5. The effects of a magnetic field on glasses as insulators similarly involves energy absorptions and their decay with time. Problems relating macroscopic magnetic quantities to microscopic structures of glasses are no less difficult than their electrical counterparts.

8.1 DEFINITIONS AND NOMENCLATURE

The macroscopic magnetic quantities can be defined analogously to electrical quantities.

Magnetic		*Electrical*	
Magnetic field	H	Electrical field	E
Magnetic induction	B	Electric displacement	D
Permeability	μ	Dielectric constant	k
Magnetization	M	Polarization	p
Susceptibility	γ	Susceptibility	ϵ
$M = \gamma H$		$P = \epsilon E$	
$B = \mu H$		$D = k E$	

Imposition of a magnetic field causes changes in the magnetic fields of materials. The material is designated paramagnetic if there results an increase in the number of lines of force in the same direction of those in the imposed field. In this case the magnetic susceptibility γ has positive values. Diamagnetic materials reduce the density of lines of force and produce a flux opposed to that of the external field. These materials have negative values of γ.

The effect of magnetization M on structural constituents of a material may be categorically indicated as follows:

$$M = M_e + M_i + M_O \qquad (8.1)$$

where M_e, M_i, and M_O relate to separate effects of electrons, ions, and orientation of ionic groups in the lattice or substructure. Sorting out these separate effects in relation to M and other measurable magnetic quantities is difficult because they are interdependant. Distinctions can be drawn between paramagnetic and diamagnetic elements from our knowledge of atomic physics. Permanent magnetic moments of the atoms result from the following contributions: (1) the intrinsic or spin moment of the electrons, (2) the orbital motion of the electrons, and (3) the nuclear magnetic moment. Nuclear magnetic moments are three orders of magnitude smaller than electron moments and can be neglected in this discussion. Distinctions can be drawn between the two types of element in terms of electron quantum numbers n, ℓ, $m\ell$, and m_S, where

n = principal quantum number ($n = 1, 2, 3, \ldots$)
ℓ = orbital angular momentum quantum number ($= 0. 1. 2 \ldots$)
$m\ell$ = magnetic quantum number ($m\ell = 0, \pm2, \pm2, \ldots$)
m_S = spin quantum number ($m_S = \pm\frac{1}{2}$)

The ground state of an atom is usually the state with the lowest set of quantum numbers. The electronic shells of atoms, where n is the shell number, can be described in terms of Pauli's exclusion principle, which states that no two electrons in an atom can have identical sets of quantum numbers.

Table 8.1 gives electron states in the first two atomic shells. The $n = 1$ shell can have at most two electrons because $\ell = 0$, $m\ell = 0$, and $m_S = \pm\frac{1}{2}$. A third electron has to go into the second shell, where $n = 2$. Thus the first shell is filled at the element He, indicated in Table 8.2. Let us consider the second shell, which starts with Li and ends as a filled shell at Ne. For the filled shell we have

TABLE 8.1   Electron States in the First Two Atomic Shells

Shell	n	ℓ	$m\ell$	m_S	Number of electrons
K	1	0	0	$+\frac{1}{2}$	2(1S)
		0	0	$-\frac{1}{2}$	
L	2	0	0	$+\frac{1}{2}$	2(2S)
		0	0	$-\frac{1}{2}$	
		1	0	$+\frac{1}{2}$	
		1	0	$-\frac{1}{2}$	
		1	+1	$+\frac{1}{2}$	6(2P)
		1	+1	$-\frac{1}{2}$	
		1	-1	$+\frac{1}{2}$	
		1	-1	$-\frac{1}{2}$	

The 2S and 2P states together total 8.

TABLE 8.2 Electron Shell Structures of Elements

| Element | | K ($n=1$) | L ($n=2$) | | M ($n=3$) | | | N ($n=4$) | | | |
Symbol	Number	1S	2S	2p	3S	3p	3d	4S	4p	4d	4f
H	1	1									
He	2	2									
Li	3	2	1								
Be	4	2	2								
B	5	2	2	1							
C	6	2	2	2							
N	7	2	2	3							
O	8	2	2	4							
F	9	2	2	5							
Ne	10	2	2	6							
Na	11	2	2	6	1						
Mg	12	2	2	6	2						
Al	13	2	2	6	2	1					
Si	14	2	2	6	2	2					
P	15	2	2	6	2	3					
S	16	2	2	6	3	4					
Cl	17	2	2	6	2	5					
Ar	18	2	2	6	2	6					
K	19	2	2	6	2	6		1			
Ca	20	2	2	6	2	6		2			
Sc	21	2	2	6	2	6	1	2			
Ti	22	2	2	6	2	6	2	2			
V	23	2	2	6	2	6	3	2			
Cr	24	2	2	6	2	6	5	1			
Mn	25	2	2	6	2	6	5	2			
Fe	26	2	2	6	2	6	6	2			
Co	27	2	2	6	2	6	7	2			
Ni	28	2	2	6	2	6	8	2			
Cu	29	2	2	6	2	6	10	1			
Zn	30	2	2	6	2	6	10	2			
Ga	31	2	2	6	2	6	10	2	1		
Ge	32	2	2	6	2	6	10	2	2		
As	33	2	2	6	2	6	10	2	3		
Se	34	2	2	6	2	6	10	2	4		
Br	35	2	2	6	2	6	10	2	5		
Kr	36	2	2	6	2	6	10	2	6		

$\ell = 0$ and $\ell = 1$. For $\ell = 0$, $m\ell = 0$, and $m_S = \pm\frac{1}{2}$. At $\ell = 0$ there is no orbital moment, and the two spin moments cancel. For $\ell = 1$ we have $m\ell = -1, 0, 1$, which cancel, and for each we have $m_S = +\frac{1}{2}$ and $m_S = -\frac{1}{2}$. Thus the spin moments again all cancel. giving zero total moment. The elements He and Ne with filled shells have zero magnetic moments and are diamagnetic with negative values of magnetic susceptibility γ. This is also true for the other rare gas atoms Ar, Kr, and Xe, which is why inert gases are so suitable for studying diamagnetic susceptibility. Tables 8.1 and 8.2 in conjunction with Figure 8.1 give complete

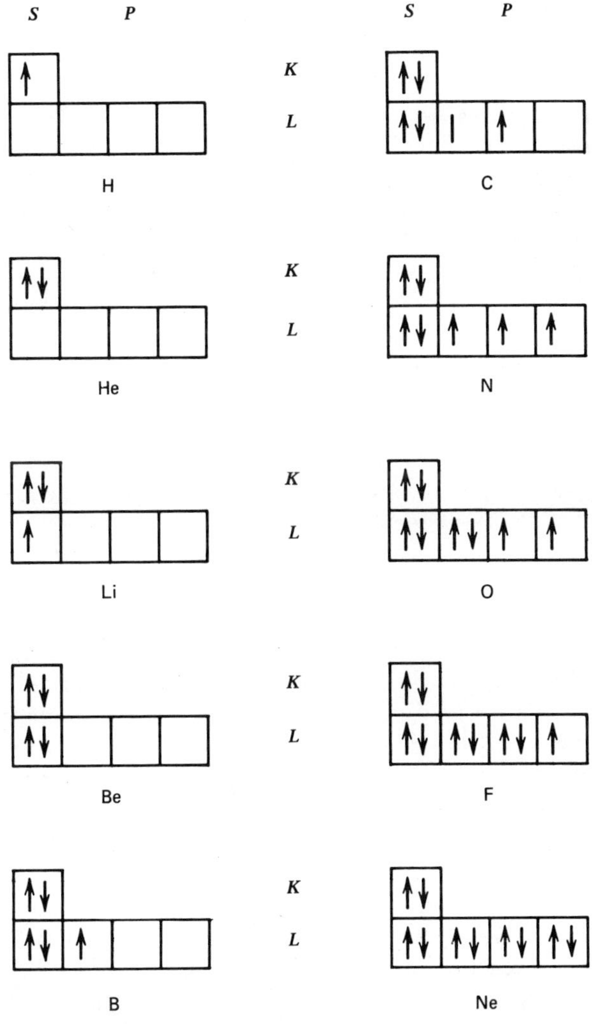

Figure 8.1 Filling of first two shells of atomic electrons.

information on electronic structures of elements with atomic numbers Z from 1 to 10. The first column in Figure 8.1 shows the S states ($1S$ and $2S$) and the P states in the next three columns. The boxes represent magnetic substates, each of which contains two spin states shown by arrows. Generally speaking, elements having single unpaired spins are paramagnetic and those with paired spins in opposite directions are diamagnetic. However two electrons may pair up to form diatomic molecules as in H_2, Cl_2, and so on. Also atoms such as Li and B may form covalent bonds with oxygen by sharing electrons. Such combinations may be diamagnetic. In other words the magnetic characters of such complexes must be evaluated in terms of bonds and structures of the given material. Thus relating macroscopic magnetic measurements to microstructures presents the materials scientist with serious difficulties.

The K and L shells are filled in a normal manner as indicated. Filling of the M shell involves electronic shielding and relative energies of electronic states. The order of filling of electron subshells is $1S$, $2S$, $2P$, $3S$, $3P$, $4S$, $3D$, $4P$, $5S$, $4D$, $5P$, $6S$, $4P$, $5D$, $6P$, $7S$, $5F$, and $6D$. The transition elements with Z in the range 21 to 28 have unfilled inner shells, which give them paramagnetic properties that they retain in combination with other elements, as in the oxides. The unfilled shells are next to the valence shells, and their properties are influenced by the structural environment. Elements with ranges of Z from 57 to 72, and from 91 to 102 also have unfilled inner shells. However the electrons in their inner shells are less affected than those of the transition elements, because there is a filled shell between the valence shell and the unfilled shell. Table 8.2 gives shell structures for most of the elements of interest in the magnetic properties of glasses. The reader can locate tables showing structures for the remaining elements.

8.2 MAGNETIC PROPERTIES OF GLASSES

Published data on magnetic properties of glasses are not extensive. Interests in glass properties are strongly influenced by the technological importance of such properties in glass manufacturing. The study of magnetic properties has been enhanced by development of new instrumentation and techniques in recent years. Efforts have been directed toward an understanding of relations between macroscopic properties and glass microstructures. Glasses containing the transition elements are the most fruitful and interesting for magnetic property investigations. The contribution of electrons to magnetic properties are strongly influenced by surrounding electrostatic fields. This approach, known as the crystal field theory, was first developed for ions in a crystal lattice. Because the major effects result from nearest neighbors, use of this model does not require an extended crystal lattice. Crystal field theory has been extended to glasses and is generally referred to as the ligand field theory. The analysis of interaction between ionic electron orbits and fields of neighboring ions can be simplified somewhat by considering the symmetry characteristics of the field. Selwood (1956), Figgis (1966), Earnshaw (1968), and similar texts supply details on the use of crystal and ligand field theory.

The presence of paramagnetism in glasses containing transition elements, whether these glasses absorb light in or near the visible spectrum, are properties that depend on the existence of incomplete electron shells and accompanying energy levels. These phenomena are closely related and cannot be distinctly separated. However for discussion purposes spectral absorptions are covered separately in Chapter 9. Recent research indicates that paramagnetism and spectral absorption can best be understood in terms of ligand field theory.

Morey (1954) reviewed the meager contemporary data on magnetic suseptibility of glasses. Included were measurements by Cole (1950) on 21 Chance optical glasses and vitreous silica. These representative data show that glasses free from the presence of transition elements are weakly diamagnetic. However the addition of relatively small amounts of these paramagnetic oxides cause the glass itself to become paramagnetic. Cole (1951) covers theoretical background and measurement method in connection with magnetic susceptibility data on glasses containing iron oxide. He added varying amounts of Fe_2O_3 to a base glass containing the following approximate weights percent: SiO_2, 72; Al_2O_3, 1; CaO, 12; Na_2O, 15. Magnetic moments were expressed in terms of Bohr Magnetons, which are equivalent to 0.9273×10^{20} erg/gauss. The magnetic moment of the glass containing 0.6% Fe_2O_3, melted under oxidizing conditions, was 5.83 Bohr Magnetons. When reducing agents were introduced into the glass batches in increasing amounts, the magnetic moment decreased regularly from 5.83 to 5.30. In the same sequence the glass colors changed from a yellow-green through blue-green, blue, and dark amber. These data relate to effects of changes in surrounding fields on paramagnetic orbital electrons. Cole concluded that color of the oxidized glasses could be attributed to the presence of Fe^{3+} ions or $(FeO_4)^{3+}$ groups, that reduced iron glasses contained Fe^{2+} ions, free or coordinated in the network, and that sulfur-amber glasses contained a chromophoric group made up of Fe, O, and S. Bamford (1960) studied the magnetic susceptibility of sodium borate glasses containing iron. He relates coloring actions of three types of iron ions and ionic groups with unpaired electron spins and ligand field influences.

Investigators have turned to magnetic resonance pheonomena for clarification of microstructure details in glasses. Resonance results from magnetic dipole transitions in a magnetic field and can be caused by both electrons and nuclei. Electron paramagnetic resonance (EPR) is produced by interaction between the microwave magnetic field and the electron magnetic moment. In nuclear magnetic resonance (NMR) the interaction occurs between the magnetic field and the nuclear magnetic moment. Methods of measuring magnetic resonance and other magnetic properties are described by Lark-Horovitz and Johnson (1959) and by the authors of other solid state texts. The following outlines representative publications on EPR and NMR in glasses.

Yafaev (1966) made EPR studies on sodium borate, sodium silicate, and sodium phosphate glasses containing one mole % Cr_2O_3, with Na_2O contents from 11 to 50 mole %. Measurements were made at 9280 X 10^6 Hz and at temperatures of 77 and 295°K. There were detailed differences between the three types of glasses, but the Cr^{3+} ions were usually observed in octahedral environments of six oxygen ions. The data were related in terms of crystal field theory. Toyuki and Akagi (1972, 1974) studied relations between EPR and optical spectra of vanadium in alkali borates containing Li_2O, Na_2O. K_2O, and Cs_2O in the range 5 to 20 mole %. EPR measurements were made at the microwave frequency of 9.1 X 10^9 Hz. Major interest was centered on determining the valence states of vanadium in the glasses. Various workers have indicated the presence of the three valence states V^{3+}, V^{4+}, and V^{5+} in glasses, and some have considered that the VO^{2+} ion is also present. Toyuki and Akagi indicate that vanadium is present in the V^{4+} state in the form of the vanadyl ion VO^{2+} in a ligand field of 4-coordinated symmetry in B_2O_3 glasses.

P. J. Bray and his colleagues have done extensive work on nuclear magnetic resonance in glasses, particularly those containing B_2O_3. Silver and Bray (1958) describe measurements on nuclear quadrupole effects in B_2O_3, Na_2O-B_2O_3, and borosilicate glasses. The quadrupole interaction shifts the four equally spaced Zeeman energy levels, splitting the single magnetic resonance frequency into three lines. Silver and Bray were concerned with spin transitions of ^{11}B and ^{23}Na isotopes, which had resonant frequencies at 7.177 X 10^9 and 5.91 X 10^9 Hz, respectively. Their data on Na_2O-B_2O_3 glasses with Na_2O ranging from 0 to 33 mole % showed the presence of two different boron environments. They assigned a 3-coordinated boron to 100% B_2O_3 glass. The coordination changed to 4 as Na_2O replaced B_2O_3. The NMR data on boron contained in two borosilicate glasses did not yield definite conclusions.

Bray and Silver (1960) published a comprehensive review of NMR studies in glasses, stating:

> The nuclei serve as sensitive detectors of the internal magnetic and inhomogeneous electric fields. Static or time-independent local fields can produce shifts in the resonant frequencies and changes in the spectra. Time-dependent local fields can cause energy exchange between the nuclei and their environment. The rate of this energy transfer is associated with a characteristic time called the spin-lattice relaxation time.

These authors outlined the theory of quadrupole splitting in nuclear resonance and related it to NMR studies of the common isotopes of lithium, beryllium, boron, sodium, and aluminum. They note that ^{29}Si is not adaptable to this particular technique because it has no nuclear quadrupole moment. However they suggest that information on the silicon-oxygen network might be accomplished by the use of glasses enriched in the oxygen isotope ^{17}O, which has a small quadrupole moment. One of their interests was to relate NMR findings to the nature of glass structures. Are glasses made up of random networks, or are the anions and cations in glasses arranged in some orderly manner? Though NMR studies are not conclusive in answering this question, they do furnish valuable information on bonding and oxygen coordination of the metal cations. Kriz, Park, and Bray (1971), following their more recent research, reexamined NMR studies on the ^{11}B isotope. They found that BO_3 groups with one or two nonbridging oxygens could be distinguished from BO_3 groups in which either all three oxygens are bridging or all three oxygens are nonbridging. Kim and Bray (1974) in a study of Na_2O-MgO-B_2O_3 glasses concluded that both Na_2O and MgO can behave as network modifiers in terms of random network teachings. However as the MgO content of the glasses increased, some of the MgO took positions in the glass network, presumably in the form of MgO_4 tetrahedra.

EPR and NMR studies in glasses continue to be actively researched. Only a few published studies have been outlined in this chapter for purposes of illustration, but the regularly published journals listed in the Appendix provide comprehensive and continuing coverage of the subject.

9

Spectral Absorption

Interactions between electromagnetic radiation and glasses may be essentially passive, as in the passage of visible light through lenses and prisms, or they may be active, as in selective absorption in all regions of the spectrum. This chapter outlines representative examples of research directed toward understanding relations between spectral absorption in glasses and their composition and microstructures.

Our knowledge of spectral absorption in materials is based on developments in physics and chemistry made particularly during the first quarter of the present century. Planck in 1900 announced the quantum theory of radiation to explain the energy distribution among the wavelengths of continuous spectra given out by a hot black body. In 1905 Einstein extended the idea of quanta to the radiation itself, thus explaining the photoelectric effect. Bohr in 1913 applied the quantum theory to the individual atoms and derived the equation for the Balmer series of the hydrogen spectrum. Energy exchanges between electrons of the atom and incoming radiation were expressed in terms of energy level diagrams for simple atoms like hydrogen. In 1924 de Broglie developed a scheme of waves and quanta that formed the basis of Schrodinger's invention of wave mechanics in 1926. Pauli in 1925 published the exclusion principle, which stated that no two electrons in an atom can have identical sets of quantum numbers. A major step was taken by Goudsmit and Uhlenbeck in 1926, with their concept that the electron itself possesses a magnetic moment and a mechanical angular momentum. In 1927 Davisson and Germer shot electrons into a nickel crystal and observed an array of spots similar to those discovered by von Laue when he bombarded crystals with x-rays.

Theoretical developments and improved experimental apparatus since the early 1900s have provided the investigator with the means of research on spectral phenomena. These tools have confirmed that light consists of waves and particles. The statement on this problem by Marion (1971) has much to recommend it:

> In the quantum mechanical view of nature, an experimenter's apparatus and the object under study together constitute a system. The only mean-

ingful way to discuss the behavior of the object studied is in terms of the results of measurement. Therefore, whether an electron or a photon appears as a wave or a particle depends on the nature of the measurement that is made. The wavelike or particlelike character of an electron or a photon therefore lies only in the eye of the beholder.

9.1 PASSIVE INTERACTIONS IN THE VISIBLE REGION

When light at normal incidence interacts with glass, some of it is reflected, some is absorbed by the glass, and the remainder is transmitted through the glass. The fraction of light reflected at each glass-air surface increases with refractive index and is given by

$$R = (n - 1)^2 / (n + 1)^2 \tag{9.1}$$

If the refractive index is 1.5, then $R = 0.04$ or 4%. The amount of light absorbed by the glass follows the equation

$$I = I_0 e^{-\epsilon c \ell} \tag{9.2}$$

where I_0 is the intensity of light entering at the first boundary, I is the intensity of light arriving at the second boundary, ϵ is the extinction coefficient, c is the concentration of absorbing species, and ℓ is the path length. Table 9.1 gives extinction coefficient values for four Chance-Pilkington optical glasses. Each value is equivalent to the absorption coefficient per centimeter path length; extinction coefficients are unique for each glass, and they vary with wavelength. Extinction coefficients are dependent on glass composition and the thermal history of the glass in terms of the annealing processes that occur during its manufacture. Composition effects on extinction coefficient are of two kinds: (1) extinction coefficients for silicate glasses differ from those of borate, phosphate, and other glass types; (2) extinction coefficients depend on specific absorbing species and their amounts. The optical glasses in Table 9.1 contain very low levels of absorbing species and are essentially colorless.

Chapter 6 explains that refractive index is a complex quantity, as defined by equation (6.2) and illustrated in Figure 6.1. The real part is the usual refractive index and the imaginary part is related to spectral absorption in terms of the extinction coefficient. Changes in refractive index with wavelength in the visible region are small, and the dispersion is designated as normal. For example, compare data on glass BSC 510644 given in Tables 6.3 and 9.1. Refractive indices of glasses attain higher values as wavelengths move toward the ultraviolet and reach so-called anomalous dispersion regions. For a given thickness, each glass has a characteristic ultraviolet cutoff or band edge where absorption becomes quite high. Glasses also have characteristic infrared cutoffs or band

TABLE 9.1   Extinction Coefficients of Optical Glasses (cm^{-1})

Wavelength Å	BSC, 510644	MBC, 572577	LF, 576434	DEDF, 748278
3,200	0.900		2.275	
3,600	0.035	0.220	0.099	
4,000	0.009	0.025	0.025	0.300
4,500	0.005	0.013	0.017	0.040
5,000	0.003	0.007	0.012	0.010
5,500	0.002	0.005	0.007	0.005
6,000	0.004	0.004	0.009	0.006
6,500	0.006	0.005	0.010	0.007
7,000	0.006	0.004	0.008	0.005
8,000	0.007	0.005	0.006	0.003
9,000	0.008	0.006	0.006	0.004
10,000	0.010	0.007	0.007	0.004

Source.   Chance-Pilkington catalog.

edges. Wavelengths where band edges occur depend on glass composition and the integrated effects of the material's structural constituents. Glass technology controls in optical glasses would be greatly enhanced if our knowledge of glass composition and microstructure included quantitatively related information on refractive index, dispersion, and absorption limits.

9.2   ABSORPTIONS IN THE VISIBLE AND ULTRAVIOLET REGIONS

The absorption and transmission properties of silicate glasses approach more and more closely the curve for vitreous silica (Figure 9.1) as the amount of SiO_2 increases. The solid line represents the transmission of vitreous silica made by the melting of quartz crystals. Absorption regions in the SiO_2 curve depend on impurities in the quartz crystals from which it is made. Principal absorption bands at 1.4, 2.75, and 4.25 are caused respectively by FeO, free OH ions, and tightly bound OH ions. Vitreous silica made by the aqueous $SiCl_4$ process contains few impurities but has increased absorption, particularly at the 2.75 band. This band can be reduced in intensity if the free OH ions are removed by heating. The wavelength positions of ultraviolet and infrared absorption limits or cutoffs of vitreous silica depend on sample thickness and impurities. They can be changed by replacing SiO_2 with other oxides and glass constituents. Such substitutions usually result in moving the ultraviolet cutoff to longer wavelengths. Changes in the position of the infrared cutoff are less sensitive to

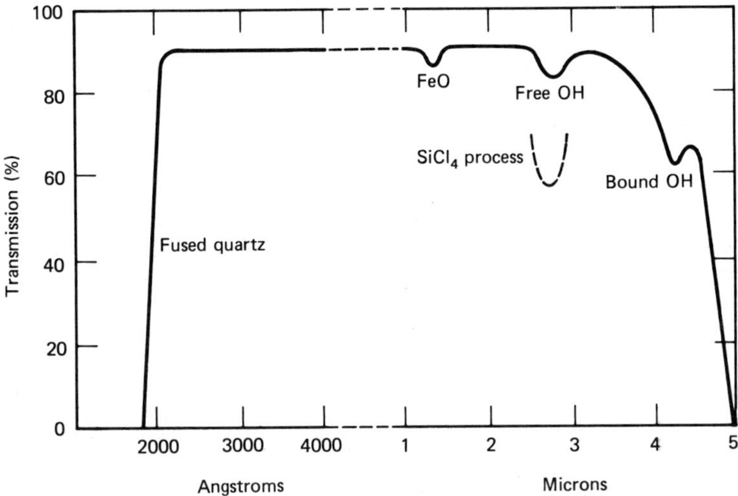

Figure 9.1 Spectral transmission of vitreous silica.

composition changes. Absorptions in the ultraviolet and visible regions strongly depend on electronic structures; but as wavelengths move more and more into the infrared, the absorptions are determined by glass lattices or substructures.

Spectral absorption in visible and ultraviolet regions involves interactions between the incoming radiation and electronic states of the glasses under consideration. These absorptions range in wavelength from the ultraviolet cutoff, where anomalous dispersion occurs, into the near infrared. Electronic states depend on glass compositions, and each one has its own microstructure governed by ions and ionic groups. Absorption of incoming radiation depends on how tightly the electrons are bound in the microstructure. Before interaction with radiation, the electrons occupy their lowest energy levels. Upon absorbing a quantum of energy, an electron moves to the next higher energy level. This excited absorption may result in color formation, fluorescence, and related phenomena. A quantum of energy or photon is released when an electron drops into a lower energy level. The sharpest spectral lines are obtained with gases and vapors and are produced by electrons that drop to the lowest energy levels within a given quantum shell. These electron transitions are essentially unhindered by constraints of neighboring electromagnetic fields in their immediate environment. Transitions of electrons in materials in the condensed state (e.g., glasses) do not give rise to line spectra but to broad bands or regions of absorption. In this case the transitions are strongly influenced by short-range field forces.

Numerous published studies on the absorption of transition elements in glasses indicate the complexity of electronic spectra. Electronic structures of these elements are described in Tables 8.1 and 8.2 and Figure 8.1. Absorptions relate to unfilled $3d$ shells that have from one to eight electrons. In these neutral atoms two electrons go into the $4s$- subshell before the $3d$- subshell is filled, as indicated in Table 8.2. Upon ionization, the $4s$- electrons are removed first. Iron has a valence of 2 when it loses its two $4s$- electrons and a valence of 3 when it loses two $4s$- electrons and one $3d$- electron. Bonds where electronic exchange or transfer occurs may be called ionic, those involving electron sharing are known as covalent bonds. As noted in Chapter 1, the bonds in glasses do not clearly belong to one type or the other; they are partly ionic and partly covalent in these terms. Knowing that iron has different valences and may be surrounded by either four or six oxygens, we see at once the difficulty of describing how this element affects spectral absorption. The arrangements of electrons among different states of an incomplete shell in the transition elements result in characteristic absorptions. The d-d transitions are strongly influenced by short-range electromagnetic fields of neighboring ions and ionic groups. The f-f transitions of the rare earths occur in inner incomplete shells and are less influenced by short-range force fields.

Data on absorption spectra of glasses have been obtained by the use of several techniques, including spectrometric research, the Raman effect, the Mossbauer effect, EPR, NMR, and fluorescence. The results of these studies on different types of oxide glass are too extensive to review here. Postulates have been advanced to relate absorptions in the ultraviolet and visible regions to glass microstructures. Absorption characteristics have been related to covalent and ionic bond types, bridging and nonbridging oxygens, coordination numbers, and other aspects of the random network theory. Crystal field theory, which is purely an electrostatic approach, was originally developed for ions in crystal lattices. This treatment has been extended to glasses because the local ligand fields do not have to be considered as parts of an extended crystal lattice. Earnshaw (1968) provides details on crystal or ligand field theory and its uses. Bates (1962) has noted that complex ligands are formed from the central ion, together with its surrounding atoms, ions, and ionic groups. He observes that absorption spectra of the transition elements generally fall into two classes— weak and strong absorptions. The weak bands are caused by internal transitions between the d-electron levels of the central ion as modified by the ligand field. Such bands occur in the low evergy spectral regions. He considers that the strong bands arise from transitions that involve a transfer of charge between ligands and the central ion. These electron charge transfer bands occur mainly in the high energy regions of the spectrum. Though most of the available data deal with absorptions related to transition elements, other types of absorption have been studied. Our discussion is limited to representative experimental data on transition element absorptions, with only minor attention to their relations with glass microstructures.

Turner (1967) studied effects of transition and rare earth elements on ultraviolet absorption limits in a number of base glass compositions. Table 9.2 shows compositions of eight glasses made from relatively pure materials that were melted under oxidizing conditions. If we neglect the presence of 0.001 to 0.004% Fe_2O_3 in these glasses, their absorptions can be considered to be intrinsic for the given base glasses. Intrinsic absorption curves for three base compositions are illustrated in Figure 9.2. The spectral data are given in terms of absorbance per centimeter. That is, absorbance = $\log_{10} 1/T$, where T does not include reflection losses. When absorbance = 1, then T = 10%, when absorbance = 2, T = 1%, and so on. The wavelength scale in Figure 9.2 is converted to a linear energy scale in terms of electron-volts per photon. The intrinsic absorption curves of the three base glasses derive from the presence of their individual oxide combinations, not from their very low Fe_2O_3 contents. Postulated explanations of absorption limits must be able to account for such intrinsic absorption limits. Shifts in absorbance limits caused by additions of substantial amounts of transition and rare earth oxides are given in Table 9.2. Figure 9.3 shows effects of adding 0.25% Fe_2O_3 to the three base glasses of Figure 9.2. The effects of adding 1% of the rare earth CeO_2 are illustrated for three base glasses in Figure 9.4. The effects of adding 0.5% of CuO are indicated in Table 9.2.

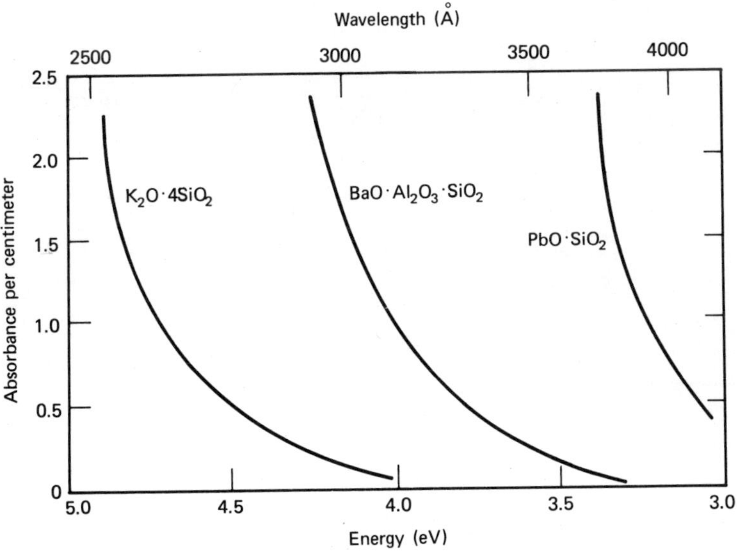

Figure 9.2 Absorbance in base glasses. Adapted from Turner (1967).

TABLE 9.2 Absorbance Limits in Experimental Glasses

Oxide	Compositions (wt %) of eight glasses							
	Na_2O, $4SiO_2$	K_2O, $4SiO_2$	Cs_2O, $4SiO_2$	MgO, Al_2O_3, SiO_2	BaO, Al_2O_3, SiO_2	PbO, SiO_2	PbO, Al_2O_3, SiO_2	PbO, K_2O, SiO_2
SiO_2	79.47	71.83	46.02	61.00	39.29	21.11	19.00	27.48
Al_2O_3				18.50	10.72		10.00	
Na_2O	20.53							
K_2O		28.17						20.00
Cs_2O			53.98					
MgO				20.50				
BaO					49.99			
PbO						78.89	71.00	52.52
Oxide (wt %)	Shifts in absorbance limits (e V) caused by added oxides							
Fe_2O_3, 0.25	2.2	1.2	0.9	0.7	0.7	0.6		
CuO, 0.50	1.85		0.8	1.3	0.8		0.9	
TiO_2, 1.00	1.7	0.8	0.35	1.0	0.25		0.5	0.4
CeO_2, 1.00	2.2		1.00	1.5	1.00		0.7	

Source. Turner (1967).

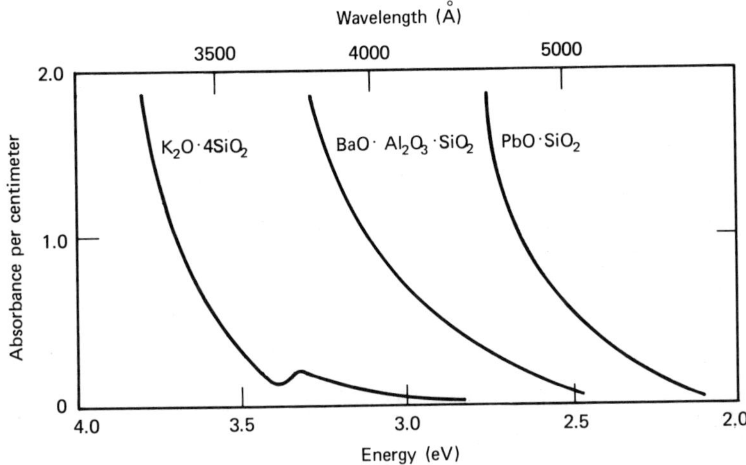

Figure 9.3 Absorbance in glasses containing 0.25% Fe_2O_3. Adapted from Turner (1967).

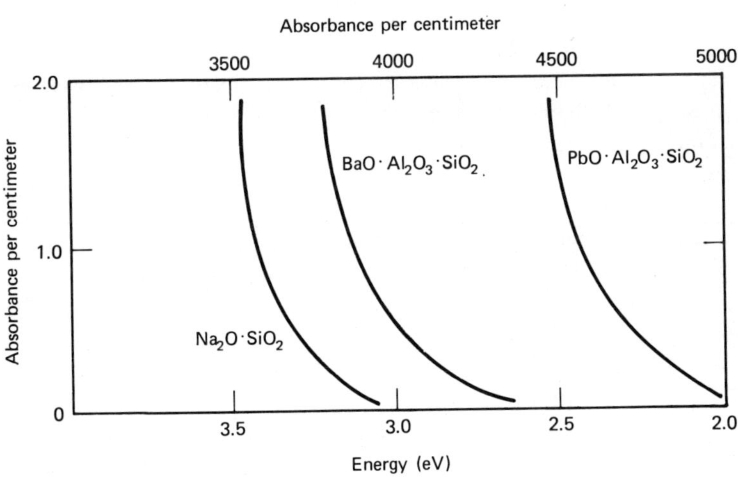

Figure 9.4 Absorbance in glasses containing 1.0% CeO_2. Adapted from Turner (1967).

TABLE 9.3   Molar Absorptivities of Transition Ions

Ion	Absorptivity	Wavelengths (Å)	
		d-d Transitions	Electron transfer
Mn^{2+}	0.05-0.44	4,150 (1)	
Cu^{2+}	15-20	8,000 (2)	
Co^{2+}	5-300	6,000 (2)	
Fe^{3+}	0.5	4,350 (2)	
Fe^{3+}	7,000		2,300 (3) and (4)
Fe^{2+}	25	10,600 (2)	
Fe^{2+}	3,000		2,000 (3)
Cr^{6+}	4,200		3,700 (5)
Cr^{3+}	20	4,500 and 7,000 (2)	

(1) Bingham and Parks (1965).
(2) Bates (1962).
(3) Schulman and Compton (1962).
(4) Swarts and Cook (1965).
(5) Nath, Paul, and Douglas (1965).

Data source. Turner (1967).

All these additions cause the absorbance limits to shift toward longer wavelengths. Numerical values of the shifts derive from specific elements with unfilled inner shells and their amounts, together with influences of given base glass compositions. Molar absorptivities in Table 9.3 are from Turner (1967) who comments that when located outside the visible region, some strong absorptivities from electron transfer spectra may contribute to visible coloration. Ferric iron (Fe^{3+}) is an example.

The foregoing data and those of other investigators demonstrate that the onset of electron transfer in terms of ultraviolet absorption limits depends on base glass composition and specific transition ions. Smith and Cohen (1963) determined ultraviolet absorption limits for a variety of cations when added to $R_2O \cdot 3SiO_2$ glasses, where R = Li, Na, or K. Data were obtained on transition, rare earth, and other cations such as Mg, Ca, Sr, Ba, Cd, and U. Quantitative data cannot be derived because the samples varied in thickness and the data are shown only in graphical plots. Wavelengths of some of the absorbing species in the ultraviolet and visible regions are given. Schultz (1974) studied optical absorption of transition elements in vitreous silica. He used variations of the flame hydrolysis method in preparing samples. For example, vanadium was added to SiO_2 by using $SiCl_4$ and VCl_4 vapors, which were deposited as a boule in a high temperature furnace. Absorptivity spectra were obtained for V,

Cr, Mn, Fe, Co, Ni, and Cu, ranging in amounts from 0.01 to 0.1 weight percent. Ultraviolet cutoffs and absorption peaks were in reasonable agreement with data on other silicate glasses, although they were generally shifted to shorter wavelengths.

Hensler and Lell (1969) studied ultraviolet absorption in a number of silicate glasses. Polished samples 2 mm thick were measured and the data given in terms of absorbance (optical density). The ultraviolet absorption limit or cutoff shifted to longer wavelengths as Na_2O and other alkalies replaced SiO_2. As Mg, Ca, Sr, and Ba replaced SiO_2, the cutoffs shifted to increasingly longer wavelengths. Hensler and Lell correlated their data with the number of nonbridging oxygen ions and to the bond strengths between the nonbridging oxygens and ions other than SiO_2. An increase in the number of nonbridging oxygens shifts the absorption limit to longer wavelengths.

Bates (1962), some of whose data are included in Table 9.3, applied ligand field theory to absorption spectra in transition elements Ti^{2+}, V^{2+} and V^{3+}, Cr^{2+} and Cr^{3+}, Mn^{2+} and Mn^{3+}, Fe^{2+} and Fe^{3+}, Co^{2+} and Co^{3+}, and Ni^{2+} and Ni^{3+}. The absorption spectrum of Cr^{3+} was accounted for in some detail, and it was concluded that the Cr^{3+} ion in the glasses studied had six oxygen ligands arranged in octahedral symmetry with little departure from regularity. Tentative interpretations of other transition ion spectra were attempted, and there was reasonable agreement with ligand field predictions. The spectra of glasses containing iron are especially complex and Bates recommends further research on the subject. He observes that "Additional difficulties are the occurrence of charge transfer bands in the near ultraviolet region for Fe^{3+} in both octahedral and tetrahedral symmetry, the equilibrium between Fe^{3+} and Fe^{2+} and a possible interaction between these two valence states." His use of ligand field theory confirmed that Co^{2+} and Ni^{2+} can change from octahedral to tetrahedral symmetry.

Turner and Turner (1972) studied the spectra of nickel in silicate and aluminosilicate glasses with additions of alkali and alkaline-earth oxides. Application of ligand field theory showed that several of the glasses contained at least two spectral types that corresponded to previously proposed 4- and 6- coordinated Ni^{2+} species. Both tetrahedral and octahedral ligand calculations were made from directly observed bands free of overlapping. The calculations of the investigators suggested that the postulated tetrahedral species may instead be cubic with 8-coordinated Ni^{2+} species. They indicated that packing requirements of the alkali-oxygen, alkaline ertha-oxygen, and silicate polyhedra may preclude Ni^{2+} sites other than those which are either 6- or 8-coordinated. Their research led them to conclude that "silicate glasses are interpreted as disordered but highly structured, even to the extent of determining the nature of a local site such as that occupied by Ni^{2+}." This paper gives detailed Ni^{2+} spectra and absorbance per centimeter as these values are affected by base composition, annealing temperature, concentration of Ni^{2+}, and oxidation-reduction conditions during melting. It also reviews prior work on borate glasses.

The foregoing résumé of typical spectra in silicate glasses illustrates problems involved in relating measured data to details of glass microstructures. Published data on silicate, borate, and phosphate glasses indicate that absorption band locations and absorbance levels differ from one glass type to another. Moreover these properties vary measurably with changes in base compositions in silicate-type glasses. The problems encountered in all types of oxide glass are similar in kind, although they may differ in detail. Published data indicate that absorptions related to specific ions depend on their electronic structures and the lattice environment in which they are situated. Proposed interpretations of the data have been based on the random network theory in terms of coordination numbers, bridging and nonbridging oxygens, ionic field strengths, and so on. The author's postulated primary phase substructures in silicate glasses involve only ions and ionic groups that characterize glass lattices or substructures. In its present form the postulate cannot be used to evaluate glass spectra that depend so strongly on electronic structures of ions. However it does differentiate between substructure environments that also have strong influences on the base compositions of spectra. Glass compositions in presently available spectral data are not arranged in terms of primary phase fields. This precludes testing the validity of the substructure postulate. It seems reasonable to suppose that all glasses located in the tridymite (SiO_2) primary phase field would have similar substructure environments. The same would be true for glasses lying in the sodium disilicate $(Na_2O \cdot 2SiO_2)$ phase field. For example, research outlined in Section 4.3 indicates that lattice or substructure environments in these two contiguous phase fields are different as interpreted in terms of refractive index and specific volume data. The possible usefulness of the substructure postulate in clarifying glass spectra could be tested by adding one of the transition ions, say Cr^{3+}, to groups of glasses located in the tridymite and sodium disilicate phase fields. Such systematic research might disclose whether substructures in the two adjacent phase fields had differences that might be related to glass spectra. This new method might lead to clarification of our understanding of glass spectra in the ultraviolet and visible regions.

9.3 ABSORPTION IN THE INFRARED REGION

Experimental methods for determining optical constants of glasses vary with spectral absorption levels involved. Prism methods can be used to determine refractive index in the visible region, where absorption can usually be neglected. Measurements in regions of intermediate absorption can be made on thin film samples of different thicknesses from which the absorption coefficient can be calculated. These transmission methods cannot be used in regions of high absorption, and reflection techniques must be followed. Infrared absorptions in silicate glasses range from intermediate to strong depending on the wavelength region of interest. Simon (1960) developed equations relating optical parameters for both types of measurement. Both methods involve the complex refractive index [see equation (6.2), $n = n' - in''$, where n' and n'' are the real and imaginary components, respectively]. The real part is the refractive index as ordinarily defined, and the imaginary part relates to absorption in terms of the extinction coefficient. Cleek (1966) made measurements of the optical constants n' and n'' for vitreous silica, three silicate glasses, and a phosphate glass, using reflection methods in the region between 7 and 13 microns. It is noted that the nomenclatures of Simon and Cleek differ from those used in our equation (6.2), which were made to conform to our equation (5.9), $k = k' - ik''$ for the corresponding complex dielectric constant. It is noted that structures and properties of glasses in the form of thin films may differ from those of the bulk glass. Compositions of the glass in these two forms may differ. If measurements are made on a series of films of increasing thickness, the composition of the films can be expected to approach that of the bulk glass. Experimenters must be cognizant of this problem when using thin glass films for infrared and other kinds of glass measurement.

Cleek (1966) made both transmission and reflection measurements on vitreous silica (Corning 7940) from which he derived curves showing the relation of n' and n'' to wavelength. Both curves had sharp maxima (at 9.52 microns for n' and at 9.05 microns for n''). Cleek determined from his data that the principal absorption peak of vitreous silica was at 9.24 microns which is in good agreement with the results of other investigators. This sharp absorption is considered to be caused by the Si-O stretching vibration in the lattice or substructure. Cleek's data show that absorptions of the other glasses are high in the region investigated. The three silicate glasses had single broad reflection bands in the 8 to 13 micron region. The phosphate glass had three definite bands in the region. The broad absorption maxima were attributed to a range of vibration frequencies, and no detailed assignments could be made in terms of microstructures.

Simon and McMahon (1953a) made infrared reflection studies on quartz, cristobalite, and vitreous silica in the range 700 to 1400 cm^{-1} wave numbers, corresponding to the wavelength range, 14 to 7 microns. Sharp absorption maxima were found at 9.43, 9.17, and 9.10 microns for quartz, cristobalite, and vitreous silica, respectively. These absorption were found to be sharpest at $4°K$, becoming less sharp as the temperature was increased. The sharp absorption peaks of cristobalite and vitreous silica occur at closely comparable wavelengths, and it is noted that cristobalite is the primary phase in vitreous silica. Absorption peaks in all three modifications of SiO_2 were related to Si-O stretching vibrations.

Using the method previously reported (Simon and McMahon, 1953a), the same workers (1953b) studied the infrared absorptions of a number of binary silicate glasses. Compositions ranged from 0 to 0.46 mole fraction Na_2O; other glasses containing Li, K, Ca, Sr, and Ba oxides were also investigated. The method involved measurements of the reflecting power at the two angles of incidence of 20° and 70° followed by graphical analysis of the data. They determined the optical constants n' and n'', which they and Cleek (1966) designate as n and k, respectively. They chose to plot the imaginary part of the dielectric constant k'' whose maxima indicate the natural vibrational frequencies. Figure 9.5 gives their plot of reflecting power of the Na_2O-SiO_2 glass series at a 20° angle of incidence. Note that peak reflections move toward

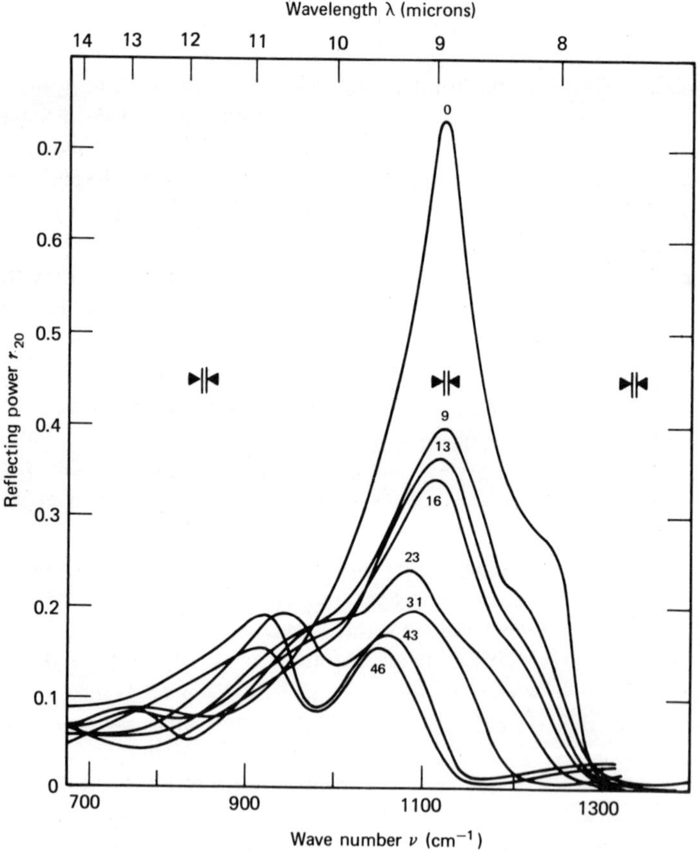

Figure 9.5 Reflecting powers at 20° angle of incidence for a series of soda-silica glasses with 0 to 46 mole percent Na_2O. Figures on curves indicate mole percent Na_2O. Slit width is indicated by arrows. From Simon and McMahon (1953b).

longer wavelengths and that the curves become more complex as Na_2O replaces SiO_2. The imaginary part of the complex dielectric constant also changes as SiO_2 is replaced by Na_2O (Figure 9.6). The authors' ϵ'' is equivalent to our k''. They suggest that the width of the absorption band indicates the frequency range over which the vibrating constituents are spread out, and this in turn indicates the spread of the force constants of the ionic bonds in the glass lattice or substructure. Simon (1960) reviews these data in conjunction with Raman spectra measured by Wilmot (1954) on Na_2O-SiO_2 glasses containing from 0 to 0.43 mole fraction Na_2O. Simon comments that "a new band begins to emerge

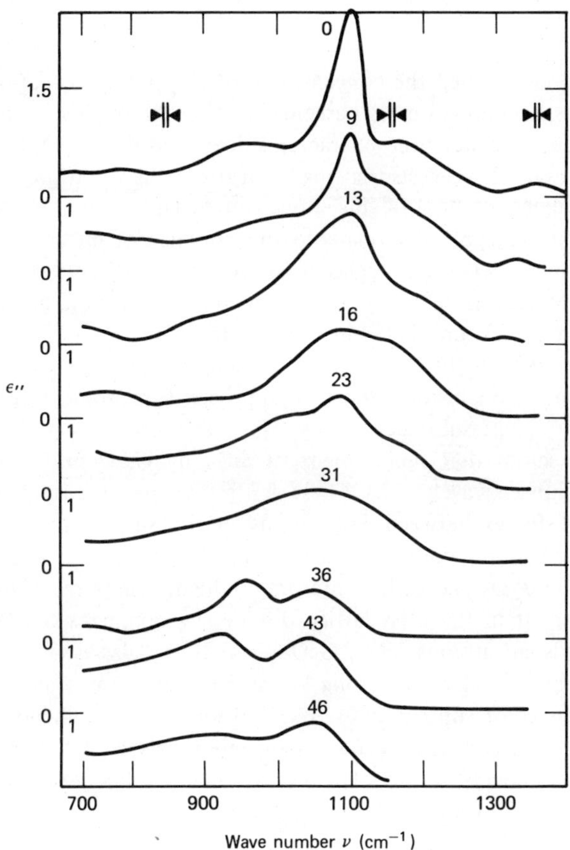

Figure 9.6 Imaginary part ϵ'' of the complex dielectric constant for the series of soda-silica glasses with 0 to 46 mole percent Na_2O. Figures on curves indicate mole percent of Na_2O. From Simon and McMahon (1953b).

TABLE 9.4   Compositions of Primary Phases in Na_2O-SiO_2 Binary Glasses

Primary phase	Na_2O ranges (mole fractions)
SiO_2 (cristobalite)	0-0.11
SiO_2 (tridymite)	0.11-0.24
SiO_2 (quartz)	0.24-0.26
$Na_2O \cdot 2SiO_2$	0.26-0.37

Data source. Lavin, Robbins, and McMurdie (1964).

at about 950 cm^{-1} when the concentraion of Na_2O exceeds approximately 25 mole percent. Beyond 33 mole percent of Na_2O the original structure appears to be replaced by a new one characterized by two bands (approximately 940 and 1075 cm^{-1})." He suggests that the bond-stretching vibrations may be related to either bridging or nonbridging groups of Si and O. It is possible, as well, that these data categorically relate to the presence of different substructures existing in Na_2O-SiO_2 binary glasses as indicated in Table 9.4. The data of Simon and McMahon (1953b) also show that absorption maxima for 0.22 mole fraction Na_2O and Li_2O in binary alkali silicates occur at closely the same frequency; both these glasses are located in the SiO_2 (tridymite) primary phase field. It would be of interest to approach this problem in terms of postulated primary phase substructures.

Figure 9.5 shows that replacement of SiO_2 by Na_2O shifts the absorption peak toward longer wavelengths. These shifts toward lower frequencies imply a weakening of forces between Si-O bonds in the substructure. Jellyman and Proctor (1955) found similar shifts in peak absorptions when Na_2O replaced B_2O_3 in a pure B_2O_3 glass. They interpreted the data in terms of B-O bonds and their arrangement in the glass lattice. Figure 9.7 compares strong infrared absorption bands in vitreous SiO_2, GaO_2, and B_2O_3 glasses. Positions of these bands undoubtedly relate to strong bonds between cations and anions in the respective lattices or substructures. Explanations of these phenomena in terms of the random network theory are not entirely satisfying. Whether more reasonable explanations can be given in terms of postulated primary phase substructures in glasses is a question for future research.

9.4   LUMINESCENT PHENOMENA

The foregoing outline of absorption in glasses leads to consideration of what happens to the absorbed energy. Luminescence of materials may result from a variety of energy-absorbing processes and phenomena, including electromagnetic

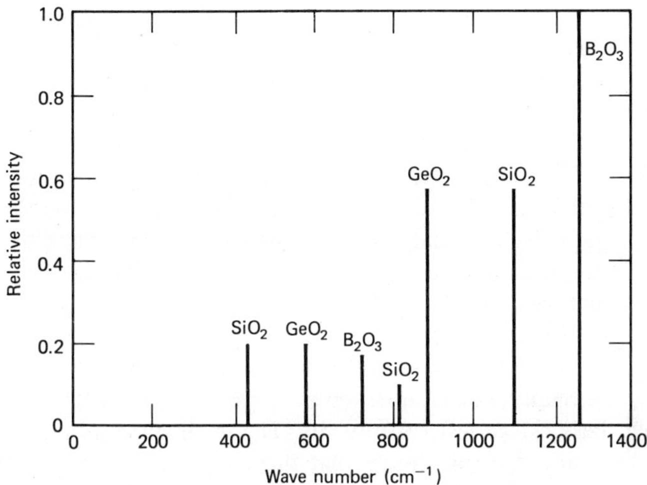

Figure 9.7 Comparison of strong infrared absorption lines in vitreous SiO$_2$, GeO$_2$, and B$_2$O$_3$. From Doremus (1973).

radiation, electrons or cathode rays, electrical fields, heat, mechanical grinding, and chemical reactions. Our interest is confined to emission of electromagnetic radiation in excess of thermal radiation. The process may yield ultraviolet, visible, or near infrared emissions depending on the wavelength of the incoming radiation and the electronic and ionic structures of the glasses. Descriptions of luminescent research should include full information on the glass sample, the incoming radiation, and the resulting excited luminescent radiation. Luminescent phenomena may be classified according to the duration of the emission after removal of the exciting radiation. Materials that have an exponentially decaying afterglow can be categorized as flueorescent. Materials that have an additional component of afterglow that decays more slowly are termed phosphorescent. The decay of the phosphorescent materials is more complex and may involve absorption of thermal energy. Glasses may exhibit either type of emission depending on their electronic and lattice structures. Luminescence can be caused by the presence of identifiable phosphors and by more complex luminescent centers that may not yield easily to description. Structural defects or imperfections may also trap electrons and release them only after additional energy has been supplied in the form of heat. Luminescent emission results in optical absorption bands on the long wavelength side of the fundamental absorption edge or cutoff of glasses. Luminescent emission generally lies at even longer wavelengths and is characterized as bands covering a more or less broad spectral range. Thus the emission lies in a spectral region where the glass in nonabsorbing or transparent. The radiative lifetime of the excited electronic states or decay time for fluorescence vary from 10^{10} to 10^{-1} sec, whereas the period for visible radiation is approximately 10^{-14} sec. Williams (1966) gives a comprehensive treatment of the theoretical basis of solid state luminescence.

Foregoing discussions demonstrate the complex character of spectral absorption in glasses. These absorptions give rise to equally complex emission spectra. Research efforts in these areas are directed toward finding reasonable relations between experimental data and postulated glass microstructures. Out attention is centered on experimental fluorescence data on glasses. Although their relations with microstructures are still under development, the data may be discussed in terms of the presence of three types of glass constituent: (1) ions, (2) complex groups of sulfides, selenides, and tellurides, and (3) metallic elements. For purposes of continuity, color formation in glasses (Chapter 10) is related to the presence of these categorical constituents. This is simply a means of outlining our discussions and does not imply that these three complex constituents are thoroughly understood. Comprehensive reviews by Weyl (1951) and by Rindone (1966), together with cited individual papers, have been most useful in the following discussion.

Ions

Individual ions with their characteristic groups of surrounding oxygens may act as luminescent centers. The role of the lattice or substructure is important in the case of transition elements with their outer $3d$ electrons. These binding forces influence the characteristic luminescent emissions of the ions. The luminescence of rare earth elements caused by excitation of the inner $4f$ electrons is less sensitive to influences of lattice binding forces.

Karapetyan (1961) studied copper-activated glasses and attributed the luminescence to the cuprous ion or to copper atoms. His EPR measurement indicated that divalent copper was associated with the band beginning at about 6000 Å and peaking at 8000 to 8800 Å. This band was suppressed by adding reducing agents to the glasses, which were then excited by ultraviolet radiation of 3130 Å wavelength. Luminescent bands were formed in the region 4300 to 5200 Å. Peaks of these bands for phosphate, borate, and silicate glasses were respectively at 4800, 5000, and 5400 Å. Karapetyan observed an emission peak in a phosphate glass at 4400 Å under 2540 Å radiation and one at 5800 Å under 3650 Å radiation. Parke and Webb (1972) made fluorescent measurements on copper in sodium silicate, sodium borate, and calcium phosphate base compositions. Fluorescence was excited by xenon and deuterium ultraviolet radiation. They found that fluorescent emission radiation peaked at different wavelengths depending on base composition, in agreement with other investigators on fluorescence of transition ions. Decay times of silicate and phosphate glasses at room temperature ranged from 23 to 33 μsec. The decay time of a calcium phosphate glass increased from 31 μsec at room temperature to 84 μsec at $100°$K. They relate the fluorescence of Cu^+ to changes in the environments of the different base compositions.

The luminescence of manganese in crystals and glasses has received considerable attention. Manganese glasses contain both Mn^{2+} and Mn^{3+} ions, depending on oxidation-reduction conditions. Under usual melting conditions both ions are present in the glass. The Mn^{3+} ion under oxidizing conditions colors the glass purple. The Mn^{2+} ion under reducing conditions causes weak yellow to brown colorations. The luminescence of manganese glasses caused by divalent manganese is green to orange to red, depending on base glass composition. Wilke (1962) found that adding 7.2 and 1.3% to a metaphosphate glass caused red and yellow luminescent colors, respectively. The green fluorescence of Mn^{2+} ion was related by Linwood and Weyl (1942) to the presence of MnO_4 tetrahedra of 4-coordinated Mn^{2+}. They suggested that the red or orange luminescence could be caused by less symmetrical groups having oxygen coordinations of 6. Turner and Turner (1970) reviewed previous publications on the fluorescence of manganese and presented new data on manganese in K_2O-$4SiO_2$ and soda-lime-silica glasses. The specially designed commercial apparatus used in their studies is recommended to those doing research on this subject. Figure 9.8 gives concentration dependence of luminescence spectra of Mn^{2+} in the soda lime glass. The emission spectra of both sets of glasses change from yellow-green to red as the Mn^{2+} concentrations increase. Differences in spectral details caused by base compositions are apparent. Increases in Mn^{2+} concentration causes decreases in green emissions and increases in red emissions. Comparison of absorption and emission spectra of Mn^{2+} revealed only one type of absorption curve but two concentration dependent emission bands. Figure 9.9 presents nonexponential data on decay times. The authors concluded from their data that green and red luminescence of Mn^{2+} cannot be assigned to 4- and 6-coordination, respectively, as indicated by Linwood and Weyl (1942). Their improved resolution of bands and theoretical calculations indicate that a covalent-octahedral Mn^{2+} species can account for luminescence spectra in silicate glasses.

The U^{6+} uranium ion causes very strong luminescence when it is present as the uranyl group $(UO_2)^{2+}$. Luminescent colors range from bluish green to yellow, and locations of emission bands depend on base glass composition. Rodriguez, Parmelee, and Badger (1943) studied the luminescence of U in silicate, borate, and phosphate glasses. They found two emission bands at 5300 and 5630 Å in soda-lime-silica glasses which increased in intensity with increasing SiO_2 content and with replacement of Na_2O by K_2O. Three strong luminescent bands were found at 5100, 5300, and 5600 Å when SiO_2 was completely replaced by P_2O_5. Sharp emissions bands appeared at 4600, 4950, 5180, 5420, 5720, and 6010 Å in a glass containing 90% B_2O_3 and 10% Na_2O.

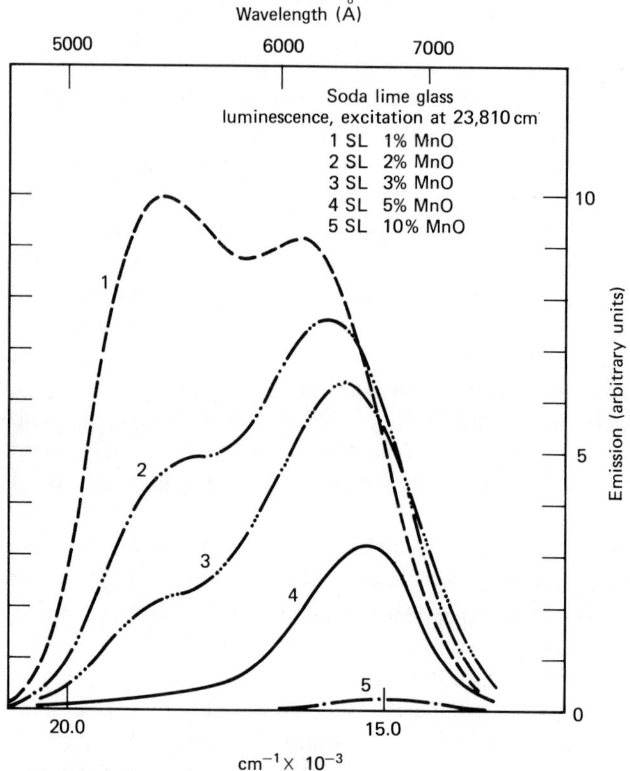

Figure 9.8 Concentration dependence of luminescence spectra of Mn^{2+} in Na_2O-CaO-SiO_2 glasses. From Turner and Turner (1970).

Luminescence of rare earth ions in glasses has received wide attention, especially since the development of neodymium-doped glass lasers. These ions contribute rather sharp absorption and emission bands in glasses because of their shielded $4f$ inner electron transitions (see Figure 10.5, the transmission curve of a neodymium glass). Shifts in absorption and emission bands caused by changes in base composition in rare earth glasses are less pronounced than those in glasses containing transition ions. However wavelength positions of emission bands of some rare earth ions are dependent on oxidation-reduction conditions, temperature, and ionic concentration. Rindone (1966) comprehensively reviews extensive research on this subject. A few examples of this work indicate the nature of rare earth luminescence.

Snitzer (1961) added Nd_2O_3 in amounts from 0.13 to 2.0 weight percent to a base glass containing 59% SiO_2, 25% BaO, 15% K_2O, and 1% Sb_2O_3. He

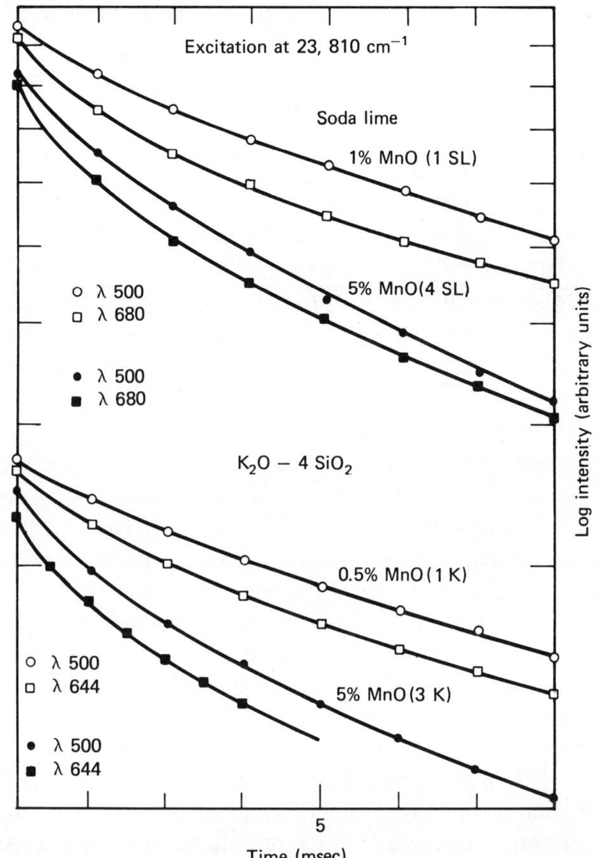

Figure 9.9 Red and green luminescence decay times in Na₂O-CaO-SiO₂ glasses. Wavelengths λ are given in nanometers. From Turner and Turner (1970).

found luminescent emmisions at 8800, 10,600, and 13,000 Å and related them to the energy level diagram of Nd^{3+} by reference to the spectrum of $NdCl_3$. Position of the well-known band at 10,600 Å and assigned energy levels has been confirmed by other investigators.

Karapetyan (1963) measured absorption and emission spectra arising from Pr, Sm, Eu, Tb, Dy, and Tm in the base glass $Na_2O \cdot CaO \cdot 5SiO_2$. Praseodymium caused emission bands at 6000, 7000, and 8600 Å . Luminescent intensities at the same wavelengths were weakened in borate and phosphate glasses. The luminescence was not sensitive to changes in oxidation-reduction conditions.

The emission spectrum of trivalent europium consisted of bands at 5870, 6090, 6530, 6980, and 7460 Å and was sensitive to oxidation-reduction conditions, temperature, and ionic concentration. Dysprosium caused intense yellow luminescence, with bands peaking at 4830, 5700, 6670, 7560, and 8440 Å. Variations in oxidation-reduction conditions, temperature, and ionic concentraion caused little change in the luminescence.

Rindone and Sproull (1964) studies luminescence of rare earth ions in a base glass of weight composition 50% B_2O_3, 40% La_2O_3, and 10% CaO. According to their data, luminescent bands caused by Eu, Sm, Dy, Tb, and Tm were at about the same wavelengths as those found by Karapetyan (1963) in his soda-lime-silica base composition. However the comparable spectra differed in detail. Rindone and Sproull (1964) found no luminescent bands in the visible region when 0.1% Er_2O_3, Gd_2O_3, Lu_2O_3, Pr_2O_3, Nd_2O_3, or Yb_2O_3 was added to the calcium-lanthanum -borate glass.

Different investigators have shown that essentially any fluorescent color can be obtained by using mixtures of rare earth ions. For example, Nelson, King, and Barber (1964) found that mixtures of Tb and Eu in vitreous silica exhibited 17 luminescent emissions between 3770 and 6530 Å. The fluorescent color of a glass containing a mixture of rare earths generally follows an additive relationship in terms of individual ions. However the fluorescent color cannot be predicted accurately from the emission curves of the single ions. Interactions between mixtures of rare earths often cause quenching of the fluorescence. Possible interactions involved have been studied by Van Uitert and Iida (1962), Pearson and Peterson (1965), Cabezes and DeShazer (1964), and Shionoya and Nakazawa (1965). For example, the fluorescence of Eu plus Nd in some glasses almost disappears. Alternatively, mixtures of some rare earths enhance fluorescent intensities. Either quenching or enhancement of fluorescence may result from mixtures of rare earths and other ions, such as those of the transition elements.

Complex Groups of Sulfides, Selenides, and Tellurides

The luminescence of CdS and CdSe groups has been attributed to their highly covalent nature. Glasses of this type exhibit little visible color when made by usual glassmaking processes. Development of visible color or fluorescence requires a reheating of the glass to a temperature within the annealing range, during which time color and fluorescent groups such as CdS may form. Buzhinski and Bodrova (1962) compared luminescent bands of CdS, CdSe, and CdS·CdSe in crystals with those in glasses. Excitation by 3650 Å radiation caused luminescent bands at 7000 and 7500 Å in both crystals and glass; bands in the glass had greater half-widths. Luminescent glass spectral characteristics were strongly dependent on heat- treatment time and temperature. A shift of absorption and luminescent bands to shorter wavelengths with increasing treatment temperature was observed in CdS·CdSe glasses with deficiency of Cd and an excess of S over Se. A shift of absorption and emission to longer wavelengths was found in glasses containing an excess of Cd or Se. The wavelength position of luminescence was in all cases related to the position of the absorption edge or limit; changes in one caused corresponding changes in the other.

Metallic Elements

Free neutral metal atoms may be produced in glass by reductions of the ions, thermal dissociation, and ionizing radiation. In small amounts and under proper conditions, the neutral atoms Cu, Ag, Au, Pb, Sb, Se, Sn, and Pt produce luminescence in glasses. The luminescence of copper has been studied by Karapetyan (1961). The luminescence of silver has been described by Weyl, Schulman, Ginther, and Evans (1949), Schulman and Etzel (1953), and others. Schulman and Etzel found that the luminescent intensity of silver was proportional to the dosage of x-rays and gamma-rays and could be used to determine these radiations dosimetrically. A dosimeter for this application was patented by Schulman, Ginther, and Evans (1950).

9.5 VACUUM ULTRAVIOLET PHOTOLUMINESCENCE

Our discussions on absorption and luminescence in glasses have been based on measurements made at atmospheric pressure. Such measurements can be made down to wavelengths of about 2000 Å. The absorption of air at lower ultraviolet wavelengths becomes very high, primarily because of the presence of oxygen. Thus it is necessary to evacuate spectrographs and auxiliary apparatus before making measurements in the vacuum ultraviolet region. Samson (1967) describes techniques of vacuum ultraviolet spectroscopy.

Lange (1973) measured vacuum ultraviolet photoluminescence on a number of commercial samples of vitreous silica. Samples of Dynasil, Suprasil, Suprasil-W, and Corning 7940 (all with about 100% SiO_2) exhibited no luminescence when subjected to the radiations of krypton and xenon sources, which peaked at about 1500 and 1700 Å, respectively. Except for Corning 7910, the samples were produced by the aqueous $SiCl_4$ process, and all contained very small amounts of metallic impurities. Measurements showed that ultraviolet transmissions at 1 cm sample thicknesses went to zero in the range from 1630 to 1700 Å. Measurements were made at pressures of 20 x 10^{-3} torr.

Samples of Optosil, Ultrasil, and Infrasil produced by the melting of quartz crystals all exhibited luminescence when radiated by the xenon source. The ultraviolet transmissions dropped to zero at about 1600 Å. Excitation spectra of these samples peaked in the range 1600 to 1800 Å and differed in detail. The excitation spectrum of Optosil appears in Figure 9.10. Emission spectra of Optosil, Ultrasil, and Infrasil peaked at about 4000, 5000, and 5000 Å, res-

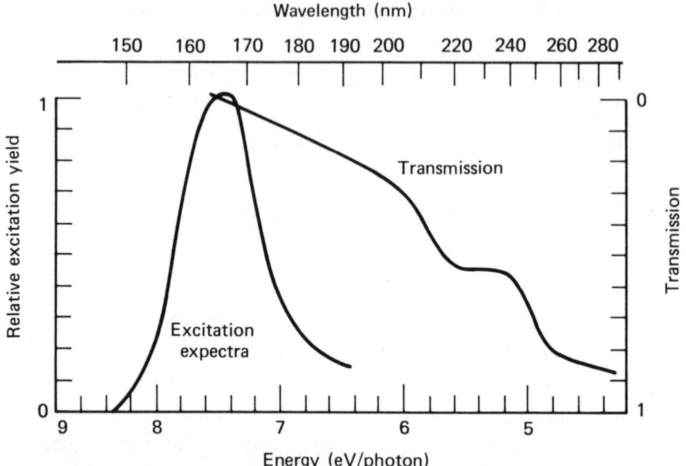

Figure 9.10 Excitation and transmission spectra of Optosil vitreous silica. From Lange (1973).

pectively, and differed in detail. Figure 9.11 presents the emisson spectrum of Optosil. Information on manufacturing differences and impurity contents of the two types of vitreous silica can account qualititatively for differences in their spectra.

During the course of this study it was found that Suprasil-W exhibited photochemistry effects. Figure 9.12 plots decline in transmission and intensity as a function of the time the sample was in the xenon source beam. Transmission loss is rapid at the 2600 Å absorption band. The absorption band and absorption down to the cutoff could be bleached out in a period of 24 hours by excitation with the 2537 Å line of mercury. Commenting on these data, Lange and Turner (1973) note that only a trace of the 2537 Å band has been previously reported. Nicoletta and Eubanks (1972) found a similar transmission loss in Suprasil-W using rather large dosages of electrons. They also observed transmission losses in Dynasil, Corning 7910, and Infrasil. Lange and Turner emphasize the importance of the nature of the exciting radiation in evaluating these effects. The use of a wide band of exciting energies could result in simultaneous bleaching losses of these effects.

Jones (1973) made vacuum ultraviolet photoluminescence measurements on some silicate glasses, and Table 9.5 lists glass compositions together with absorption limits measured in air. The study was made in conjunction with concurrent research on optical properties of these glasses in air. Glasses were prepared from silica sands containing 0.001 and 0.005% Fe_2O_3 and reagent grade chemicals.

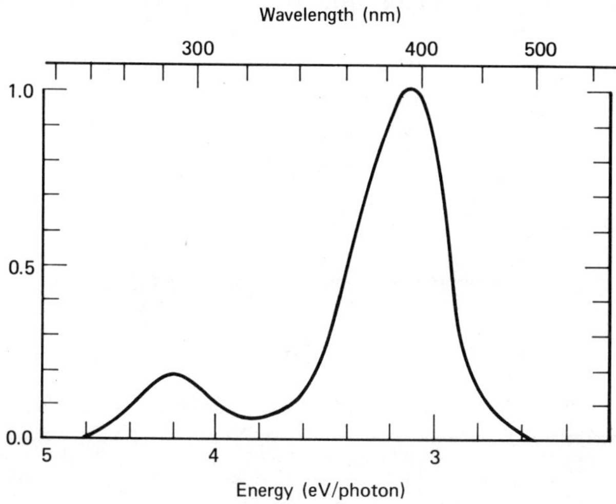

Figure 9.11 Emission spectrum of Optosil vitreous silica. From Lange (1973).

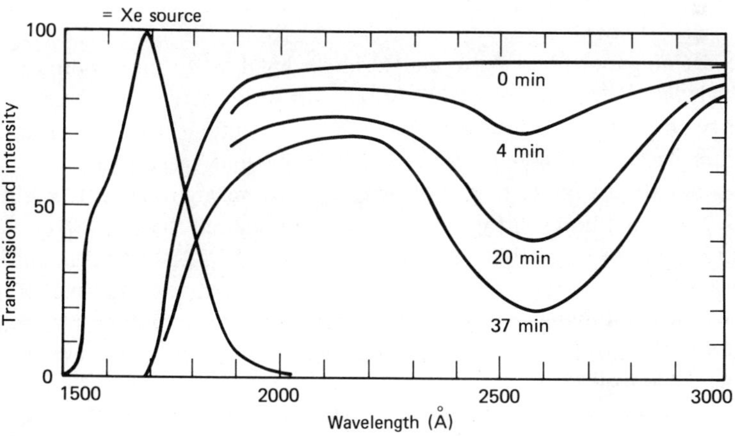

Figure 9.12 Relative source intensity and transmission loss of Supersil-vitreous silica. From Lange (1973).

TABLE 9.5   Glass Compositions with Ultraviolet Absorption Limits

Oxide	Base compositions (wt %) for three glasses		
	1	2	3
SiO_2	74.14	77.24	59.25
Na_2O	17.57	22.56	
K_2O			15.00
CaO	10.29		
BaO			23.50
Al_2O_3			1.50

Sample	Base glass	Impurity contents (wt %)	Absorption Limit (Å)
72-12-6R	1	0.005 Fe_2O_3	2300
72-17-1R	1	0.001 Fe_2O_3	2200
72-17-2R	1	0.001 Fe_2O_3 + 0.5 Mn	2600
72-17-3R	1	0.001 Fe_2O_3 + 0.5 TiO_2	3000
72-12-2	1	0.005 Fe_2O_3 + 0.5 Mn	2600
73-4-2R	2	0.001 Fe_2O_3 + 0.5 Mn	2600
73-4-3R	2	0.001 Fe_2O_3	2200
BAK 71-7	3	0.005 Fe_2O_3 + 0.75 U_3O_8	3500

Absorption limits measured on sample thickness of 1 cm.
Source. Jones (1973).

Glasses containing manganese were melted under highly reducing conditions to achieve high Mn^{2+} ion contents. Absorption limit wavelengths were dependent on impurity contents, as indicated in Table 9.5. Absorption bands in maganese and uranium glasses are caused by Mn^{2+} and $(UO_2)^{2+}$ ions, respectively.

Excitation data for the eight glasses covered the wavelength range from 1300 to 1900 Å. Relative fluorescent yield values were normalized to unity at 1600 Å. Figure 9.13 gives the excitation spectrum of sample 72-17-2R; which is similar to excitation curves of the other seven glasses. Accuracy of the excitation spectra obtained is unknown. However multiple runs with various samples demonstrated that the spectra are reproducible to within ±10% for any wavelength in the range 1300 to 1900 Å.

It was found that radiations from the xenon source and the 2537 Å line of mercury resulted in similar fluorescent spectra. The spectral fluorescence of sample 72-17-2R appears in Figure 9.14, and the spectral fluorescence of sample

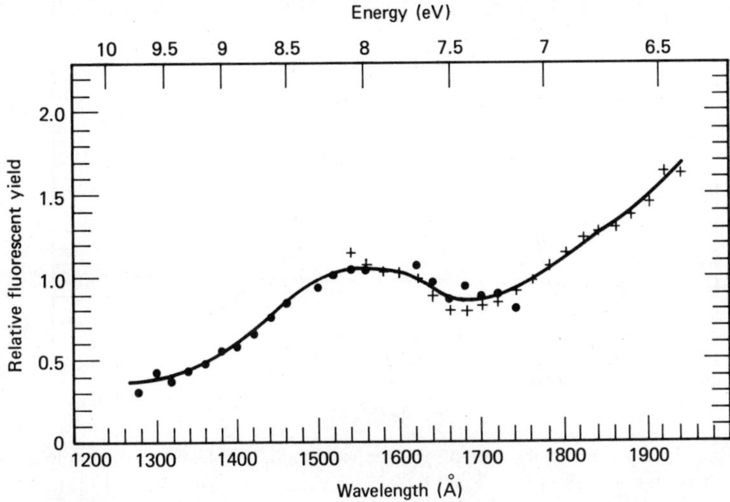

Figure 9.13 Excitation spectrum of glass sample 72-17-2R Solid points and plus signs are Krypton and xenon source data points, respectively. From Jones (1973).

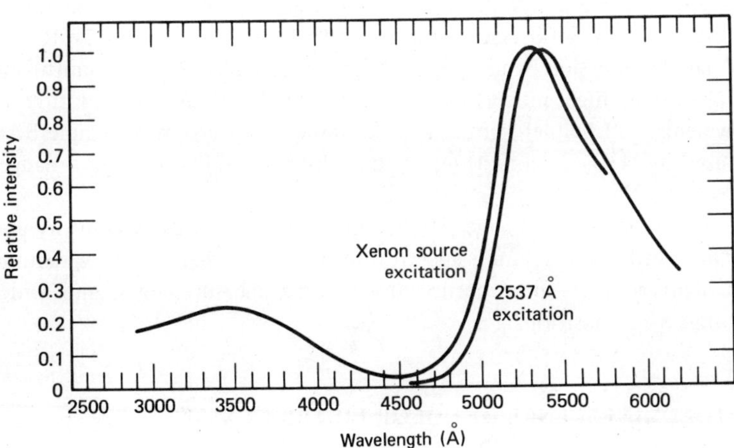

Figure 9.14 Spectral distribution of fluorescent radiation from glass sample 72-17-2R. From Jones (1973).

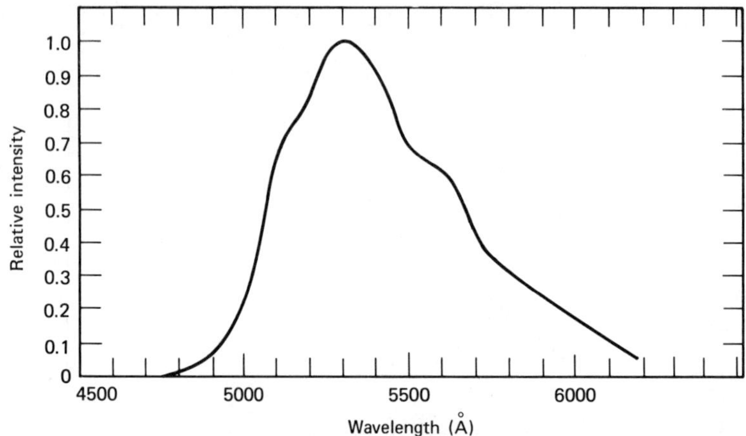

Figure 9.15 Spectral distribution of fluorescence radiation from glass sample BAK 71-7 excited by xenon source and by 2537 Å radiation. From Jones (1973).

BAK 71-7 can be seen to be the same for both radiation sources (Figure 9.15). Excitation of manganese containing samples 72-17-2R and 72-12-2 with 2537 Å radiation results in a strong fluorescent emission, caused by the Mn^{2+} ion, at about 5400 Å. Excitation by the xenon source shifts this peak to a slightly shorter wavelength of 5300 Å. Xenon excitation also caused a broad band centered around 3400 to 3500 Å in samples 72-17-1R, 72-17-2R, 72-12-6R, and 72-12-2. A similar emission band at 3500 Å was reported by Mackey, Smith, and Halperin (1966) in high purity sodium disilicate glasses during exposure to "short wavelength ultraviolet light." The fluorescence spectrum of sample BAK 71-7 is caused by $(UO_2)^{2+}$ ions as reported by Rodriguez, Parmelee, and Badger (1943).

The research reported by Lange (1973) and Jones (1973) was exploratory and suggested the need for a vacuum spectrofluorometer, higher intensity vacuum ultraviolet light sources, fluorescence decay time measurements, and wider coverage of glass compositions.

9.6 PHOTOLUMINESCENCE OF COLOR FILTER GLASSES

It is not too well known that most glasses fluoresce to some extent when irradiated with energy in the ultraviolet region of the spectrum. Turner (1973) made measurements of photoluminescence on a complete set of Corning color filters and reported data on excitation spectra, emission spectra, decay times, and estimated quantum yields. Investigators using fluorescent glasses in research should be cognizant of their possible effects on optical data. Turner states:

> To avoid the effects of filter fluorescence, it is necessary to estimate (1) the shape and locations of any emissions, (2) the range of photon energies capable of inducing fluorescence, (3) quantum yield or efficiency, and (4) decay times (it is possible to discriminate against the filter fluorescence on a time basis in some cases).

Turner presents data on excitation and emission spectra, decay time in microseconds, relative yield, and relative peak height for the filter glasses. The yields and peak heights are relative in terms of the yellow uranium glass 0-54, which was used as a standard. Discussions of the various types of filter include activating fluorescent species and other details on the observed data. Although long usage of. color filters has resulted in available filter types by different manufacturers, it is not certain that they have the same fluorescent characteristics. Color filters from the same manufacturer may have differences in fluorescence that may affect results in precision optical research. Users of color filters should read the paper by Turner.

10

Color Formation in Glasses

The first study of the effects of glassmaking constituents on color in commercial glasses was made by Zsigmondy (1901), in connection with the interests of the Jena Glass Works in colored filter glasses. Hovestadt (1902) gives a brief outline of this early work. The book by Weyl (1951) provides comprehensive coverage of the historical and technical aspects of colored glasses. Progress in the science and technology of colored glasses can be followed in technical journals and catalogs of glass manufacturers. This chapter deals with the technology of colored glasses and outlines methods of specification and control in commercial glass products.

10.1 DEFINITIONS AND NOMENCLATURE

The physical basis of the light transmitted by glasses rests on the characteristics of their spectral absorptions discussed in Chapter 9. The spectral details of the light transmitted through glass, diminished by losses caused by surface reflections and internal absorption, can be objectively measured by photodetectors of various types. These techniques are well known and need not be described here. The most used detector of color is the optical system of the human eye. The eye is not an analytical instrument for detection of spectral colors because it senses only the integrated effects of color. Moreover color sensitivity of the eye varies among individuals. The interpretation of color by different people involves the optical system of the eye, the nervous system, and the brain. Thus such color interpretations are subjective and unreliable in terms of measurement methods.

The widespread use of color in our technology requires development of systems for color specification that will ensure uniformity in the color of commercial products. The Commission Internationale de l'Echairage (CIE—International Commission on Illumination) has published the most used system for the specification of color in transparent glasses. For example, Owens-Illinois manufactures amber containers for packaging light-sensitive pharmaceuticals, whiskey, and beer. The color requirements for these three types of product differ slightly. Owens-Illinois must use color control methods that ensure that the color of containers

166

manufactured in California glass plants are the same as those made in New Jersey and other locations. The CIE system, described by Hardy (1936), Billmeyer and Saltzman (1966), and others, is used by both manufacturers and customers to verify that glass meets color specifications. Some customers wish to judge the color by ascertaining how the complete package may appear to the final consumer. Figure 10.1 illustrates the problem involved. The eye sensitivity curve shows the parts of the spectrum by which the observer judges the color. The amber appears brown, and the emerald green is judged as a brilliant green. However both types of glass may be used for absorbing short wavelengths that are deleterious to the packaged product. Thus an objective photodetection method incorporating the sensitivity curve of the so called normal eye must be used to completely specify the total response in the visible range.

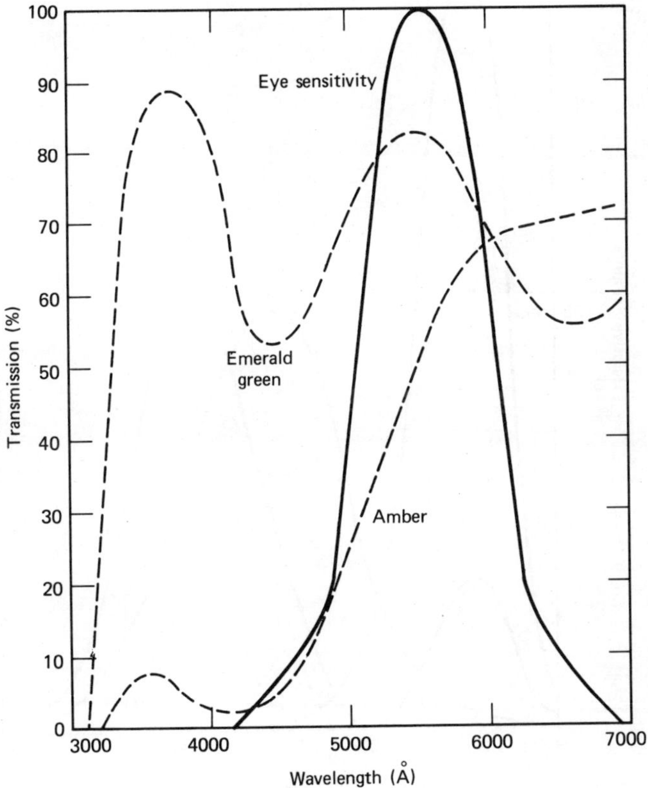

Figure 10.1 Transmission of glasses relative to eye sensitivity.

 COLOR FORMATION IN GLASSES

The essential steps in the use of the CIE system can be briefly outlined as follows.

1. The transmission curve of the glass is obtained on a spectrophotometer such as that for emerald green glass illustrated in Figure 10.1.

2. Data on the transmission curve are subjected to analysis by the CIE system in terms of specified illuminants A, B, or C. Illuminant C, typifying white light, having a color temperature of about $6500°K$ is usually used for transparent glass.

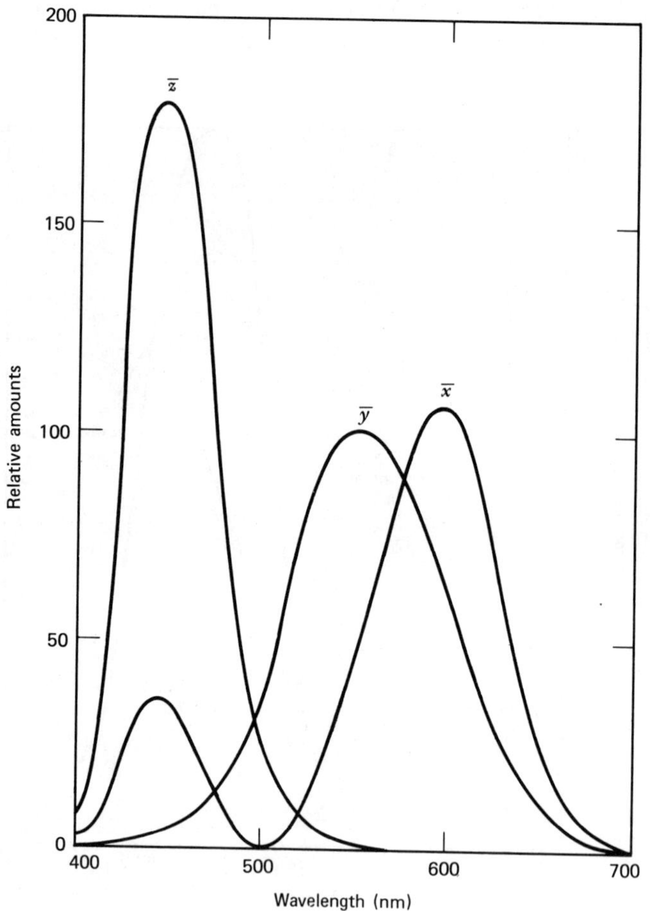

Figure 10.2 CIE color-matching functions x, y, and z for the equal energy spectrum are also the tristimulus values of the equal energy spectrum colors. From Billmeyer and Saltzman (1966, p. 36).

3. The transmission data are analyzed in terms of three color-matching functions (Figure 10.2), in conjunction with tabular numerical values of the three functions for all wavelengths of the spectrum. Values are usually given in steps of 50 Å. The y function in Figure 10.2 is immediately recognizable as the sensitivity curve of the eye.

4. Analysis by the CIE system finally provides numerical values of brightness, purity, and dominant wavelength, which completely specify the color. The value for y gives the brightness or integrated transmisssion directly. The values of purity and dominant wavelength can be displayed in a two-dimensional plot (Figure 10.3), in which all colors specified by the CIE system are located. Figure 10.4 shows the locations of common names of all the colors.

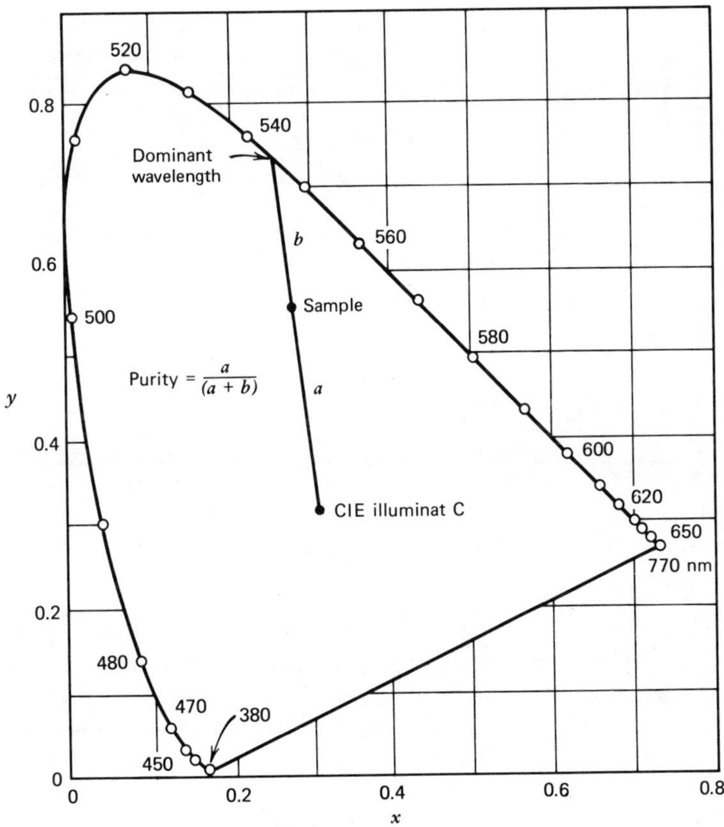

Figure 10.3 Definitions of dominant wavelength and purity. From Billmeyer and Saltzman (1966, p. 42).

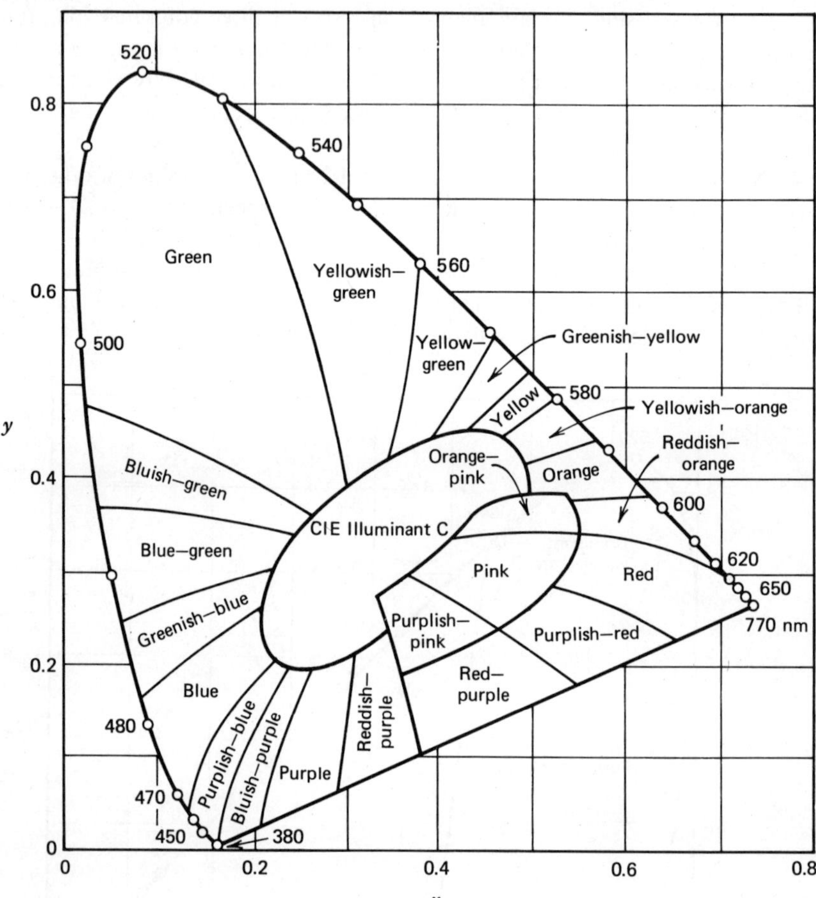

Figure 10.4 CIE chromaticity diagram with names of colors. From Billmeyer and Saltzman (1966, p. 43).

Table 10.1 gives approximate CIE color values for selected container glasses. These values are averages and may vary appreciably depending on customer requirements, which usually specify ranges of the color values. The "colorless" glass is called "flint" in the trade, and its values of purity and dominant wavelength tend to vary appreciably. Colors of art glasses may be evaluated on the CIE system but are usually evaluated by visual appearance. Colors of filter glasses can be specified on CIE terms but are usually evaluated in terms of transmission wavelength cutoffs. A standard set of filters for checking spectrophotometer-tristimulus values is available from the National Bureau of Standards.

TABLE 10.1   Approximate CIE Color Values for Container Glasses at 10 mm Thickness

Glass	Brightness (%)	Purity (%)	Dominant wavelength (Å)
"Colorless"	90	1	5700
Blue	87	1.5	5000
Green (1)	75	2.5	5300
Green (2)	72	8	5600
Green (3)	35	70	5500
Amber	30	85	5800

10.2   TYPES OF COLORANTS

Three categorical types of color formation in glasses can be distinguished. Attention is drawn to color in silicate glasses, remembering that colors in glasses based on SiO_2, B_2O_3, P_2O_5, PdO, and GeO_2 are measurably different. Color formation in these different systems derives from different arrangements of electrons, ions, and ionic groups. Integrated effects of these structural constituents cause specific absorptions that are reflected in their glass colors. The colors of some glasses also depend on thermal history and oxidation-reduction conditions during manufacture. The extent of these effects depends on types of color formation.

Ionic Colors

These colors result primarily from electronic transitions in the transition and rare earth elements discussed in Chapters 8 and 9. The curve for emerald green glass in Figure 10.1 is typical for glasses containing transition elements. The principal colorant is chromium, with a valence of 3, which has marked absorptions at about 4500 and 6500Å. The small amount of iron in the glass is dominated by chromium. As noted in Chapter 9, the Cr^{6+} ion absorbs in the ultraviolet near 3700Å. The amber glass in Figure 10.1 absorbs in the ultraviolet and is used commercially to package light-sensitive products. By control of the oxidation state, the ratio of Cr^{6+} to Cr^{3+} can be raised to the point where emerald green glass can also be used for ultraviolet protection. Emerald green glass also has a transmission band at 5500Å near the peak of the eye sensitivity curve and has high purity in terms of CIE nomenclature. The effects of the iron ion are somewhat similar to those of chromium. The Fe^{3+} ion absorbs in the near ultraviolet, whereas the Fe^{2+} ion absorbs in the infrared near 10,000Å. Here again both ions are present in glasses, and their ratio depends on the oxidation state of the glass. Both manganese and copper ions have more than one valence, and the colors

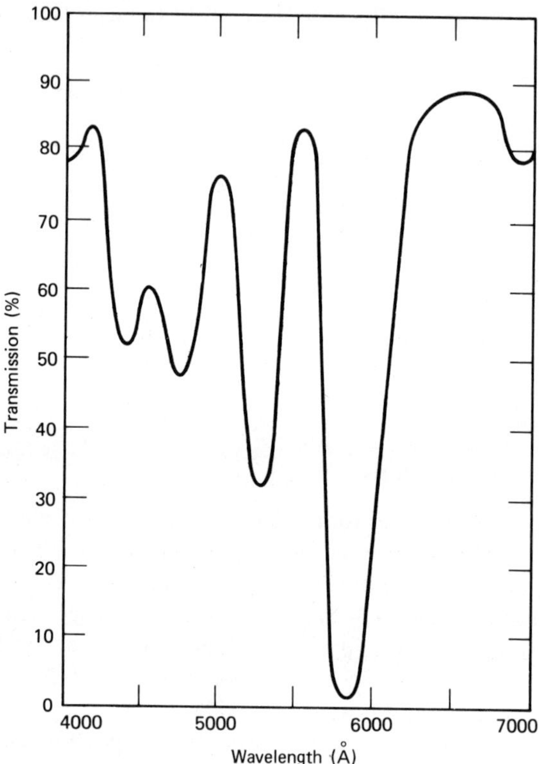

Figure 10.5 Spectral transmission of neodymium glass. From Corning color filter catalog.

produced depend on oxidation state. High proportions of Mn^{3+} produce purple colors, but glasses made under reducing conditions containing Mn^{2+} can be almost colorless. Oxidized glasses containing Cu^{2+} are blue-green. Extreme reduction can result in the presence of elemental copper, which can be classed as a different type of color formation. Nickel and cobalt have only one valence, and their colors are not changed appreciably by oxidation level. In silicate glasses nickel produces brown glasses, and cobalt glasses are blue.

The rare earth ions have protected, incomplete electronic shells, and the resulting colors are not greatly affected by their surroundings in terms of oxidation states. Because of these structures, the absorption and transmission curves of rare earth ions differ in character from those of the transition elements. The transmission curve for a glass containing neodymium (Figure 10.5) illustrates this point. The curve exhibits rather sharp absorptions, which approach those characteristic of gases and vapors. The absorptions in neodymium glasses are not much affected by states of oxidation or by other electrostatic fields of adjacent ions.

Titanium and cerium in glass absorb primarily in the ultraviolet, and their absorptions in the visible region are small except in high concentrations. However mixtures of titanium and cerium produce brilliant yellows in the visible. Small amounts of titanium and iron combine to form strong browns. Mixtures of titanium and copper result in green glasses, and manganese-titanium glasses are deep brown. Combinations of titanium with the above-named elements are used commercially for coloring borosilicate glasses.

All colors produced by transiton and rare earth elements are dependent on base glass composition, especially in terms of the R_2O/SiO_2 ratio, where R_2O generically means Li_2O, Na_2O, K_2O, and so on. Those elements with more than one valence produce colors that depend on oxidation-reduction conditions. Oxidation states are determined by glassmaking raw materials and the furnace atmosphere during the melting process. The melting temperature is of importance for variable-valence elements. For example, at high melting temperatures Fe_2O_3 dissociates into FeO; the higher the temperature, the greater the degree of dissociation. Table 10.2 lists colors resulting from glass constituents for the three types of color production. Color production in nonsilicate glasses is indicated for comparison.

TABLE 10.2   Color Formation in Glasses Based on SiO_2; Colors Based on Other Oxide Systems Are Indicated

Color	Colorants
Purple	Manganese-neodymium-nickel in K_2O glasses
Blue	Cobalt-copper-sulpher in B_2O_3 glasses;
Green	chromium-copper with Ti, Cr, Fe; iron with chromium-uranium-vanadium-molybdenum in P_2O_5 glasses
Yellow	Uranium-cadmium sulfide*-cerium and titanium-silver
Orange	Cadmium sulfide plus cadmium selenide*
Amber	Iron and sulfur*-manganese and sulfur*
Brown	Manganese and iron-titanium and iron-nickel in Na_2O glasses; iron and selenium-manganese and titanium-manganese and cerium
Red	Cadmium sulfide plus selenium*-gold-copper-uranium in PbO glasses
Black	Combinations of cobalt, manganese, nickel, iron, copper, and chromium-iron sulfide*-manganese and cobalt in PbO glasses

*Glasses are melted under reducing conditions.

Color by Sulfides, Selenides, and Tellurides

These colors are caused by the presence of compounds or ionic groups in glasses. The relation of this type of coloration to glass microstructures is not completely understood. Different mechanisms have been proposed to explain the color data. A variety of colors results from the use of these materials, sulfides and selenides being most important commercially. These colors depend on base glass composition, oxidation-reduction conditions, and melting temperature. Reducing materials added to the batch include carbon, sugar, and starch. The furnace atmosphere must be reducing. In some cases it is necessary to give the glasses added heat treatment to develop desired colors.

The amber color in Figure 10.1 is made in large tonnages for use in containers and other products; it requires additional batch ingredients of iron, sulfur, and carbon, usually in the form of high grade sea coal. Pyrite (FeS_2), sometimes augmented by additional sulfur, is the principal coloring constituent. The color results from iron sulfide plus polysulfides of the other elements present. Bacon and Billian (1954) found that CIE values of brightness and purity were closely related to light transmission of amber glasses at 5500Å. They noted that transmission at this wavelength decreased with increase in either iron or sulfur and that the glass became redder as the ratio of sulfur to iron increased. Spix and Bacon (1953) demonstrated that the amber color is affected reversibly by the heat treatment used in the annealing process. The time required to establish equilibrium color conditions increased as the annealing temperature was decreased. Refer to the annealing curve in Figure 2.4.

Although a variety of colors can be obtained by using sulfides, selenides, and tellurides, either singly or in combination, we confine our attention to optical filter glasses. The colors produced by these materials are dependent on the same factors that control the amber colors as outlined earlier. Figure 10.6 illustrates the effect of adding CdSe to glasses containing CdS for commercial production of a series of optical color filters. Optical manufacturers must use very pure batch materials to achieve the sharp cutoffs needed in filter glasses. The yellow glass is made from CdS. If the batch contains CdS plus free sulfur, the transmission curve will approach that of the amber shown in Figure 10.1 and the glass will not be suitable for use as a filter. The color of the red glass is caused by CdSe. Optical glass manufacturers offer a series of some 15 glasses ranging from yellow to red, with cutoffs moving toward longer wavelengths. The color and cutoffs of these glasses can be changed appreciably by the heat treatment used in their manufacture. Manufacturers refer to these filters as "temperature-colored glasses."

Coloration by Metallic Elements

Glasses can be colored by the selective absorption of finely divided metals. It is thought that the colors result from the presence of rather large groups of metallic atoms or ions incorporated in the lattice or substructure. Metallic gold and cop-

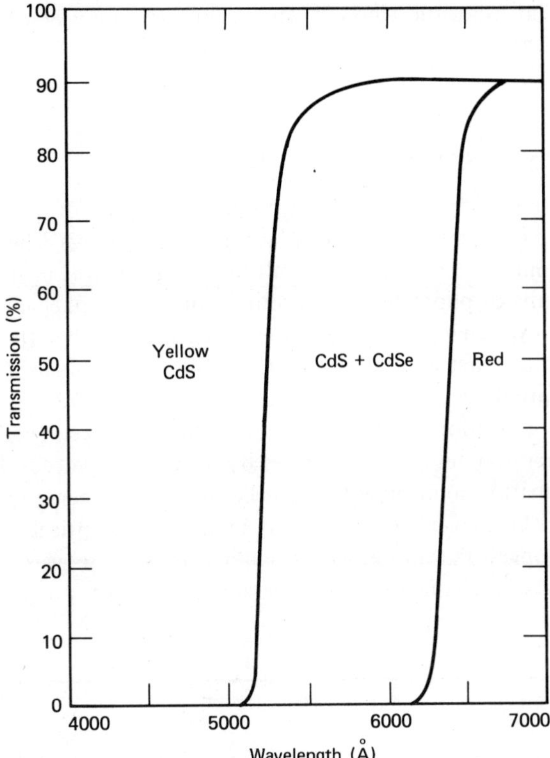

Figure 10.6 Spectral transmission of glasses containing CdS and CdS + CdSe. (Commercial filter glasses.)

per augmented by the use of small amounts of lead, tin, and bismuth give rise to the most common examples of this type of coloration. Small amounts of gold are soluble in glasses, the solubility being dependent on base glass composition. These glasses are colorless when made by usual glassmaking processes and become red only after reheating. This process is known as "flashing" in the glass trade. The CIE purity of the red color is determined by constituents in the glass batch. These glasses require batches that contain reducing materials and are melted under reducing furnace conditions. Ruby glasses derive their coloration from either elemental gold or copper.

Describing the colors in three categorical types is an oversimplification. Actually the three types can be considered as forming a rather continuous series ranging from electronic transitions (as in the transition elements), to compound colorants (as in the sulfides and selenides), to effects due to metallic elements incorporated in the lattice or substructure. The technology and production of colored glasses involves a number of parameters too extensive to be covered here in any detail.

10.3 EFFECTS OF RADIATION ON COLOR

The colors of glasses may be changed by different types of high energy radiations, including solar-rays, x-rays, beta-rays, and gamma-rays. The absorption of light quanta from these sources can result in lasting changes in the glass microstructure in terms of photochemical reactions. Indeed, these changes can be permanent unless reversed by annealing processes. Historically the effects of sunlight on glasses were the first to receive attention. The raw materials used in making large tonnage products, like containers and plate glass have small amounts of iron and other transition elements that color the glass green (Figure 10.7). To make the glass appear colorless, it is necessary to add so-called decolorizing materials, which impart a complementary purple tint. Both green and purple colorants are present in the glass, but the eye is an integrating optical system and sees the glass as colorless. Pyrolusite, which contains manganese, was one of the first decolorizing glass batch materials used. Faraday (1825) was probably the first to

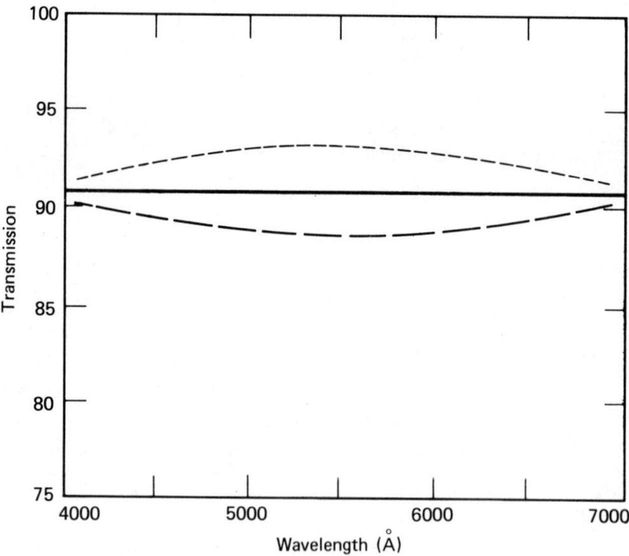

Figure 10.7 Decolorized glass (idealized curves). Solid curve, with decolorizers; dotted curve; without decolorizers (green); dashed curve, color caused by decolorizers (complimentary purple).

describe deepening of the purple color of manganese-containing glasses caused by exposure to sunlight. Splittgerber (1839) described an interesting experiment:

> I am in possession of a plate of glass which has been used as a windowpane for more than twenty years, and on which was an inscription in gold letters This inscription was taken off by grinding the plate on both sides, and polishing so as to have a new surface. When the glass had been polished, the inscription could again be clearly seen. The parts which had been under the letters remained white, while the remainder of the plate had assumed a violet tint, in consequence of the manganese it contained, a colouring which permeates the whole mass, as the grinding of the surface proved.

This account clearly describes the action of light on photochemical processes in glasses. In this case a light quantum is absorbed and the energy is used to remove an electron from the Mn^{2+} ion, which then becomes Mn^{3+}. The electron is trapped probably by an Fe^{3+} ion, which then becomes Fe^{2+}. The Mn^{2+} ion imparts little color to the glass, but the Mn^{3+} ion contributes purple. The action of the sunlight deepens the color as the ratio of Mn^{3+} to Mn^{2+} increases. The action of the light quanta is localized, leading to the use of photosensitive glasses for photographic purposes.

The use of manganese as a decolorizer was discontinued several years ago because of its light-sensitive properties. The purple color is now obtained by the use of cobalt plus selenium, which are much less sensitive to ultraviolet radiation. White and Silverman (1950) described studies on the effects of solar and high pressure mercury arc radiation on a typical $Na_2O-CaO-SiO_2$ glass containing small amounts of Se, CoO, CaF_2, $BaSO_4$, As_2O_3, CeO_2, MnO_2, and Sb_2O_3. Controlled radiation from a mercury arc was used to evaluate suitable added materials and their amounts, which could be used to develop glasses that were essentially insensitive to solar radiation. Some photochemical reactions can be minimized or prevented by adding materials that act as electron traps, thus hindering the formation of color centers. Cerium performs this function in a number of commercial glasses. Thus the coloration of manganese glasses may be reduced by the addition of cerium oxide.

Transition elements easily react to radiation by releasing electrons. The resulting absorption bands are specific and often so stable that they may act as measures of the level of radiation. Radiation causes color changes in vitreous silica made from fusion of quartz crystals and in silicate glasses made from sand. The coloration depends on the presence of small amounts of impurities and on the character and energy level of the radiation. Vitreous SiO_2 made by hydrolysis of $SiCl_4$ does not discolor if impurity levels are low.

The Corning Glass Works developed a series of photosensitive glasses in which the formation and growth of colloidal particles are controlled by the action of light and heat. Gold, silver, and copper were the photosensitive elements in these glasses. Stookey (1949) described the process for making photographs in glass. More recent developments led to the use of photosensitive properties for intricate machining of desired patterns in glasses. Corning personnel have done considerable research in this area, leading to a number of commercial glass products. Armistead and Stookey (1964) described glasses containing silver halides, which darken in sunlight and bleach in the dark. These plus later developments resulted in the commercial use of photochromic glasses for eye spectacles. These photochromic eye spectacles, which darken under the action of sunlight and rather quickly become more transparent when the light level is reduced, are now very popular.

Berthelot (1901) first discovered that a glass containing manganese was colored violet by x-rays. Subsequent advancements in modern technology have led to increasing commercial and scientific interest in the subject. High energy electromagnetic radiation primarily involves absorption of quanta by electrons. High energy particle bombardment may involve displacement of ions in the glass lattice or substructure, resulting in changes in density and other glass properties. Secondary radiation from particle bombardment may also cause changes in glass color. Commercial glasses are available for a wide range of glass usages that require protection against both types of radiation. Radiation shields to protect against x-rays, beta-rays, and gamma-rays are made from glasses containing heavy elements such as lead and barium. The shields eventually turn brown but can be restored to their original color by annealing. Glass for television screens contain the same elements in amounts that protect the viewer from harmful radiation. These glasses usually contain cerium oxide, which prevents discoloration of the viewing screen. Incidentally, the color of television viewing screens is closely specified and controlled with respect to CIE values of brightness, purity, and dominant wavelength.

Glass technology information is generally sufficient to cope with problems of radiation encountered in the earth's atmosphere. It is usually possible to achieve protection against the levels of radiation reaching the earth. However the advent of space science involves the use of glasses in hostile environments in the presence of a vacuum. Glass developments in this area are currently active subjects of research.

11

Electrolytic Conduction

Glasses behave simultaneously as dielectrics and electrolytic conductors, depending on the imposed electrical fields, their oxide compositions, and the temperature. Chapter 5 described the properties of glasses as dielectrics that are predominant at and near room temperature. Dielectric loss caused by movement of ions under the action of electric fields always exists and increases rapidly with temperature. As temperatures are increased, the glasses gradually lose their dielectric characteristics and become essentially electrolytic conductors. Electrolytic conduction becomes predominant when glasses are in the molten condition. This dual nature of glasses poses difficult problems to those interested in relations between the materials, macroscopic properties and their microstructures. This chapter discusses representative data on electrolytic conductivity or its inverse, electrical resistivity, in glasses and molten salts. The data can be interpreted in terms of the presence of substructures in silicate glasses.

11.1 METHODS OF MEASUREMENTS

Electrical conduction in silicate glasses is caused by the movement of ions and predominates over minor effects caused by electronic conduction. However some nonsilicate glasses exhibit substantial electronic conduction. Our discussion is limited to ionic conduction in silicate glasses. Ionic conduction in glasses has long been a subject of research and has been covered in some detail by Morey (1954). The process follows Faraday's law of electrolysis; (1) the mass of any substance liberated by electrolytic action is proportional to the quantity of electricity passing through the cell, and (2) masses of different substances liberated in electrolytic conduction of equal quantities of electricity are proportional to the combining equivalents of those substances.

The ionic resistivity of silicate glasses from room temperature to, say, $1200°C$ may range from 10^{20} down to 1 ohm/cm. Different methods of measurement must be used in covering this enormous range. As with other glass properties, we are interested here in the volume or bulk conductivity. Measurement of volume conductivity at or near room temperature is complicated by the presence of surface conductivity, and such effects must be evaluated separately in volume conductivity measurements. Our interest is in the conductivity of glasses at higher temperatures, where the electrodes are immersed in the body of the glass and surface conductivity can be ignored.

Electrical conduction in glasses is caused by cations that have the weakest bonds with oxygen, and these are the alkali ions such as Li, Na, and K. The conduction process depends on the character of the applied electric field and on the kind of electrodes immersed in the glass. If a static or direct current field is applied to platinum electrodes, the alkali cations pile up near the surface of the cathode and the concentration of oxygen is increased at the anode. These space charge effects at the electrodes insulate against the flow of current and greatly reduce it. Depending on conditions, the current may cease to flow (in which case the direction of the electric field must be reversed to establish current flow). If the electrodes do not furnish or accept the alkali cations, as in the case of platinum, they are known as blocking electrodes. If the electrodes are made of materials that can accept the mobile ions, they are called nonblocking electrodes and the conduction obeys Ohm's law. Reference is made to the definitive paper by Carlson, Hang, and Stockdale (1972), which covers space charge or polarization effects at electrodes in alkali-containing glasses. Conduction measurements on glasses at elevated temperatures usually employ platinum electrodes and cyclic or alternating electric current. The glass is usually contained in a platinum crucible, and the sample with its electrical connections is placed in one arm of a bridge circuit. The glass cell has a measurable capacity and must be placed in parallel with a variable condenser. The frequency of the ac source is of importance. Space charge effects are observable if 60 Hz ac is used, and the balance point of the bridge circuit is not sharp. Space charge effects rapidly decrease with increased ac frequency, and the balance point becomes much sharper. For example, Babcock (1934) arranged suitable measurement conditions by using platinum electrodes and 500 Hz cycle ac from the output of a motor-generator set with a sine wave voltage curve. Tickle (1967) explored the effect of frequency in detail. He found that electrical resistance varied with frequency and made measurements on glasses in the range 10^3 to 10^5 Hz. Those interested in correlating data from different investigators should pay close attention to the methods of measurement used, in addition to property-composition conclusions.

11.2 ELECTROLYTIC CONDUCTIVITY IN SILICATE GLASSES

The electrolytic conductivity of glasses increases rapidly with increases in temperature. Glass data are usually expressed in terms of the logarithm of resistivity in ohm-centimeters, the inverse of conductivity. Figure 11.1 shows the typical shape of resistivity-temperature curves for glasses from near room temperature up to, say, 1400°C. Babcock (1934) demonstrated that such relations could be

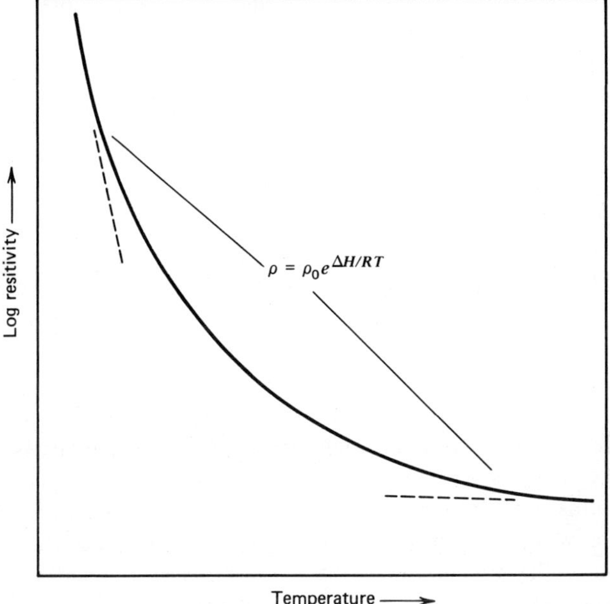

Figure 11.1   Log resistivity of glass versus temperature.

represented by equations of the form

$$\rho = \rho_0 e^{A/(T-T^0)} \tag{11.1}$$

where ρ = resistivity measured at a given equilibrated temperature T, and ρ_0, A, and T_0 are constants. This equation is of the same form as that found by Fulcher (1925) for representing viscosity-temperature data

$$\eta = \eta_0 e^{B/(T-T_0)} \tag{11.2}$$

where η = viscosity measured at a given equilibrated temperature T, and η_0, B, and T_0 are constants. The author found that the numerical values of the constant T_0 were the same for both resistivity and viscosity. This led him to the equation that relates resistivity and viscosity

$$\eta = \alpha \rho^\beta \tag{11.3}$$

where α and β are constants. It follows that there is a linear relation between the logarithms of resistivity and viscosity: A and B in the foregoing equations can be identified with the work function or activation energy of the processes, and $B/A = \beta$. Preston and Seddon (1937), Stanworth (1950), and Weyl and Marboe (1967) have commented on the significance of the Babcock equation. By comparing the temperature coefficient of resistivity A with that of viscosity B, the effect of temperature on volume is eliminated and the value of β becomes the ratio of the levels of the activation energies of the two processes. Electrolytic conduction involves the movement of ions through available spaces in the glass lattice or substructure, and the activation energy is a measure of the resistance encountered in maintaining current flow. Viscosity is a property of fluids by virtue of which they offer a resistance to flow, and it involves their rate of yielding to a shearing stress. Thus electrolytic resistance and viscosity both involve rate processes. Further research, particularly on such thermal properties of glasses as specific heat and absolute coefficient of volume expansion, will have to be done before the meaning of the Babcock equation can be clarified. Weyl and Marboe suggest that such research would be worthwhile. However this empirical equation forms the basis of controlling viscosity in manufacturing processes in terms of relatively simple measurements of electrical resistance.

Many authors have made use of the following type of equation in evaluating activation energies involved in resistivity and viscosity:

$$\rho = \rho_0 e^{\Delta H/RT} \tag{11.4}$$

where ΔH is the activation energy, R is the gas constant, and T is temperature. A similar equation is used to represent viscosity. It is evident from the equations given that neither resistivity nor viscosity varies linearly with temperature. Therefore equations of the form (11.4) fit the resistivity and viscosity data only at low and high temperatures, where property-temperature relations are closely linear, as illustrated in Figure 11.1. Thus relative comparisons can be made between activation energies at high and low temperatures. This method of obtaining activation energies has been widely used, and the data have been interpreted in terms of microstructure constituents in glasses. Most published interpretations relate to ionic radii, oxygen coordination, free volume, and other parameters characteristic of the random network theory. Considerable attention has been given to glasses containing more than one alkali oxide, commonly called mixed alkali glasses. All data show maximum values of resistivity as one alkali oxide is substituted for another. Representative publications are outlined here.

Tickle (1967) measured electrical resistivity on glasses in the binary systems $Li_2O\text{-}SiO_2$, $Na_2O\text{-}SiO_2$, and $K_2O\text{-}SiO_2$, and the ternary systems $Li_2O\text{-}Na_2O\text{-}SiO_2$, $Li_2O\text{-}K_2O\text{-}SiO_2$, and $Na_2O\text{-}K_2O\text{-}SiO_2$. He covered the method of measurement, data obtained, and interpretations of the data. After examining data on glasses in the systems named, he concluded that the availability of holes or free volume seems to be more reasonable than that of the activation jump energy approach. He states: "A comparison of the activation energy and free volume theories of transport in liquids, suggests that the free volume term is equivalent to the activation volume term in the activation energy theory." The activation volume is dependent on temperature in terms of the volume coefficient of expansion as noted in our discussion of equation (11.3).

Isard (1969) published a comprehensive review of the mixed alkali effect on electrical resistivity and other glass properties. Chaper 4 noted that many glass properties can generally be related linearly to amounts of constituent oxides over limited composition ranges. However exceptions to this approximate linearity occur in some properties when one alkali oxide in progressively substituted for another. In this case property values may reach maximum or minimum values. Although Isard reviews data on a number of properties, our discussion is limited to cases in which the mixed alkali effect seems most definite.

Isard notes that the molar volume in series of mixed alkali glasses exhibits small departures from linearity. These departures are sometimes positive and sometimes negative. Some authors have noted nonlinear relations with density as a criterion of the mixed alkali effect. Babcock points out that this is not a test of the mixed alkali effect because the progressive replacement of any glass oxide for another results in nonlinear density relations. However changes in molar volume per formula weight are tests of this effect.

Gehlhoff and Thomas (1925) first described the effect of mixed alkalies on electrical resistance in glasses. This subject has received much attention, and Isard has reviewed data obtained by a number of investigators. Data on most glasses show definite minima in conductivity and maxima in resistivity as one alkali progressively replaces another. Definite maxima are also observed in activation energy for most glasses. However the mixed alkali effect was absent in $Na_2O\text{-}K_2O\text{-}GeO_2$ and $Na_2O\text{-}K_2O\text{-}SiO_2$ glasses containing 10% or less total Na_2O plus K_2O. Present postulates on the mixed alkali effect cannot account for such data. The effect becomes more pronounced as the difference in ionic radii of the two cations increases. The resistivity reaches a maximum at certain proportions of all three alkalies, and this maximum is greater that that when only two alkalies are present. The concept that free volume space in the interstices of the substructure permits movement of cations according to size is reasonable. Isard noted that most postulates have been developed to explain the effect on electrical conductivity. However, not all the effects can be explained in term of cationic mobilities. There is no doubt that electrolytic conduction involves cationic mobility, which results in absorption of energy. The measurement of volume at an equilibrated temperature involves no energy exchange in the system. All constituent oxides in the glass are and remain in positions of minimal potential energy that are undisturbed during volume measurements. Under these conditions there is no reason to expect the behavior of monovalent alkali cations to differ from that of the other oxides. Therefore measurements of volume should not indicate mixed alkali effects. This expectation is confirmed by published volume data. It is suggested that departures from linearity are caused by errors in volume measurements. Isard (1969) covers a number of explanations that have been postulated to explain this effect in terms of cationic mobilities, cationic radii, bridging and nonbridging oxygens, oxygen coordination, free volume, and the presence of polar groups and crystals. These postulates relate to parameters characteristic of the random network theory and none has been universally accepted.

Isard points out that if dielectric constant k' and dielectric loss k'' are measured at constant frequency and temperature, both properties in general show minimum values (see the discussion on dielectric properties in Chapter 5).

Isard notes that measurements on chemical durability or chemical resistance of glasses definitely indicate the presence of the mixed alkali effect. Although chemical resistence is a complex property cationic mobilities are most definitely involved. Important uses of this fact have been made in commercial glass production.

It has been known for some time that measurements of internal friction of mechanical damping exhibit the effect. Relaxation occurs in mechanical damping and at low temperatures (100 to 150°C) the relaxation times range from 1 to 10 sec. Mixed alkali effects in other mechanical properties are less definite. Mechanical damping in glasses is discussed in some detail in Part III.

Hakim and Uhlmann (1971) measured electrical conductivity of $Na_2O\text{-}SiO_2$, $K_2O\text{-}SiO_2$, and $Cs_2O\text{-}SiO_2$ glasses at low temperatures. Values of activation energies at 350°C decreases smoothly as the alkali oxides replaced silica. The authors point out that in binary alkali silicate glasses the isothermal conductivity always increases with increasing alkali concentration. However the form of the compositional dependence of conduction and activation energy is less certain. They note that different investigators have described the data either by smooth curves or by straight lines of different slopes. Some have attempted to relate the data to structural changes in the glasses. According to Hakin and Uhlmann, "Most of these structural changes are based on the assumption that as the alkali is added to a silica network, a progressive breakdown of the three-dimensional network occurs with the rupture of the Si-O bonds." They make use of the postulated relation between the mixed alkali effect and strain energy as proposed by Anderson and Stuart (1954). Based on this model, the strain energy contribution to activation energy was significantly larger than electrostatic energy for the SiO_2 glasses, and significantly smaller for the $Na_2O\text{-}SiO_2$ glasses.

Doremus (1974) studied the interdiffusion of Na and K ions in a glass having the approximate composition in weights percent 72.6 SiO_2, 15.2 Na_2O, 4.6 CaO, 3.6 MgO, and 1.7 Al_2O_3. Tubes of the sample glass were placed in a bath of molten KNO_3 and the Na in the glass was replaced by K by ionic exchange, but the Na ions were not affected by the presence of the K ions. In other words their mobilities were not functions of the relative concentration during ionic exchange.

These interdiffusion results are consistent with the assumption that ionic mobilities are independent of ionic ratio. Doremus considers it likely that the mixed alkali effect must be accompanied by structural rearrangement in the glass and that ionic interaction alone is not sufficient to explain the data. He states:

> It is difficult to propose a model for the effect of structure on ionic mobilities that might explain the mixed alkali effect, since not even a satisfactory model for ionic transport in glass has been developed. Mobilities of monovalent cations in glass are influenced by (1) their binding to negative oxygen ions and (2) the resistance of the lattice to movement of the ion through it.

11.3 MIXED ALKALI EFFECT AND GLASS STRUCTURES

The author has reviewed published data on the mixed alkali effect in glasses and suggests that it can be categorically related to the presence of primary phase substructures. These relations are qualitative to the same extent as those previously proposed in terms of the random network theory. Published data show conclusively that the phenomena involve strengths of cationic-anionic bonds, ionic radii, free volume, and ionic mobilities. The author's proposal accepts experimental data on these microstructure details but does nothing to clarify their roles in the process. However it departs radically from explanations based on the random network theory. It deals with the type of resistance offered by the lattice or substructure to movement of ions through it. The type of resistance offered is thought to be qualitatively related to the primary phase substructure of the glass under consideration. In other words the resistence encountered by ions in a glass with an Na_2O-$2SiO_2$ primary phase substructure differs from that offered by a glass with an SiO_2 substructure. It is assumed that somewhat orderly ionic groups characteristic of specific primary phase fields present different levels of resistance to ionic movement. The following examples demonstrate the validity of this assumption. Questions relative to microstructure differences between ionic substructure groups cannot be answered and must await future research efforts in glasses.

Tickle (1967) measured electrical resistivity on a number of systematically arranged alkali silicate glass compositions in the temperature range 500 to $1400°C$. Review of his data indicates that maximum values of resistivity were attained in mixed Na_2O-K_2O-SiO_2 glasses as K_2O replaced Na_2O, and according to Levin, Robbins, and McMurdie (1964, Fig. 381,) glass compositions at maximum resistivity values approximate those at primary phase boundaries. The author has recently reviewed the density data on Young, Glaze, Faick, and Finn (1939) at room temperature. Equations for specific volume of glasses in the primary phase fields involved appear in Table 11.1. These equations differ slightly from those given by Babcock (1969) and are considered to be more representative of the

TABLE 11.1   Primary Phase Boundary Compositions in Na_2O-K_2O-SiO_2 Glasses

Oxide	Phase diagram	Maximum log resistivity	Specific volume*
Electrical resistivity data from Tickle (1967), 1000°C			
Tridymite-quartz boundary: SiO_2 = 75.5%, Na_2O = 9.5%, K_2O = 15%			
SiO_2	0.8022	0.8000	0.8000
Na_2O	0.0927	0.1000	0.1044
K_2O	0.1051	0.1000	0.0956
N2S-K2S boundary: SiO_2 = 64%, Na_2O = 13%, K_2O = 23%			
SiO_2	0.7012	0.7000	0.7000
Na_2O	0.1381	0.1500	0.1414
K_2O	0.1608	0.1500	0.1586
Electrical resistivity data from Mazurin and Borisovskii (1957)			
N2S-K2S boundary: SiO_2 = 60%, Na_2O = 15%, K_2O = 25%			
SiO_2	0.6631	0.6667	0.6667
Na_2O	0.1607	0.1458	0.1529
K_2O	0.1762	0.1875	0.1804

*The following equations were used to calculate compositions from specific volume data.

$$
\begin{aligned}
\text{tridymite field:} \quad & V = 0.45285\ SiO_2 + 0.28277\ Na_2O + 0.26888\ K_2O \\
\text{quartz field:} \quad & V = 0.44629\ SiO_2 + 0.30206\ Na_2O + 0.30282\ K_2O \\
\text{N2S field:} \quad & V = 0.43627\ SiO_2 + 0.33219\ Na_2O + 0.32449\ K_2O \\
\text{K2S field:} \quad & V = 0.43756\ SiO_2 + 0.31208\ Na_2O + 0.33676\ K_2O
\end{aligned}
$$

Phase diagram data from Levin, Robbins, and McMurdie (1964).
Density data from Young, Glaze, Faick, and Finn (1939).

density data measured by Young et al. Table 11.1 compares compositions at phase boundaries from the data of Levin et al., including resistivity maxima and specific volume minima for glasses containing 0.80 and 0.70 SiO_2. There is reasonable agreement between compositions obtained from the three independent sets of measurements, and it cannot be the result of chance or coincidence. Figure 11.2 plots resistance and specific volume for glasses containing 0.70 SiO_2. It is noted that resistivity increases when volume decreases and decreases when volume increases. This is true in all cases, although changes in volume in some primary phase fields must be considered as statistical in terms of accuracy of density measurements.

Figure 11.3 plots electrical resistivity data determined by Mazurin and Borisovskii (1957) for Na_2O-K_2O-SiO_2 glasses containing 0.6667 SiO_2. Resistiv-

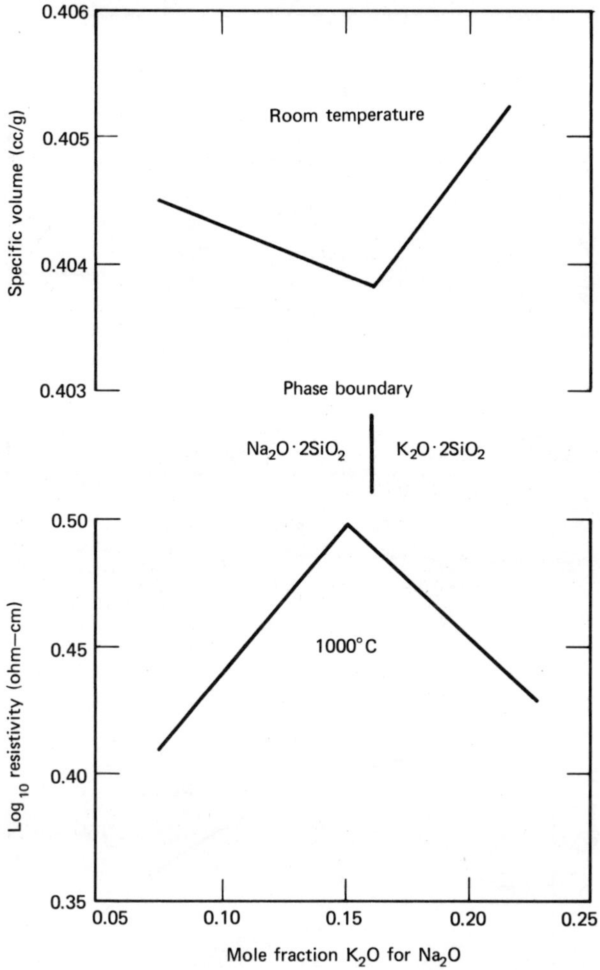

Figure 11.2 Resistivity and volume of 0.70 SiO$_2$ glass. Tabular data from Tickle (1967).

ity values at 150°C are orders of magnitude higher than those at 300°C, and the resistivity changes more rapidly with composition. Compositions derived from three sets of measurements (Table 11.1) are in reasonably good agreement.

Figure 11.4 gives data obtained by Tickle (1967) on Li$_2$O-Na$_2$O-SiO$_2$ glasses containing 0.80 SiO$_2$. Resistivity decreases regularly as Li$_2$O replaces Na$_2$O, and the resistivity maximum is absent. These glasses are located in the SiO$_2$ (tridyite) primary phase field as indicated by Levin, Robbins, and McMurdie (1964,

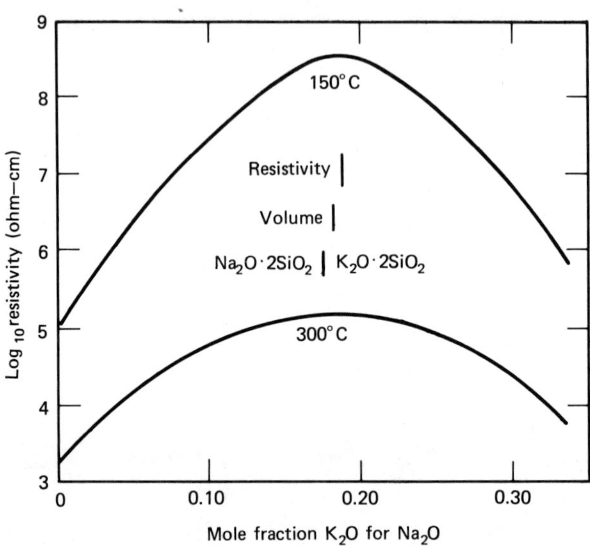

Figure 11.3   Log resistivity for 0.67 SiO_2 glasses. Tabular data from Mazurin and Borisovskii (1957).

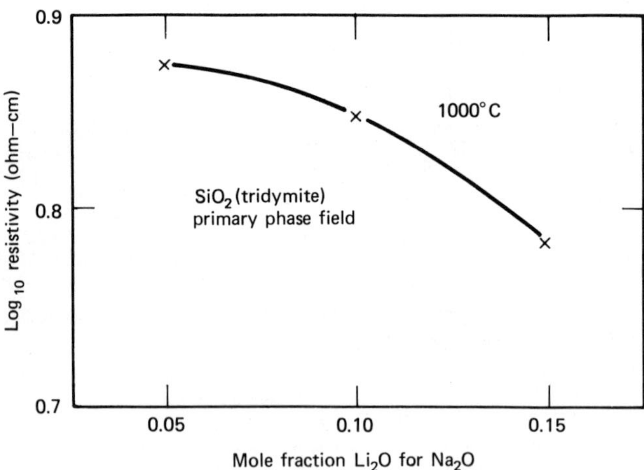

Figure 11.4   Log resistivity of 0.80 SiO_2 glasses. Tabular data from Tickle (1957).

188

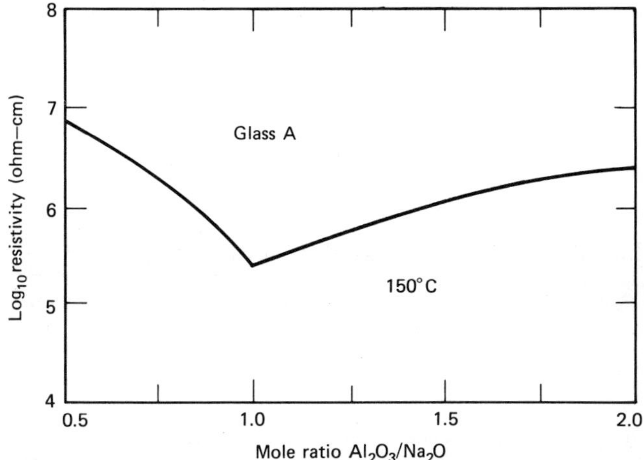

Figure 11.5 Log resistivity of Na_2O-Al_2O_3-SiO_2 glass. Tabular data from Mazurin (1965).

Fig. 430. These data are understandable in terms of the substructure postulate.

Mazurin (1965, Fig. 30) presents data on Na_2O-Al_2O_3-SiO_2 glasses at 150°C. Figure 11.5 illustrates the effect on electrical resistivity of replacement of Na_2O by Al_2O_3. It is noted that resistivity reaches a minumum value in glass A at the composition where the ratio $Al_2O_3/Na_2O=1$. Mazurin's plot shows that electrical resistivity minima for glasses A, B, and C (Table 11.2) occur where $Al_2O_3/Na_2O=1$. Levin et al. (1964, Fig. 501) show these compositions to be close to the respective phase boundary compositions. These data are in agreement with the substructure postulate. Interestingly, other glass property values change

TABLE 11.2 Eutectic Compositions of Na_2O-Al_2O_3-SiO_2 Glasses

Oxide	A	B	C
SiO_2	0.74	0.670	0.60
Al_2O_3	0.13	0.165	0.20
Na_2O	0.13	0.165	0.20
Eutectic	NA6S-3A2S Albite-mullite	NA2S-A Nepheline-corundum	NA2S-A Carnegieite-corundum

$N = Na_2O$, $A = Al_2O_3$, and $S = SiO_2$.

Source. Levin, Robbins, and McMurdie (1964, Fig. 501).

sharply in Na_2O-Al_2O_3-SiO_2 glasses at compositions where the ratio $Al_2O_3/$ Na_2O=1. Reference is made to the following papers: refractive index by Babcock (1968), microindentation hardness by Georoff and Babcock (1973), and stress-optical coefficient by Nissle and Babcock (1973). Here again the several sets of experimental data are in agreement with the presence of primary phase substructures in silicates glasses.

11.4 ELECTROLYTIC CONDUCTION IN MOLTEN SALTS

The science and technology of molten or fused salts is well known. The characteristics of molten salts are similar in some respects to those of silicate glasses. The melting of crystalline salts into molten liquids and the melting of crystalline

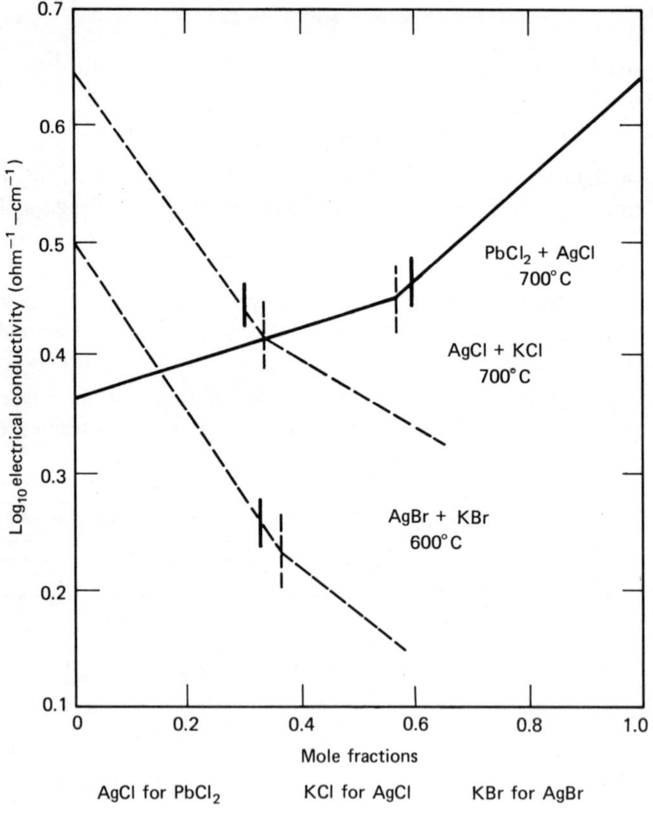

Figure 11.6 Conductivity of eutectic molten salt mixtures. Solid vertical lines: primary phase boundaries *Broken vertical lines: log conductivity slope changes*

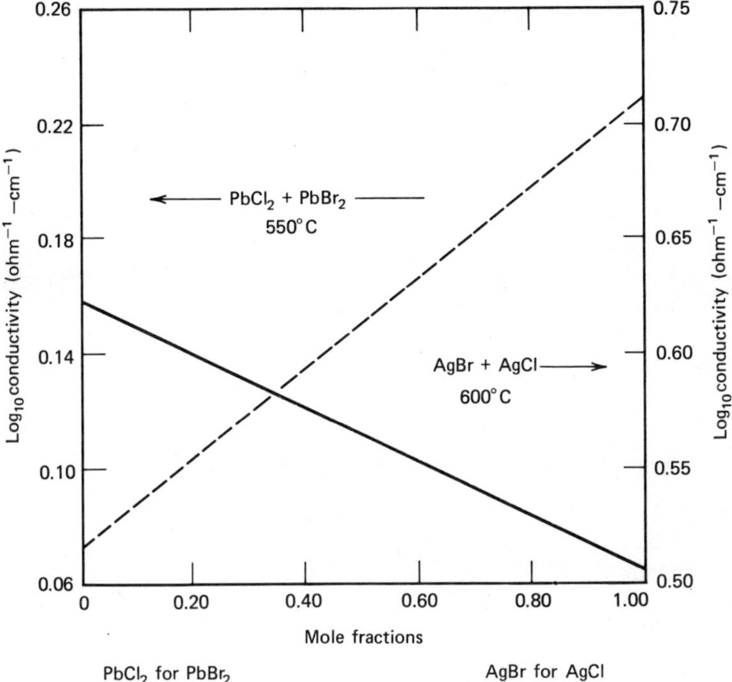

Figure 11.7 Conductivity of noneutectic molten salt mixtures.

silicates into molten glasses both involve thermodynamic parameters, as indicated by equation (2.1). As discussed in Chapter 2, the degree of randomness can be measured numerically in terms of the entropy S, and at the melting point there is a balance between the internal energy and the thermal disorganizing force. The drive toward minimum potential energy is balanced by the drive toward maximum randomness. The microstructures of molten salts and molten and solid glasses are active subjects of research. Investigators working on the two types of material have a common problem in deciding whether the ions are randomly arranged or structured in some manner.

Our interest here is in the electrolyticity of molten salts as this property may relate to similar phenomena in silicate glasses. In this connection Rawson (1967) discusses halide glasses and glasses in the system KNO_3 - $Ca(NO_3)_2$. Weyl and Marboe (1967) also treat the ionic properties of molten salts.

Herrap and Heymann (1955) measured electrical conductivity of a number of mixtures of systematically arranged molten salts in the temperature range 400 to 700°C. These data can be reasonably explained in terms of the presence of substructures in the molten salt mixture. The method is similar to that used for glasses, and the results from its use are given in Figures 11.6 and 11.7 and Table

TABLE 11.3   Primary Phase Boundary Compositions in Molten Salt Mixtures

Mixture	Temperature (°C)	Phase diagram	Conductivity data*
Eutectic mixtures			
PbCl$_2$ + AgCl	700	1223	(1) and (2)
PbCl$_2$		0.40	0.43
AgCl		0.60	0.57
AgCl + KCl	700	1215	(3) and (4)
AgCl		0.70	0.66
KCl		0.30	0.34
AgBr + KBr	600	1175	(5) and (6)
AgBr		0.67	0.63
KBr		0.33	0.37
Noneutectic mixtures			
AgCl + AgBr	600	1593	(7)
PbCl$_2$ + PbBr$_2$	550	1604	(8)

*The following equations were used to calculate compositions from electrical conductivity data:

(1) $\log^{10}$ conductivity $= 0.36871$ PbCl$_2$ $+ 0.51801$ AgCl.

(2) $\log^{10}$ conductivity $= 0.20655$ PbCl$_2$ $+ 0.64238$ AgCl.

(3) $\log^{10}$ conductivity $= 0.65093$ AgCl $- 0.02255$ KCl.

(4) $\log^{10}$ conductivity $= 0.51468$ AgCl $+ 0.23900$ KCl.

(5) $\log^{10}$ conductivity $= 0.50247$ AgBr $- 0.23174$ KBr.

(6) $\log^{10}$ conductivity $= 0.36278$ AgBr $+ 0.01094$ KBr.

(7) $\log^{10}$ conductivity $= 0.50518$ AgBr $+ 0.62391$ AgCl.

(8) $\log^{10}$ conductivity $= 0.22391$ PbCl$_2$ $+ 0.07179$ PbBr$_2$.

Electrical conductivity data from Harrap and Heymann (1955).

Phase diagrams from Levin, Robbins, and McMurdie (1964).

11.3 Figure 11.6 reveals that $\log_{10}$ conductivity values are linear within given phase fields for eutectic mixtures and that the linearity changes at their respective phase boundaries. Figure 11.7 presents linear relations between log conductivity and composition for noneutectic compositions where only one phase field is involved. Equations derived for all mixtures (Table 11.3) represent log conductivitiy to within 2%. Table 11.3 also indicates that compositions calculated from log conductivity data are in reasonable agreement with those given in the phase diagrams for eutectic mixtures.

The foregoing interpretation indicates that mobile ions in molten salts do not move through a randomly arranged network of other ions but rather between ionic groups that are characteristic of given primary phase fields. As in the case of glasses, the substructure postulate can only categorize in terms of different types of substructure. Microstructure details will be revealed following future research.

III

Mechanical Properties

> It is difficult even to attach a precise meaning to the term "scientific truth".
> Thus the meaning of the word "truth" varies according to whether we deal
> with a fact of experience, a mathematical proposition, or a scientific theory.
>
> Albert Einstein (1879-1955)

The reactions of glasses to imposed mechanical and electrical forces are similar in a number of ways, and some can be described in terms of analogous mathematical expressions. As noted in previous discussions, glass microstructures are considered to consist of electrons, ions, and ionic groups, which execute forced oscillations under the action of imposed force fields. Though it is customary to speak separately of the actions of electrical, magnetic, mechanical, and thermal forces, we know that the effects of these forces are interpendent and cannot be clearly isolated from one another. For example, all types of imposed forces cause changes in polarization that can be described categorically in terms of equation (5.1).

In all cases the energy absorbed by the glass is distributed among the coupled structural oscillators in ways that depend on the character and intensity of the imposed forces. Part III outlines the direct effects of imposed mechanical forces on a number of glass properties. Distinctions are drawn between pressures of ordinary intensity and high pressures of the order of 10^4 atm and higher. Ordinary pressures leave the glass microstructures unchanged and the properties are generally reversible. Very high pressures cause modifications in the lattice or substructure that result in permanent property changes in some glasses.

12

Definitions and Nomenclature

Similarities of definitions and nomenclature used to describe the actions of mechanical and electrical forces are well known. Principal effects of electrical forces are described in Chapter 5. Mathematical expressions of the actions of mechanical forces are similar in form to those for electrical forces; it is unnecessary to set them down in detail in this chapter. Readily available texts describe relations between mechanical and electrical forces imposed on isotropic materials such as glasses. Efforts have been made to use accepted symbols for the properties involved, but in some cases symbols have been changed to preserve uniformity in the present discussion.

12.1 EFFECTS OF STATIC MECHANICAL FORCES

Discussions on the nature of glasses outlined in Chaper 2 emphasized that glass properties change in a continuous manner from a liquid at high temperatures to a solid at room temperatures. In terms of property measurements, glass is a viscous liquid at high temperatures and gradually becomes an elastic solid at room temperature. The effects of temperature on Young's modulus and viscosity of an $Na_2O\text{-}CaO\text{-}SiO_2$ glass are illustrated in Figure 12.1. The temperature spread is arbitrarily divided into three regions. It is to be understood that all silicate glasses follow this same patterrn of property changes; the temperature ranges are dependent on glass composition. The Young's modulus data are from McGraw (1952), and the viscosity data are supplied through the courtesy of Owens Illinois, Inc.

Figure 12.1 shows that the glass in region A is a viscous liquid and that its elastic characteristics are vanishingly small. In this region the viscosity is dependent only on composition and temperature. When temperatures move into region B, generally termed the transformation range, the viscosity increases rapidly with temperature and the elastic modulus increases rapidly. Properties in this region depend on composition, temperature, and time. When temperatures move into region C, the elastic modulus increases and viscous flow becomes very small. Properties in this region are dependent only on composition and temperature. These designations are in terms of technological uses and are approximate. For

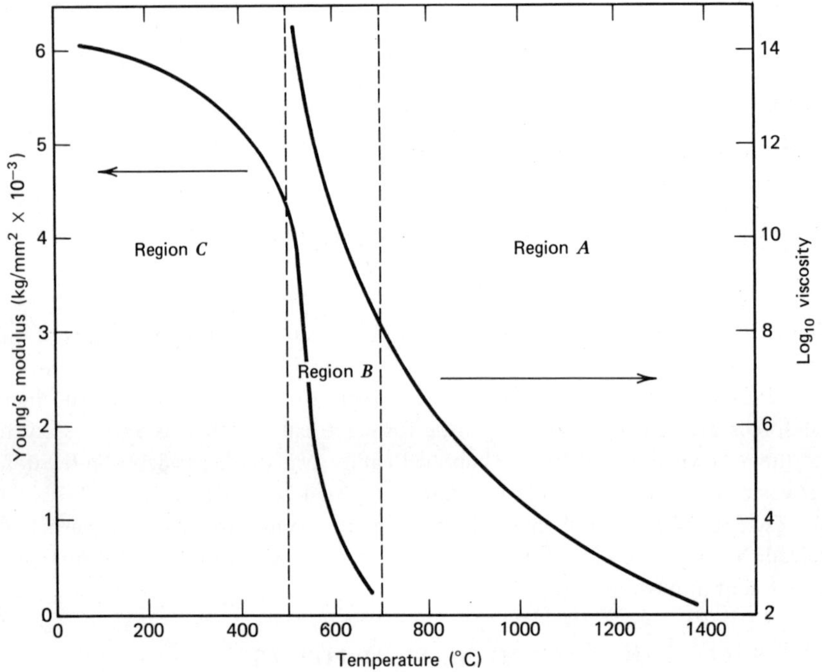

Figure 12.1   Young's modulus and viscosity versus temperature.

example, it can be demonstrated that glass exhibits elasticity in region A and viscous flow in region C.

Consider the case of weight suddenly applied to a fiber of glass suspended vertically in a furnace at some intermediate temperature (Figure 12.2). Sudden application of the weight at time t_1 results first in an instantaneous extension, followed by a region of delayed elasticity during which time the structural elements adjust to the force. This occurs during the time period known as the relaxation time λ. At time t_2, when relaxation is completed, the glass will flow until the force is removed at time t_3. During this period the extension is linearly related to time, and the slope of the line is equivalent to the viscosity coefficient. At time t_3 there occurs an elastic contraction, which is followed by delayed elastic recovery.

It is evident that the chameleonlike nature of glass is further dependent on the character of the imposed force. Two separate aspects of applied forces can be noted.

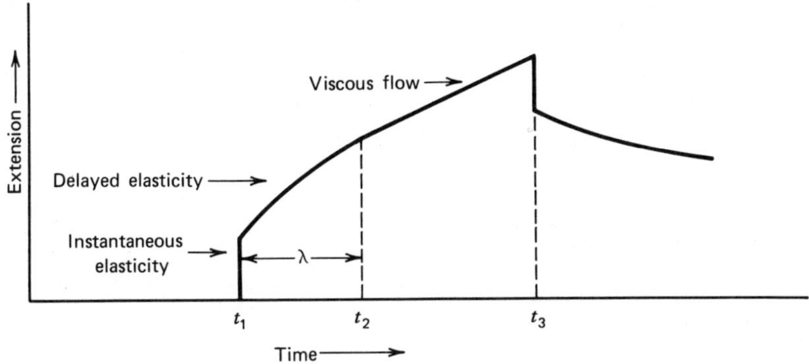

Figure 12.2 Effects of force applied to glass at intermediate temperatures.

1. Forced in which the time duration of application is much longer than the relaxation time. Such forces result in flow at high temperatures and will produce fracture at low temperatures if the force exceeds the bond strength of the glass.
2. Forces in which the time duration of application is much shorter than the relaxation time. Such impulsive forces result in fracture if the energy of the impulse exceeds the bond energy of the glass. Refined experiments show that glass can be fractured by such forces at temperatures well into region A of Figure 12.1. Glass manufacturers using high speed production machines are well acquainted with the action of impulsive forces.

12.2 EFFECTS OF CYCLIC MECHANICAL FORCES

Mathematical expressions of the actions of cyclic mechanical forces are similar to those for cyclic electrical forces discussed in Chapter 5. Equations resembling (5.5), (5.6), and (5.7) can be obtained readily by replacing electrical stress E by mechanical stress and electrical displacement D by mechanical strain. The meanings of reversibility of polarization, time delays, angular velocity, and loss angle given in Chapter 5 apply equally well to the action of mechanical forces. The mechanical relaxation time λ corresponds to the dielectric relaxation time τ and is the time required to relax all but $1/e$ of the strain.

Mechanical power losses are related to the frequency of vibrations and relaxation time λ. Figure 12.3 illustrates conditions when power loss is high. The area of the hystersis loop (Figure 12.3a) is a measure of the power loss. High loss occurs when the frequency equals $1/\lambda$ (Figure 12.3b). Energy is absorbed by the glass when impressed frequencies are in resonance with vibrations of structural constituents. The energy absorbed during damping causes the glass to be strained.

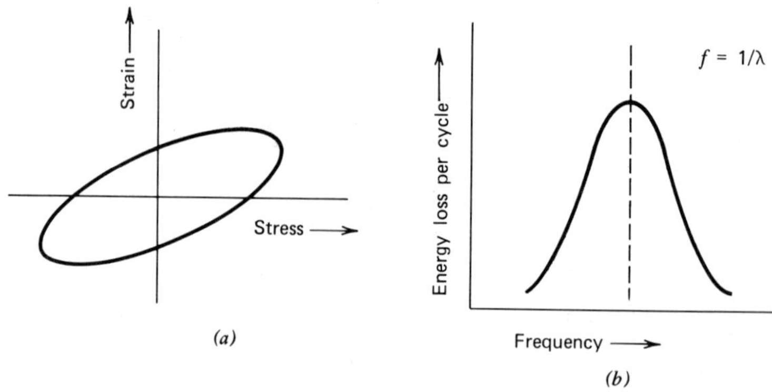

Figure 12.3 Conditions for high mechanical loss.

This absorption can result in heat absorption, which may in turn cause chemical diffusion, changes in volume, and so on. Review of experimental data on electrical and mechanical behavior during relaxation times in the processes indicate that the elastic aftereffect is caused by several mechanisms with varying relaxation times. Since elastic flow and viscous flow take place simultaneously, this region of adjustment of structural constituents has been referred to in the literature as "viscoelastic" and "elasticoviscous."

12.3 RELATIONS BETWEEN ELASTICITY AND VISCOSITY

The application of a shearing stress S to an elastic solid results in elastic deformation. The shear strain is defined as the tangent of the shear angle θ (Figure 12.4). The amount of the deformation depends on the shear modulus G.

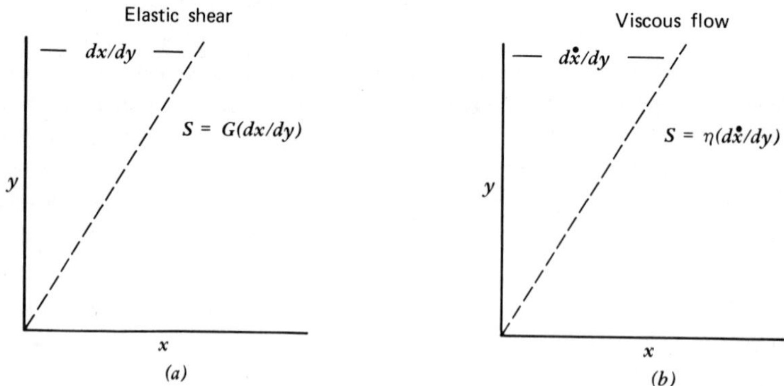

Figure 12.4 Effects of shearing stress on solid and liquid glass.

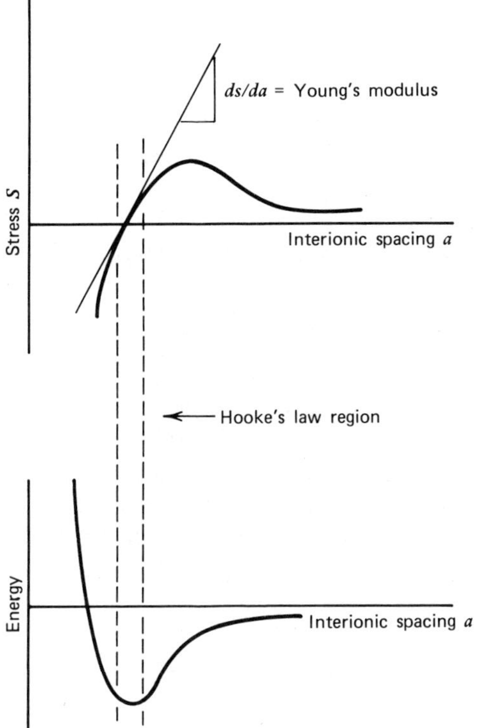

Figure 12.5 Relation of Hooke's law to interionic spacing.

$$S = G\theta = G\,dx/dy \qquad (12.1)$$

The shear modulus is a constant for levels of strain within the range where Hook's law is obeyed, as illustrated in Figure 12.5, which applies directly to Young's elastic modulus. All elastic moduli refer to stress/strain relations where the interionic spacing is changed by quite small amounts from their equilibrium distance. Though it is usual to assume that elastic moduli are the same for both tensile (+) and compressive (−) stresses. the curves in Figure 12.5 indicate that elastic moduli in compression should be larger than those for tensile stresses. In this connection remember the demonstration by Van Zee and Noritake (1958) that the stress-optical coefficient for compressive stress was larger than that for tensile stress. It is emphasized that in all cases of elastic strain the ions maintain their original positions in the structure. The action of a tenile stress is to extend dimensions of the sample in the direction of the stress and to contract them in perpendicular directions. The magnitude of the ratio of contraction/extension is known as Poisson's ratio and can be shown to be 0.5 for ideal materials that maintain constant volume when elastically stressed. This ratio for real materials is always less than 0.5; Poisson's ratio for glasses is more nearly equal to 0.25. The classical theory of elasticity deals with the effects of macroscopically applied stresses on reactions of the material in terms of the following elastic moduli: Young's modulus E, shear modulus G, Poisson's ratio μ, bulk modulus B, and $1/B$, termed compressibility β. All elastic moduli can be expressed in terms of E and G or any two of the others. For example,

$$G = 3BE/9B\text{-}E = E/2(1+\mu) \tag{12.2}$$

Poisson's ratio is a pure number, and the others are expressed in force per unit area (e. g., $1b/in.^2$, kg/cm^2, $dynes/cm^2$).

The application of a shearing stress to molten glass results in viscous flow. The magnitude of the flow depends on the viscosity modulus or coefficient and is constant for Newtonian or laminar flow. The relation between shear stress and flow can be expressed as

$$S = \eta(d\dot{x}/dt)/dy \tag{12.3}$$

where η is the viscosity coefficient, which is the shearing stress involved in maintaining the uniform flow of a liquid per unit rate of shear. The viscosity coefficient is called the poise in honor of Poiseuille, who did extensive research on the flow of blood in veins and capillaries. The poise is expressed in grams per centimeter-second or dyne-seconds per square centimenter. Figure 12.4 illustrates relations in viscous flow. It is evident that elastic shear modulus determines the amount of strain set up by an applied stress, whereas the viscosity modulus or coefficient determines the rate at which the substance yields to the applied stress. It is noted that the velocity of flow is directly proportional to the applied stress. This is a condition for laminar flow in fluids. Under turbulent or hydraulic conditions the stress is proportional to the second or third powers of the velocity. In viscous flow the ions no longer maintain their original neighbors. As mentioned earlier, glass exhibits elastic and viscous properties simultaneously at intermediate temperatures. Sorting out their separate effects in such cases presents major problems for the materials scientist.

Published data on elasticity, relaxation processes, and viscosity in glasses is covered in some detail in the following chapters.

13

Elastic Behavior

Morey (1954) reviews early work on the elastic behavior of glasses. Most of the early data on glasses were determined by static methods in which the sample was subjected to tension, compression, or bending. More recent data have been obtained by the use of sound waves and related dynamic elastic behaviors. We have seen in Chapter 12 that certain elastic displacements have measurable delays in terms of the relaxation time. Therefore a dynamically measured strain may be less and the modulus higher than those determined by static methods. However the static and dynamic values are approximately the same for glasses if measurements are made at room temperature. McSkimin (1961) published a comprehensive review of dynamic methods for measurement of elastic moduli and wave propagation in solids. This definitive paper, which covers measurements at normal temperatures and pressures, also outlines high temperature and high pressure techniques; it lists 111 references.

13.1 ELASTIC PROPERTIES OF COMMERCIAL GLASSES

The mechanical strength of a glass product is dependent on elastic constant of the glass, the shape or architecture of the product, and the character of the force to which it is subjected. The time duration of applied force may be greater than the relaxation time of the glass. The force may be a sudden impact, in which case its time duration is much less than the relaxation time. An impact force propagates through the product elastic waves that may interfere constructively or destructively, depending on the shape of the product. Thus the point of impact becomes important in relation to the shape of the product. Evaluation of the mechanical strength of a product is a complex problem. Instantaneous and delayed elasticity have major influences on mechanical strength, but mechanical strength per se is not a property of the material.

An additional requirement must be considered if the glass is to be used for optical purposes. Wright (1921) has noted that even at room temperature lenses and prisms made of strained glass do not retain their shapes. This is also important for large astronomical mirrors in reflection optics. It is for this reason that all optical glasses must be given more careful annealing than other glass products.

Optical glasses must have very low levels of birefringence to meet requirements in various optical applications.

Table 13.1 shows values of Young's modulus, the shear modulus, and Poisson's ratio for selected Corning commercial glasses. Values of G have been estimated in terms of E and μ. Silicate glasses have relatively high values of E and G with μ ranging up to about 0.25. Data in Table 13.1 demonstrate that elastic moduli are strongly dependent on glass composition. Formulation of glasses having desired values of elastic properties requires information on the effect of each oxide used in the composition. Hovestadt (1902) gives value of constants for calculating Young's modulus in terms of weights percent of the oxides. The three types of glass listed had constants whose values depend on glass type. The glasses were considered to be mutual solutions of the constituent oxides. The formulas for calculating Young's modulus were of the form of equation (4.1) previously described. This method implies that glass microstructure constituents are randomly distributed in the glass. The approach is compatible with classical elasticity theory, which treats a solid as a continuum whose parts cannot be separately discerned. In the absence of external forces, the solid is in a condition of internal equilibrium and minimum potential energy. Imposition of external forces brings into play balancing attractive and repulsive forces of

TABLE 13.1 Compositions and Elastic Properties of Commercial Glasses

| Oxide | Compositions (wt %) of five Corning glasses | | | | |
	7940	7913	9700	1990	7070
SiO_2	99.9	96.5	80.0	41.0	71.0
H_2O	0.1				
Al_2O_3		0.5	2.0		1.0
B_2O_3		3.0	13.0		26.0
Li_2O				2.0	0.5
Na_2O			5.0	5.0	0.5
K_2O				12.0	1.0
PbO				40.0	
Property					
E	10.5	9.6	9.6	8.4	7.4
$G*$	4.5	4.0	4.0	3.4	3.0
μ	0.16	0.19	0.20	0.25	0.22

$*G = \dfrac{E}{[2(1 + \mu)]}$, where E and G in lb/in.$^2 \times 10^6$.

Data source. Hutchins and Harrington (1966).

microstructure constituents. Elastic moduli relate to integrated macroscopic reactions of the solid. Therefore relations between elastic moduli data and glass microstructures must be empirical.

13.2 PROPERTIES ON BASIS OF RANDOM NETWORK THEORY

Morey (1954) reviewed elastic property data on both commercial and experimental glasses. Subsequent papers by Kozlovskaya (1960), Han Fu-si (1963), Phillips (1964), and Demkina and Kisin (1972) based an evaluation of elastic property data on the condition that glasses were mutual solutions of oxides. Demkina and Kisin measured data on Young's modulus, the shear modulus, Poisson's ratio, density, and the number of seconds it took for longitudinal and transverse elastic waves to travel one kilometer. They reviewed data of the authors just named and concluded that Young's modulus and the shear modulus were not additive quantities, since values of oxide constants for these properties differed widely. The author does not agree with this reasoning and suggests that such differences result because different types of glasses were examined by the different authors. As noted in Chapter 4, glass properties are not linear functions of glass composition if wide ranges of composition are examined. They are only linear for limited composition ranges. Results from Phillips, Demkina, and Kisin are of interest in this connection.

Phillips (1964) examined the Young's modulus data of various authors, for 36 glasses. The compositions included binary, ternary, and more complex silicate glasses expressed in mole percent. Young's modulus values are given in kilobars and are represented by equations of type (4.1). The author's evaluation showed that coefficients for SiO_2, Al_2O_3, and CaO had constant values of 7.3, 12.1, and 12.6 per mole percent, respectively, for all compositions. Coefficients for the following oxides had variable values depending on the complexity of compositions: Na_2O, K_2O, Li_2O, B_2O_3, PbO, MgO, BaO, and ZnO. Phillips' coefficients give good representation of the measured data, but this complicated method cannot be recommended for future research on glass properties.

Demkina and Kisin (1972) measured elastic property data on 32 experimental glasses whose compositions were expressed in mole percent. Three base glasses containing the following oxides were used: (1) SiO_2, Na_2O, and K_2O; (2) SiO_2, PbO, and K_2O; (3) SiO_2, B_2O_3, ZnO, BaO, Na_2O, and K_2O. Additional glasses were made by adding other oxides to the three base glasses. The authors give coefficients for 12 oxides for Young's modulus, the shear modulus, Poisson's ratio, density, and travel times for longitudinal and transverse elastic waves in seconds per kilometer. Young's and shear moduli were expressed in kilograms per square millimeter. Their oxide coefficients for Young's modulus differ considerably from those proposed by Han Fu-si and Phillips, perhaps because they tried to incorporate data from widely different glass types into one computational scheme. Their paper is principally concerned with travel times of elastic waves in glasses which are the reciprocals of the velocities in kilometers per second. They note that a similar relationship exists for light propagation. In this case the additive quantity is not the light propagation velocity but its reciprocal, the refractive index. Coefficients for times of travel of elastic waves represent the data to within 2%; accuracies of representation are not given for the other properties.

The technological importance of coefficients developed by the various authors cannot be questioned in terms of formulating glasses of the different types reviewed. Values of coefficients used in predicting elastic properties of new glasses depend on glass type and must be carefully scrutinized in each prospective application.

13.3 PROPERTIES ON BASIS OF SUBSTRUCTURE METHOD

Representation of elastic property data in terms of the random network theory is of technical importance if used properly, but such information can tell us nothing about glass micro-structures. Johnston and Babcock (1975) took a new approach to the problem in the effort to determine whether elastic data could be interpreted in terms of postulated substructures in silicate glasses. They made measurements on Young's modulus, the shear modulus, and Poisson's ratio on Na_2O-TiO_2-SiO_2 glasses located in two contiguous primary phase fields. A number of other related elastic properties were derived from their measurements.

The ASTM designation C623 - 71 (1974) procedure was followed in making measurements at $26 \pm 2°C$. The frequency counter was capable of measuring frequency to the nearest hertz. The frequency counter and apparatus were checked by repeated measurements of the resonant flexural frequency of an Al_2O_3 reference material described by Dickson and Wachtman (1971). Fundamental frequency ranges used were 896.9 to 3496.0 Hz for flexural measurements and 3253.3 to 8347.0 Hz for torsional measurements. Tables 13.2 and 13.3 give glass compositions and measured property data on glasses located in the D and E primary phase fields. Property data for each composition are averages of the indicated number of samples.

Elastic moduli data on E, G, and μ for the 15 D samples and 16 E samples were subjected to least squares analyses. Values of other moduli were calculated for each sample in terms of E, G, and μ. Representative least squares equations were determined for D and E glasses separately and then as combined groups of 31 glasses. Representing E and G moduli by seqparate equations for D and E groups of glasses was statistically advantageous. However combining

TABLE 13.2   Data for Glasses in D Primary Phase Field

Oxide	Composition (wt %) of four glasses			
	D1	D2	D3	D4
SiO_2	0.65	0.60	0.60	0.67
TiO_2	0.15	0.15	0.20	0.12
Na_2O	0.20	0.25	0.20	0.21

Property				
Number of samples:	4	4	4	3
Density (g/cc):	2.658 ±0.002	2.685 ±0.002	2.733 ±0.003	2.614 ±0.001
E (dynes/cm^2 × 10^{11}):	7.8821 ±0.0186	7.4237 ±0.0112	8.2018 ±0.0200	7.4270 ±0.0056
G (dynes/cm^2 × 10^{11}):	3.2066 ±0.0081	2.9834 ±0.0068	3.3114 ±0.0066	3.0105 ±0.0022
μ	0.2290 ±0.0007	0.2442 ±0.0012	0.2384 ±0.0010	0.2335 ±0.0016

Source. Johnston and Babcock (1975).

TABLE 13.3 Data for Glasses in E Primary Field*

Oxide	Composition (wt %) of five glasses				
	E1	E2	E3	E4	E5
SiO_2	0.50	0.45	0.40	0.55	0.50
TiO_2	0.20	0.25	0.30	0.20	0.24
Na_2O	0.30	0.30	0.30	0.25	0.26
Property					
Number of samples:	3	4	2	4	3
Density (g/cc):	2.764 ±0.001	2.831 ±0.004	2.900 ±0.002	2.755 ±0.004	2.813 ±0.003
E (dynes/cm^2 × 10^{11}):	7.3759 ±0.0017	7.8255 ±0.0449	8.1557 ±0.0000	7.8361 ±0.0412	8.1515 ±0.0274
G (dynes/cm^2 × 10^{11}):	2.9294 ±0.0031	3.1067 ±0.0192	3.2178 ±0.0020	3.1360 ±0.0178	3.2572 ±0.0034
μ	0.2589 ±0.0011	0.2595 ±0.0018	0.2673 ±0.0008	0.2494 ±0.0006	0.2513 ±0.0007

*These data on glasses are in reasonable agreement with data published by Manghani (1972) on glasses of the same compositions, prepared by the National Bureau of Standards.

Source. Johnston and Babcock (1975).

the 31 samples into one computer group gave equations that represented the measured data within experimental error

$$E = 8.27496\ SiO_2 + 16.57520\ TiO_2 - 0.17357\ Na_2O \qquad (13.1)$$
$$G = 3.49181\ SiO_2 + 6.62532\ TiO_2 - 0.43746\ Na_2O \qquad (13.2)$$

where E and G are expressed in dynes per square centimeter $\times 10^{11}$. Analyses showed that moduli that are functionally related to both E and G could best be represented by separate equations for glasses in the two phase fields. This is demonstrated in Table 13.4, which lists measured and calculated values of compressibility for glasses in the separate phase fields. Compressibility is given by $= 1/B = 3(3G - E)/EG$, where B is bulk modulus. Lines of equal compressibility and measured values appear in Figure 13.1. Note that glass E4 lies on the boundary between phases D and E and can be represented by either of the D and E equations. Simultaneous solution of D and E equations from Table 13.4 and the equation $SiO_2 + TiO_2 + Na_2O = 1$ showed that the composition of the D-E boundary could be represented by the equation $TiO_2 = -1.539\ Na_2O + 0.583$. This boundary (broken line in Figure 13.1) is in reasonable agreement with those obtained by Babcock (1973) from refractive index and specific volume data and by Nissle and Babcock (1973) from data on the stress-optical coefficient C.

TABLE 13.4   Measured and Calculated Compressibilities

		D Phase Field		E Phase Field	
Glass	Measured	Calculated*	Difference	Calculated†	Difference
D1	0.2063	0.2063	0	0.1999	0.0064
D2	0.2068	0.2068	0	0.2043	0.0025
D3	0.1914	0.1014	0	0.1875	0.0039
D4	0.2153	0.2153	0	0.2082	0.0071
E1	0.1961	0.1923	0.0038	0.1964	0.0003
E2	0.1844	0.1774	0.0070	0.1840	0.0004
E3	0.1712	0.1626	0.0086	0.1717	0.0005
E4	0.1919	0.1919	0	0.1920	0.0001
E5	0.1831	0.1800	0.0031	0.1829	0.0002

Compressibility β $(cm^2/dyne \times 10^{-11})$

*$\beta = 0.24900\ SiO_2 - 0.04890\ TiO_2 + 0.25873\ Na_2O$.
†$\beta = 0.21903\ SiO_2 - 0.02763\ TiO_2 + 0.30746\ Na_2O$.
Source. Johnston and Babcock (1975).

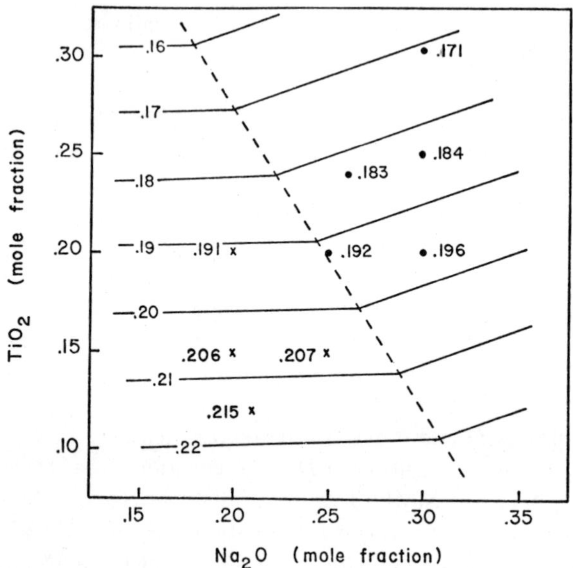

Figure 13.1 Lines of equal compressibility in plot of Na$_2$O versus TiO$_2$ (compressibility in cm^2/dyne × 10^{-11}.) From Johnston and Babcock (1975).

Analyses showed that Poisson's ratio could best be represented by separate D and E equations:

$$\mu = 0.16355 \; SiO_2 + 0.27455 \; TiO_2 + 0.42109 \; Na_2O \qquad (13.3)$$
$$\mu = 0.18728 \; SiO_2 + 0.25200 \; TiO_2 + 0.38003 \; Na_2O \qquad (13.4)$$

Least squares fitted lines were determined for the ratio G/E for glasses in the two phase fields. Figure 13.2 reveals that glasses in the two groups are distinctly separated and suggests that D and E glasses have different substructures. Zwikker (1954) states that G/E is near 0.375 for most metals. This extrapolated line for metals is shown for comparison. Substructures of metals are known to differ from those of glasses.

This exploratory research indicates that macroscopic stress-strain data on glasses can be interpreted as evidence of the presence of primary phase substructures. Since elastic properties involving both compression (or tension) and shear appear to exhibit these differences, both cation-anion distances and Si-O-Si angles may be involved. This work suggests that the substructure concept as it relates to elastic properties should be further verified by research on other types of silicate glass. In this connection Young's modulus data on three Na$_2$O-

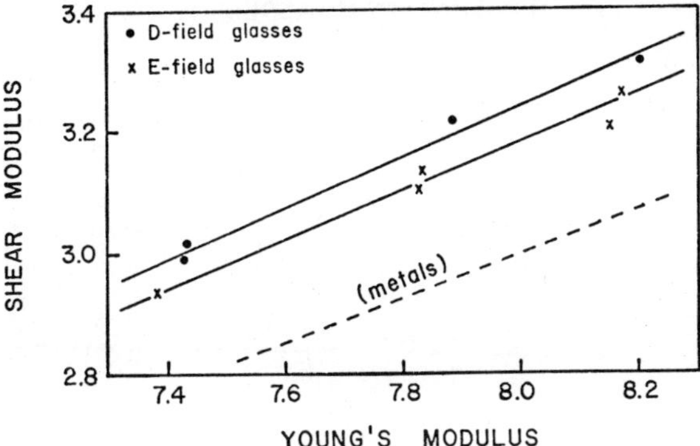

Figure 13.2 Young's modulus versus sheer modulus for glasses and metals. From Johnston and Babcock (1975).

SiO_2 glasses used by Phillips (1964) and measured by Koslovskaya (1960) can be interpreted in terms of the substructure postulate. Figure 13.3 plots these data against composition. It is evident that the effects of substituting SiO_2 for Na_2O are distinctly different in the $Na_2O \cdot 2SiO_2$ and SiO_2 phase fields. The position of the indicated phase boundary [from Morey (1954 p. 35)] lies at 73.9% SiO_2 or 0.745 mole fraction SiO_2.

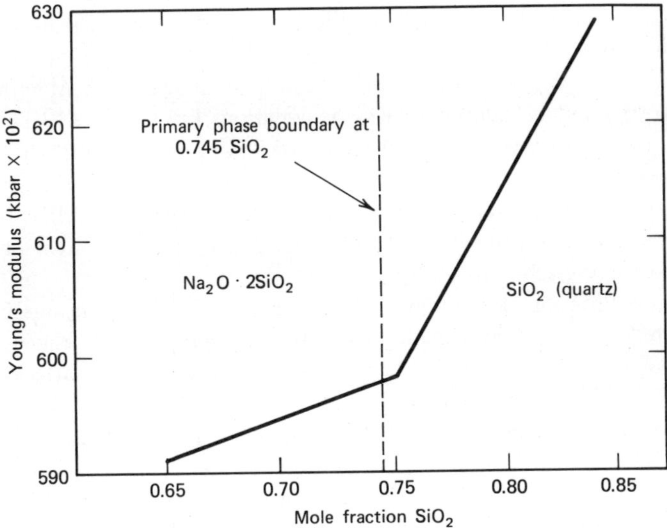

Figure 13.3 Young's modulus of $Na_2O\text{-}SiO_2$ glasses. Tabular data from Koslovskaya (1960).

13.4 THE VELOCITY OF SOUND IN GLASSES

Elastic wave velocities in glasses can be derived from elastic moduli data obtained by dynamic resonance methods. The data of Johnston and Babcock (1975; see Tables 13.2 and 13.3) were measured on samples having dimensions $12 \times 2.5 \times 0.3$ cm. Following Zwikker (1954) for samples whose transverse dimensions are small compared to their lengths, the longitudinal and transverse velocities can be expressed by

$$V_L = (E\ d)^{1/2} \tag{13.5}$$
$$V_T = (G\ d)^{1/2} \tag{13.6}$$

where d = density. Derived values of V_L, V_T, $1/V_L$, and $1/V_T$ were determined for each of the 31 samples, then subjected to least squares analyses. It was found that combining D and E glasses into one group gave the following equations, which represented the data within experimental error:

$$V_L = 5.84584\ SiO_2 + 7.30188\ TiO_2 + 2.63550\ Na_2O \tag{13.7}$$
$$V_T = 3.79543\ SiO_2 + 4.61355\ TiO_2 + 1.47037\ Na_2O \tag{13.8}$$
$$1/V_L = 0.16956\ SiO_2 + 0.11829\ TiO_2 + 0.28242\ Na_2O \tag{13.9}$$
$$1/V_T = 0.25960\ SiO_2 + 0.18791\ TiO_2 + 0.46328\ Na_2O \tag{13.10}$$

where V_L and V_T are in kilometers per second. Reciprocal velocities $1/V_L$ and $1/V_T$ are in seconds per kilometer, the times required for acoustic waves to travel 1 km.

Demkina and Kisin (1972) used equations of the forms (13.5) and (13.6) to derive values of $1/V_L$ and $1/V_T$ in their glasses. Their values for vitreous silica (0.174 and 0.265 sec/km, respectively) are in qualitative agreement with coefficients 0.16956 and 0.25960 for SiO_2 in equations (13.9) and (13.10). These investigators give average values of 0.186 and 0.284 per mole fraction of SiO_2 for their 32 glasses. Their values for $1/V_L$ and $1/V_T$ for Na_2O in their glasses were 0.22 and 0.35 per mole fraction, respectively. Here again the coefficients are in qualitative agreement with values of 0.28242 and 0.46328, but care must be taken in comparing values for different types of glass.

The propagation of elastic waves along a bar of finite dimensions must be distinguished from that in a medium of infinite extent. Apart from waves in the body of the medium, there are surface waves that follow the surfaces of two media having different elastic properties. Following Birch (1942) the velocity of longitudinal waves in a medium of infinite extent is expressed by

$$V_L* = [E(1 - \mu)\ d(1 + 2\mu)\ (1 - \mu)]^{1/2} \tag{13.11}$$

The velocity of transverse waves is given by equation (13.6). Equations for V_L* and $1/V_L*$ for the 31 samples of Johnston and Babcock are as follows:

$$V_L = 6.02175 \; SiO_2 + 8.03769 \; TiO_2 + 3.64781 \; Na_2O \qquad (13.12)$$
$$1/V_L = 0.16532 \; SiO_2 + 0.10926 \; TiO_2 + 0.23594 \; Na_2O \qquad (13.13)$$

Comparison of equations (13.7) and (13.12) shows that longitudinal waves travel faster in a glass sample of infinite extent than in a sample of finite dimensions because of boundary constraints on the latter. Transverse waves are not hindered in this way.

Manghnani (1972) measured longitudinal and transverse wave velocities in six Na_2O-TiO_2-SiO_2 glasses prepared by Hamilton and Cleek (1958). Table 13.5 contains compositions and measured velocities of these glasses. Measured data obtained by Johnston and Babcock (1975) are given for two of the glasses prepared at the University of Arizona. Calculations of velocities using equations of Johnston and Babcock are in good agreement with measured values of Manghnani. Glasses 1 to 4 are located in the D and E primary phase fields, whereas glasses 5 and 6 lie outside these two phase fields (Figure 4.5). Calculated values for these two glasses are in good agreement with Manghnani's measured values, indicating that wave velocities are not sensitive to glass substructures. These property relations are understandable on the following basis. We have assumed that sound velocities are independent of frequency of vibration. Manghnani used a frequency of 3×10^4 Hz for his measurements, and the frequencies used by Johnston and Babcock were of the order of 10^3 Hz. The corresponding wavelengths are orders of magnitude higher than the lattice constants in silicate glasses, which are usually expressed in angstroms. Thus the glass reacts as a structureless medium to such long wavelengths. In other words there is no frequency dependence or dispersion in this case. The data on sound velocities in glasses cannot be used to evaluate details of the materials' microstructures.

13.5 EFFECTS OF TEMPERATURE ON ELASTIC PROPERTIES

The foregoing discussions relate to measurements at room temperature and 1 atm pressure, conditions that pose no experimental difficulties. Measurements at elevated temperatures necessitate more care in that readings must be taken immediately after loading. In addition the whole sample must be held under constant and known temperature conditions. It will be pointed out that the thermal history of the sample is important in terms of its fictive temperature or equilibrated temperature condition.

The accurate measurements of Birch and Dow (1936) referred to in Section 4.6 demonstrated that the elastic properties of vitreous silica were unique among the silicate glasses. They found that compressibility (i.e., the ratio of the decrease in volume per unit volume to the applied pressure) increased with temperature, while that of most silicate glasses decreased. Section 4.6 indicated that most of the unique properties of vitrous silica were understandable in terms of changes in the Si-O-Si angles in the microstructure. Changes in the Si-O

TABLE 13.5 Sound Velocities in Na_2O-TiO_2-SiO_2 Glasses

Oxides	Composition (wt %) of six glasses					
	1	2	3	4	5	6
SiO_2	0.500	0.550	0.625	0.475	0.525	0.575
TiO_2	0.200	0.200	0.200	0.250	0.250	0.250
Na_2O	0.300	0.250	0.175	0.275	0.225	0.175
Velocities (km/sec)						
Longitudinal						
Manghnani (1972)	5.731	5.794	5.895	5.806	5.906	5.955
Johnston and Babcock (1975)	5.709	5.839				
Calculated, equation (13.12)	5.713	5.832	6.010	5.873	5.992	6.110
Transverse						
Manghnani (1972)	3.315	3.386	3.516	3.353	3.461	3.526
Johnston and Babcock (1975)	3.256	3.374				
Calculated, equation (13.8)	3.262	3.378	3.552	3.361	3.477	3.594

distances were considered to be minor in view of the very strong bonding forces in the SiO_4 tetrahedra. When Na_2O and other oxides are added to form silicate glasses, the bonding forces between tetrahedra are weakened. Decreases in elastic properties with temperature then become understandable in terms of changes in cation-anion distances. Changes in elastic properties caused by pressure (Section 13.6) become reasonable on the same basis. Birch and Dow found that compressibility of silicate glasses decreased with temperature and reached minimum values at temperatures dependent on composition. Compressibility of vitreous silica increased with temperature and reached a minimum value above $400°C$, which is higher than that of any of the silicate glasses. Glass microstructures cannot provide a ready explanation for this behavior. It is apparent that properies of both vitreous silica and silicate glasses must be finally explained in terms of both cation-anion distances and angles, however small these changes may be.

Manghnani (1972) measured the effects of temperature up to $300°C$ on elastic properties of the Na_2O-TiO_2-SiO_2 glasses given in Table 13.5. Shear and bulk moduli for all glasses decreased with temperature, but the Poisson ratio increased. It follows from his data that longitudinal and transverse sound velocities decrease as temperatures increase.

McGraw (1952) described a static method for measuring instantaneous values of Young's modulus for three commercial container glasses up to about $700°C$. Data on one of his glasses are plotted in Figure 12.1 and are typical of silicate glasses. Stong (1937) found similar rapid decreases of Young's modulus with temperature; however his measurements were taken while the glass samples were being heated.

Ide (1937) measured the velocity of sound in silicate glasses in the temperature range 0 to $800°C$ at a frequency of 1.43×10^4 Hz. His data on the following glass types are of interest: (1) vitreous silica, (2) chemically resistant glass containing about 81 weight percent SiO_2, and (3) Na_2O-CaO-SiO_2 glass. The longitudinal velocity in glass 1 increased almost linearly up to $800°C$, attaining a value of 6 km/sec. The velocity in glass 2 increased regularly up to a value of 5.5 km/sec at $550°C$, then decreased to 5.25 km/sec at $700°C$. The velocity in glass 3 decreased linearly from 5.3 km/sec at $0°C$ to 5.2 km/sec at $250°C$. Derived data from Ide's measurements for Young's modulus follow the same pattern of changes with temperature. It is evident that glass 2, with its high SiO_2 content, retains some of the characteristics of vitreous silica.

Characteristic effects of temperature on glass properties were outlined qualitatively in Section 2.3. The effects of composition, temperature, and time in the intermediate temperature range were discussed in terms of "transformation range," "temperature-equilibrated structure," and "fictive temperature." The definitive paper by Spinner and Napolitano (1956) describes quantitative relations between refractive index and elastic moduli in a borosilicate glass as affected by a series of heat treatments. This important paper also gives a good review of pertinent literature. Results of the authors' research can be outlined as follows.

A borosilicate crown glass BSC 517/645 having the following composition in weight percent was used in the investigation; SiO_2 66.3, PbO 0.5, B_2O_3 12.5, Na_2O 7.5, K_2O 12.8, and As_2O_3 0.4. Eight samples were cut from a disk of the glass, which had previously been given fine optical annealing. The experimental procedure included (1) measuring the refractive index and elastic moduli of each sample, (2) subjecting the samples to a series of different heat treatments, and (3) measuring and comparing these properties again at room temperature after each heat treatment. The average value of refractive index n_D at room temperature was 1.51755 with a standard deviation of 0.000017 for the eight specimens. This average value was used as the starting value for all specimens. Elastic moduli E and G were measured at frequencies within and slightly above the sonic range. The average starting value for Young's modulus was 810.5 kibar, with a standard deviation of 1.7; that for shear modulus was 335.6 kibar, with a standard deviation of 0.09. As in the refractive index measurements, these deviations were within the errors of measurement. The heat treatments consisted of raising the temperature of the sample to a given value and maintaining this temperature until the temperature of the sample and refractive index had attained constant equilibrium conditions. Figure 13.4 shows linear relations between refractive index and elastic moduli which can be expressed in terms of least squares equations

$$E = 10416 \, (n_D - 1.515) + 781.9 \qquad (13.14)$$
$$G = 4291 \, (n_D - 1.515) + 324.4 \qquad (13.15)$$

Poisson's ratio does not change with heat treatment within experimental error, which is understandable from the known relation $\mu = E/2G - 1$.

Spinner and Napolitano noted that changes in elastic moduli as affected by heat treatment were greater than those for density and refractive index. The relative sensitivities to annealing for BSC 517/645 glass were given as

$$\Delta E/E = 6 \, \Delta d/d = 21 \, \Delta n/n \qquad (13.16)$$

where E is either Young's modulus or the shear modulus, d is density, and n is refractive index. As a practical matter, it is necessary to consider the annealing sensitivity of a property and the precision with which it can be measured. The data of Spinner and Napolitano show that refractive index could be measured to within 0.002%, whereas the corresponding figure for elastic moduli was 0.1%. Thus refractive index may be a somewhat better indicator of annealing affects. Facility of measurement further recommends refractive index

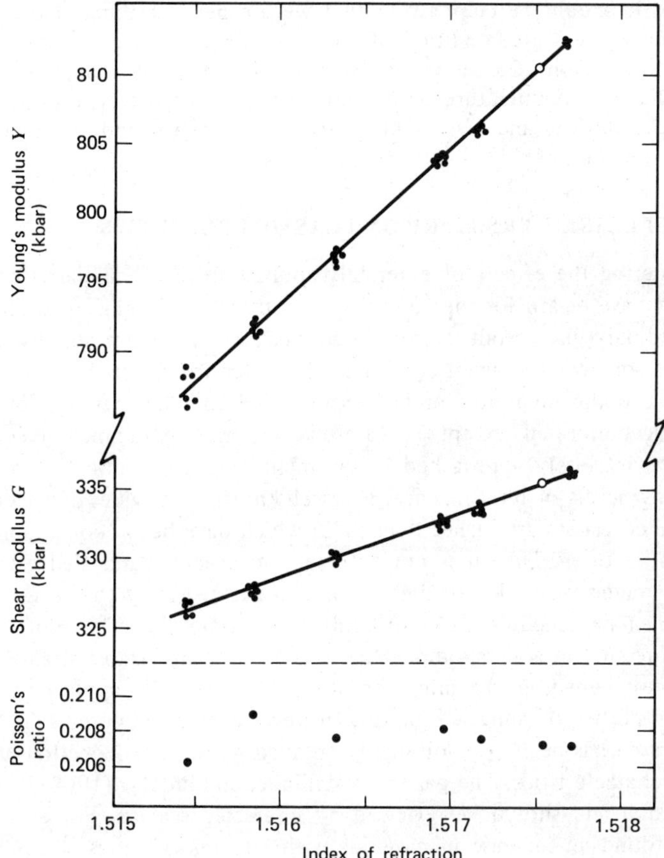

Figure 13.4 Young's modulus versus index of refraction for borosilicate crown glass 517. Open circles represent average of initial values. From Spinner and Napolitano (1956, Fig 1).

as a useful indicator of annealing affects on properties.

With respect to the nature of microstructural changes occurring during annealing Spinner and Napolitano state:

To account for the observed changes in structure-dependent properties (e.g., density, refractive index, and elastic moduli) it is proposed that the overall effect of annealing is one in which compaction or expansion on a macroscopic scale is associated with a corresponding statistical volume change in the holes in the network without any significant change in their shape. The usual picture of the glassy network structure may be found in Morey

(1954). It should be emphasized that we are here speaking of the frozen-in room temperature condition of a glass as a result of changes in the transformation region. Consequently, significant structural changes occurring with changing temperature or pressure, or both, of the type proposed by Babcock, Barber, and Fajans (1954) or by Smyth, Londeree, and Lorey (1953) are still possible.

13.6 EFFECTS OF PRESSURE ON ELASTIC PROPERTIES

In considering the effects of externally applied pressures on elastic properties, one must have regard for the condition of internal strain in glasses. The annealing of ordinary glass products such as containers and flat glass usually leaves the glass in a somewhat strained condition, the outermost layers being under compression and the innermost under tension. The complete removal of strain is seldom accomplished except in carefully arranged experimental conditions. This state is closely approached in the manufacture of optical glasses, which require low levels of birefringence. It is well known that values of elastic moduli for annealed glasses are higher than those which have been cooled rapidly from high temperatures. Thus it is not sufficient to specify "annealed glass" when reporting values of moduli of high accuracy; it is necessary to specify the complete annealing schedule. It is difficult to evaluate the effects of externally applied pressure in a glass sample that already contains sizable strain gradients. Our interest here is in outlining the effects of externally applied pressure on elastic properties of annealed glasses. However it must be noted that considerable research has been done on highly strained glasses in connection with their strength characteristics. The paper by Mallinder and Proctor (1964) is typical of this research and should be reviewed by interested readers. For example, the authors found an increase in modulus with strain in vitreous silica and a decrease in a glass containing Na_2O.

A number of authors including Vukcevich (1972) have noted that Young's modulus and the shear modulus of vitreous silica decrease with increasing pressure. Bridgman (1925, 1948) found that room temperature compressibility of vitreous silica increased up to a pressure of about 30 kibar, then decreased rapidly at higher pressures. This so-called anomalous behavior of vitreous silica has been previously referred to, particularly in Sections 4.6 and 13.5. This designation has been attached to properties of vitreous silica because they deviate from the usual behavior of most other glasses. Bridgman (1924) found that a chemically resistant glass containing about 81 weight percent SiO_2 exhibited an increase in compressibility with increase in pressure. This is understandable because its microstructure approaches that of vitreous silica.

Manghnani (1972) measured the effects of pressures up to 7 kbar on elastic properties of Na_2O-TiO_2-SiO_2 glasses given in Table 13.5. Young's modulus for all glasses increased with pressure. Shear modulus increased with pressure except in glasses 3 and 6, which showed slight decreases. The bulk modulus of all glasses increased with pressure; thus its reciprocal, the compressibility, decreased. Manghnani found that longitudinal wave velocities of all glasses increased with pressure while the transverse velocities decreased. He observes that increases in bulk modulus for the glasses is normal behavior for most crystalline oxide solids. He related property changes to SiO_2/Na_2O ratios in the glasses. The pressure and temperature derivatives of the moduli varied systematically with this ratio but not always linearly with the SiO_2/Na_2O ratio. This treatment, which follows implications of the random network theory, can be compared with data evaluations on this glass system by Johnston and Babcock (1975). Questions on property-structure relations in glasses nevertheless remain and can be settled only by further research studies.

13.7 HIGH PRESSURE EFFECTS ON GLASS PROPERTIES

The pioneering work of P. W. Bridgman on effects of high pressures on the properties of materials has been extended by subsequent investigators who concentrated on experimental procedures, materials, and properties. Wentorf (1962) discusses the use of very high pressure techniques. The comprehensive review by Sakka and Mackenzie (1969) on the effects of high pressure on glasses describes potential difficulties relative to experimental arrangements and interpretation of the data obtained. High pressure devices used by different workers vary and are outlined according to type of device, pressure-transmitting medium, pressure range, temperature range, and shape of glass sample. Application of hydrostatic pressure to the sample may be accompanied by unavoidable shear stresses. Thus exact knowledge of the nature of the applied stress is not easy to achieve. Imposition of very high pressures may bring about temperature increases, which must be taken into account.

Glasses react as reversible elastic solids at low pressures but exhibit plastic flow, strain hardening, and changes in properties beyond certain threshold pressures. Bridgman and Simon (1953) applied compressive pressures ranging up to 2×10^5 atm on vitreous SiO_2 and B_2O_3 and several silicate glasses. Glass samples were in the form of disks about 5 to 8 mm diameter and from 0.15 to 0.25 mm thick. Pressures were applied at a rate of 2×10^4 atm/sec in most tests. The effect of compacting was evaluated by measuring densities, dimensions, and x-ray diffraction patterns. Definite threshold pressures were observed for vitreous silica and silicate glasses. Measured properties were not affected below these thresholds, but collapse of the structures occurred above the given thresholds. Pressures were held constant for one minute after attaining the desired value, whereupon they were released at the same rate at which they had been applied. Rapid removal of pressure on vitreous silica and glasses with high SiO_2 contents resulted in explosive disintegration of the samples.

Vitreous silica exhibited a definite threshold at 10^5 atm and the structure seemed to collapse beyond this critical pressure. It became premanently densified at 2×10^5 atm with a density increase of 7.5% above the original value of 2.22 g/cc. One fragment of vitreous silica underwent a density increase of 17.5%. Glasses of Na_2O-SiO_2 had lower threshold pressures near 4×10^4 atm. They showed marked decreases of permanent compressibility. Permanent density changes of 8.7, 3.5, and 0.7% were observed for glasses containing 10, 23, and 31 mole percent Na_2O, respectively. This indicated that increasing the number of Na cations embedded in the interstices of the SiO_2 network hindered the collapse of the structure. Glasses containing 0.22 Li_2O, 0.23 Na_2O, and 0.23 K_2O gave permanent densifications of 6.9, 3.5, and 1.2%. Their respective ionic radii (0.68, 0.97, and 1.33 Å: Table 1.1) suggest again the influence of filling the interstices in the SiO_2 network.

Bridgman and Simon (1953) found the effects of high pressures on vitreous B_2O_3 to be quite different from those on vitreous SiO_2 and silicate glasses. Permanent changes in density of B_2O_3 occurred at all pressures and appeared to asymptotically approach a density increase of 6%. This glass showed much higher permanent density changes in low pressure regions than any of the silicate glasses. They noted that B_2O_3 and the silicate glass containing 0.31 mole fraction Na_2O behaved more like a plastic than a brittle solid. X-ray diffraction patterns taken before and after compression of SiO_2 and B_2O_3 glasses indicated no changes in short-range order in the structures. It was suggested that essential changes were in the Si-O-Si and B-O-B angles resulting in folding and distortion of the original structures. Permanent densification of all glasses could be removed by annealing, but the times required at a given temperature were much longer than those needed for ordinary annealing processes.

Sakka and Mackenzie (1969) give a comprehensive review of high pressure effects on glasses. They note that the activation energy of the annealing process is 1 to 10 kcal/mole; this low value is one-tenth the activation energy in viscous flow. These relative values may be reasonable in terms of the two processes, which respectively involve changes in structural angles and the breaking of strong Si-O bonds. It is questioned whether densification of glasses that can be restored to original low pressure values by annealing should be designated as permanent. Annealing effects are not active unless the glass is heated to temperatures in the transformation range. Vitreous silica must be heated to 1200°C or above if annealing is to be effective. In this connection experimental studies have revealed that a glass has a "pressure history" analogous to its "thermal history." Thus it is understandable that glasses previously subjected to high pressures are more difficult to anneal at given temperatures.

Infrared spectroscopy has been used to evaluate possible structure changes caused by high pressures. Following Lippincott, Van Valkenburg, Weir, and

Bunting (1958), the spectrum of vitreous silica is characterized by strong infrared bands at 9.05, 12.4, and 21.4 microns, which may be assigned respectively to Si-O stretching, Si-Si stretching, and Si-O-Si bending frequencies. Sakka and Mackenzie (1969) review studies on effects of high pressures on infrared spectra of vitreous SiO_2, GeO_2, and B_2O_3. The strong infrared bands of these glasses were related to wave number in Figure 9.7. Bands for these glasses were broadened, which indicated a weakening of bond forces, but the oxygen coordination numbers were not changed by high pressures.

Published work on the effects of high pressures on the ultraviolet and visible spectra of transition elements furnishes added structural information. Tischer and Drickamer (1962) measured optical absorption spectra on glasses containing transition elements that had been subjected to high pressures. Three silicate and two phosphate glasses were studied using V^{3+}, Cr^{3+}, Mn^{3+}, Fe^{2+}, Co^{2+}, Co^{2+}, Ni^{2+}, and Cu^{2+} as indicator ions. The locations of absorption maxima were shifted to shorter wavelengths with increasing pressures. This shift was continuous for silicate glasses up to pressures of 120 kibar. The phosphate glasses exhibited a break in the absorption curve when the pressure reached values of about 50 to 60 kbar. This was attributed to some type of reordering of the glass structure The investigators found that Co and Ni ions, which have coordination numbers of 4 at atmospheric pressure, changed to coordinations numbers of 6 at 30 to 40 kbar. This change in coordination continued up to pressures of 120 kibar. Upon release of pressure, the coordination numbers reverted to 4. Tashiro, Yamamoto, and Sakka (1967) studied high pressure effects on Co^{2+} ions containing 0.75 B_2O_3 and 0.25 K_2O. At atmospheric pressure fourfold cobalt ions in this glass showed absorption peaks at 5300, 5950, and 6330 Å. Compression caused the peaks to move to 5000, 5730, and 6130 Å. New peaks were simultaneously observed at 4400, 4800, and 5250 Å. These new peaks together with a decrease in absorption of fourfold cobalt were attributed to sixfold cobalt.

14

Viscosity

The viscosity of a glass depends on both temperature and glass composition. The rapid change of viscosity over the temperature range used in glassmaking requires the use of different methods in its measurement. As noted previously, the liquidus temperature of a glass depends only on composition. Viscosity has controlling influence on all stages of glassmaking, and its value at the liquidus temperature is a determining factor in the choice of glass process to be used. This chapter covers the relationships of these factors in Na_2O-CaO-Al_2O_3-SiO_2 glasses, which are used for large tonnage glass manufacturing. Applications of these data to commercial glassmaking are discussed in Part V.

14.1 VISCOSITY-TEMPERATURE RELATIONS

Glasses differ only with respect to the rate at which viscosity changes with temperature and with respect to the actual temperatures associated with given viscosity levels. The property of viscosity, defined in Chapter 12, derives from measurements in which the glass test sample, free from temperature gradients, is held at a constant temperature until an equilibrium value of viscosity is obtained. Viscosity in the range from 10^2 to about 10^{12} poises is dependent only on temperature and composition. Viscosity is also dependent on time in the range from 10^{12} to 10^{15} and higher. Equilibrium viscosity-structure conditions are attained rapidly at high temperatures. At lower temperatures it is necessary to attain equilibrium temperature conditions and wait until structural adjustments in the glass reach equilibrium. At this time a maximum value of viscosity is obtained which is representative of the given glass at the given temperature. The viscosity level at which time effects become important, generally near 10^{12} poises, is dependent on glass composition. Viscosity data obtained in this manner can be reproduced by different investigators and constitute a standard glass property.

Table 14.1 presents viscosity-temperature data on a typical soda-lime-silica commercial glass. The indicated temperature coefficients of viscosity, which change rapidly with viscosity level, emphasize the experimental difficulties associated with such measurements. The development of an equation or equations to

TABLE 14.1   Viscosity-Temperature Data on a Soda-Lime-Silica Glass

Viscosity (poises)	Temperature ($^\circ$C)	$\mathrm{Log}_{10}\eta$	Viscosity range (poises)	Temperature range ($^\circ$C)	Viscosity coefficient (poises/$^\circ$C)
10^2	1451	2.0			
3.16×10^2	1295	2.5	10^2-10^3	273	3.3
10^3	1178	3.0			
10^4	1013	4.0	10^4-10^5	110	8.2×10^2
10^5	903	5.0			
10^6	823	6.0	10^6-10^7	59	1.5×10^5
10^7	764	7.0			
10^8	716	8.0	10^8-10^9	42	2.3×10^7
10^9	674	9.0			
10^{10}	639	10.0	10^{10}-10^{11}	30	2.8×10^9
10^{11}	609	11.0			
10^{12}	583	12.0	10^{12}-10^{13}	24	3.8×10^{11}
10^{13}	559	13.0			
10^{14}	539	14.0	10^{14}-10^{15}	16	5.6×10^{13}
10^{15}	523	15.0			

Source. Cole and Babcock (1970).

represent viscosity-temperature relations over this wide range involves the difficult task of dealing with the rapidly and continuously changing temperature coefficients of viscosity. Careful examination of the data in Table 14.1 indicates
that any equation is a compromise in terms of the changing temperature coefficients. In practical terms it is necessary only that the equation or equations represent the data to a precision compatible with the measured data. The equation
proposed by Fulcher (1925) is of use for technology purposes

$$\log_{10}\eta = A + B/T\text{-}T_0 \qquad (14.1)$$

where η is viscosity in poises and A, B, and T_0 are empirical constants. The situation is exactly the same as the representation of resistivity-temperature relations
illustrated in Figure 11.1. The three empirical constants give the user some flexi-
bility in matching the equation to the data. In the author's experience, best representation is obtained by using two Fulcher equations to represent the data—
one for the high temperature range and another for the low temperature range.

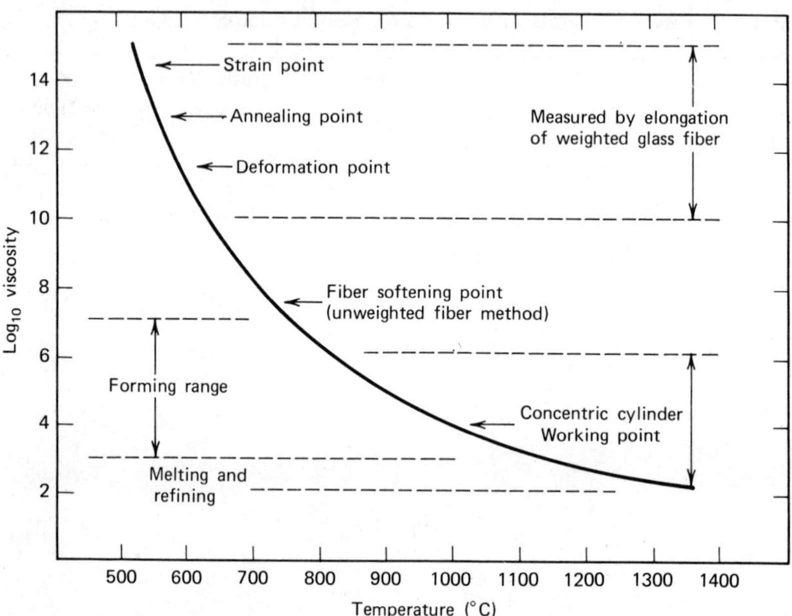

Figure 14.1 Viscosity-temperature curve of a silicate glass.

Figure 14.1 is a typical viscosity-temperature curve for a silicate glass. Two types of viscosity measurement are in general usage: tests that provide equilibrium viscosity-temperature data, and empirical tests that give data needed for control of viscosity in manufacturing processes. Roles of both tests in glass manufacturing have been developed by the American glass industry in cooperation with the American Society of Testing and Materials. These viscosity-temperature relations have gained substantial international acceptance through cooperation with the International Commission on Glass. Figure 14.1 indicates viscosity levels of fixed points together with viscosity ranges for equilibrium measurements and manufacturing processes of melting, forming, and annealing. Relations between viscosity and manufacturing processes are given in some detail in Part V.

English (1924), Lillie (1929), and Babcock (1934) were pioneers in development of apparatus and measurements of high temperature equilibrium viscosity in the range 10^2 to 10^6 poises. They all used the concentric cylinder method in which the glass was contained in a rotating platinum cylinder with measurements of torque on a platinum inside cylinder. This method is now universally employed, although methods of measuring torque exerted on the inner cylinder have undergone improvements. Early measurements involved a question of the relation between viscosity and measured torque. Babcock (1934) found a linear relation between measured torque and viscocity when times of revolution of the outside cylinder ranged from about 10 sec to 30 hours. The flow of glass and the transfer of momentum from the outer to the inner cylinder in these measurements indicated a condition of Newtonian or laminar flow. This method may be used to measure viscosities lower than 10^2 and higher than 10^6 poises, although times involved in the latter measurements may lead to tedious proceedures. Equilibrium viscosity data in the low temperature range from about 10^{10} to 10^{15} poises are obtained by measuring the elongation of a weighted glass fiber. This method, which was first described by Norton (1935), has been improved since its introduction. All types of viscosity apparatus in present use can be calibrated using standard glass samples available at the National Bureau of Standards. These samples are accompanied by certified data on equilibrium viscosities from 10^2 to 10^{12} poises together with temperatures of softening point, annealing point, and strain point. It is apparent from Table 14.1 that measurements of equilibrium viscosity and tests on fixed viscosity levels require very precise measurements of the temperatures involved.

14.2 VISCOSITY-LIQUIDUS TEMPERATURE RELATIONS

The review by Morey (1954) and subsequent publications show that viscosity is highly dependent on oxide glass composition, and this is demonstrated by the effects of composition and temperature on viscosity given in Table 14.2. Glass manufacturers use their own methods of developing glass compositions that have viscosity-temperature curves compatible with their fabrication processes. In most cases these curves are standardized and controlled in terms of the fixed viscosity points indicated in Table 14.2. The viscosity of 10^2 poises is usually associated with the approximate commercial melting temperature. Commercial compositions, usually expressed in weights percent, are not always available from the manufacturer. Available data on compositions and viscosity-temperature relations are not sufficient to develop usable analytical relations between these parameters. Development of such quantitative relations requires data obtained on systematically arranged experimental compositions. Relatively few data of this type are found in the literature, which mostly describes noncommercial binary and ternary compositions. Large glass manufacturers maintain research laboratories in which viscosity-composition-temperature relations are developed for their own purposes. Following the history of glassmaking, these relations for commercial glasses are too often held to be proprietary. In fact any modern glass laboratory can obtain these data on any glass sample, be it experimental or commercial. In the author's opinion the disclosure of these data by manufacturers would benefit the whole glass industry in the long run. Disclosure of glassmaking processes that are patentable and proprietary must be controlled by manufacturers in terms of competition in the industry. This is an entirely different matter

TABLE 14.2 Viscosity-Temperature Data on Selected Commercial Glasses

Oxide	Composition (wt %) of five glasses				
	7940	7900	1720	7740	0080
SiO_2	99.9	96.0	62.0	81.0	73.0
H_2O	0.1				
Al_2O_3		0.3	17.0	2.0	1.0
B_2O_3		3.0	5.0	13.0	
Na_2O			1.0	4.0	17.0
MgO			7.0		4.0
CaO			8.0		5.0
Condition of glass	Temperature (°C)				
Working point (10^4)	—	—	1202	1252	1005
Softening point ($10^{7.6}$)	1580	1500	915	821	696
Annealing point (10^{13})	1084	910	712	560	514
Strain point ($10^{14.5}$)	956	820	667	510	473

Sources. Hutchins and Harrington (1966) and Corning Glass Works catalog.

and is outside our present interests.

Over the years the research laboratories of Owens-Illinois, Inc., have measured a number of glass properties on systematically arranged compositions encompassing areas of their commercial interests. Some of these studies have been published in technical journals. The author is greatly indebted to the company for releasing unpublished data on chemical analyses, viscosity, and viscosity at liquidus temperatures on an extensive series of experimental Na_2O-CaO-Al_2O_3-SiO_2 glasses. Glasses in this series are widely used for large tonnage production of containers, tableware, flat glass, and so on. Oxide compositions in addition to SiO_2 cover the following ranges in weights percent:

$$Al_2O_3: \ 0.2, 2.2, 4.2, 6.2, \text{and } 8.2$$
$$CaO: \ 4, 6, 8, 10, 12, 14, \text{and } 16$$
$$Na_2O: \ 11, 12.5, 14, 15.5, 17, \text{and } 19$$

Chemical analyses normally showed 0.2% MgO and 0.2% SO_3. The MgO was added to CaO contents, the SO_3 was added to SiO_2, and the compositions normalized to totals of 100%. Compositions were converted to mole fractions, and their relations with properties determined as described in Chapter 4.

Table 14.3 allows one to find least squares equations representing temperatures at viscosity levels of 10^2, 10^3, 10^5, and 10^7 poises, as well as equations for glasses in the tridymite (SiO_2), devitrite ($Na_2O \cdot 3CaO \cdot 6SiO_2$), and wollastonite ($CaO \cdot SiO_2$) phase fields, and for all glasses combined into one group. Temperatures are obtained by multiplying the indicated oxide constants by their respective mole fraction compositions. Separating the glasses in terms of primary phase field locations follows the substructure method, while combining them into one group is in line with the teachings of the random network theory. Maximum differences between measured and calculated temperatures for combined groups are in all cases two times as large as those in separate phase groups. The contributions of the oxides toward raising and lowering of temperatures at the different viscosity levels are evident from Table 14.3. For example, Al_2O_3 is most effective in raising temperatures throughout the range from 10^2 to 10^7 poises; Na_2O lowers temperatures throughout the range, while CaO lowers temperatures only in tridymite and wollastonite glasses at the 10^2 viscosity level. These data suggest that glasses in the three phase fields have different substructures.

Silverman (1939) published liquidus temperatures of the above-mentioned series of Na_2O-CaO-Al_2O_3-SiO_2 glasses. Figures 14.2 to 14.6 are primary phase

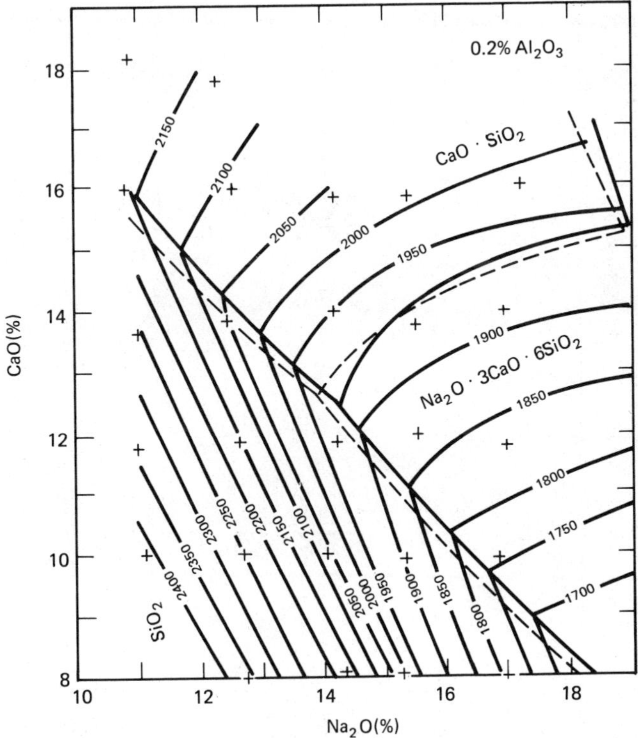

Figure 14.2 Primary phase diagram of 0.2% Al_2O_3 glasses (liquidus temperature in °F). From Silverman (1939).

TABLE 14.3   Viscosity-Temperature Relations in Na_2O-CaO-Al_2O_3-SiO_2 Glasses at Indicated Viscosity Levels

$$°C = A\ SiO_2 + B\ Al_2O_3 + C\ CaO + D\ Na_2O$$

Phase field	Ratio of number of glasses used to number available	A	B	C	D	Maximum differences between measured and calculated temperatures (°C)
Log 2 viscosity						
Tridymite	21/29	2199.40	3904.28	−466.27	−1164.48	10
Devitrite	21/23	1990.81	4037.16	87.31	−485.91	12
Wollastonite	55/75	2076.56	3356.99	−223.29	−530.60	12
Combined	97/127	2085.24	3652.41	−172.90	−695.99	22
Log 3 viscosity						
Tridymite	24/29	1713.96	3311.23	180.78	−942.39	10
Devitrite	22/23	1578.21	2999.94	418.11	−389.87	9
Wollastonite	61/75	1658.01	2814.78	274.10	−611.13	9
Combined	107/127	1660.22	2940.85	288.28	−665.81	19

Log 5 viscosity

Tridymite	28/29	1207.13	2567.76	685.83	−610.66	7
Devitrite	23/23	1139.32	2156.06	715.98	−258.96	8
Wollastonite	68/75	1187.07	2172.28	643.46	−429.84	8
Combined	119/127	1194.52	2194.71	639.11	−465.06	16

Log 7 viscosity

Tridymite	26/29	939.84	2005.16	848.94	−346.89	5
Devitrite	21/23	890.69	1720.08	875.79	−105.31	4
Wollastonite	67/75	931.13	1706.17	795.97	−230.31	5
Combined	114/127	934.00	1734.30	806.22	−259.31	13

Experimental data provided by Owens-Illinois, Inc., Toledo, Ohio.

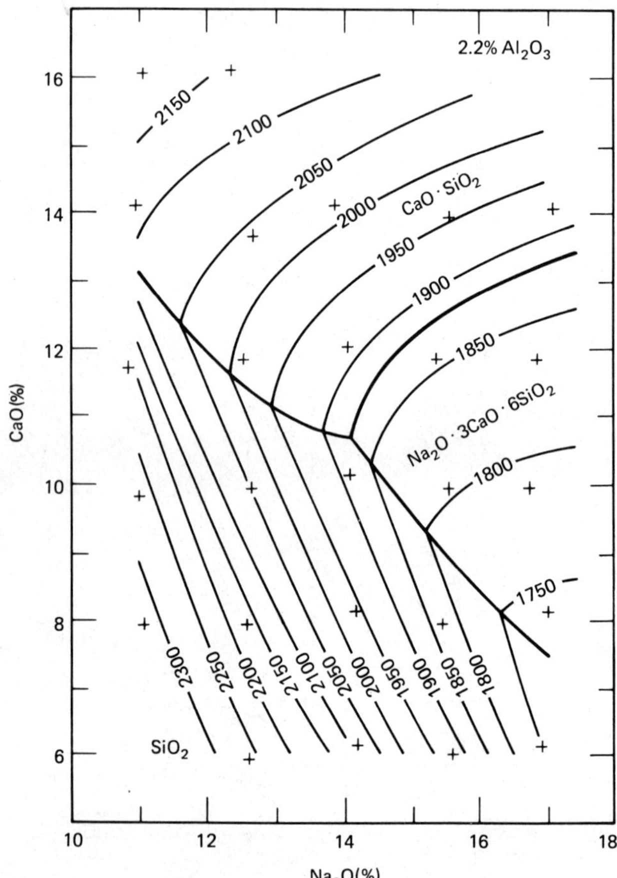

Figure 14.3 Primary phase diagram of 2.2% Al$_2$O$_3$ glasses. From Silverman (1939).

diagrams of the glasses at the five Al$_2$O$_3$ levels. Liquidus temperatures were determined by the described gradient boat method with an overall accuracy of 15°F or 8.3°C. The dotted lines in Figure 14.2 indicate phase boundary lines for Na$_2$O-CaO-SiO$_2$ glasses with no Al$_2$O$_3$ measured by Morey and Bowen (1925). Chemical analyses of the glasses are indicated by (+) in the diagrams. Review of the many published phase diagrams of oxide glasses makes it apparent that liquidus temperatures are uniquely related to glass compositions within given primary phase fields. Most diagrams exhibit nonlinear relations between liquidus temperatures and compositions, whereas some exhibit relations approaching linear-

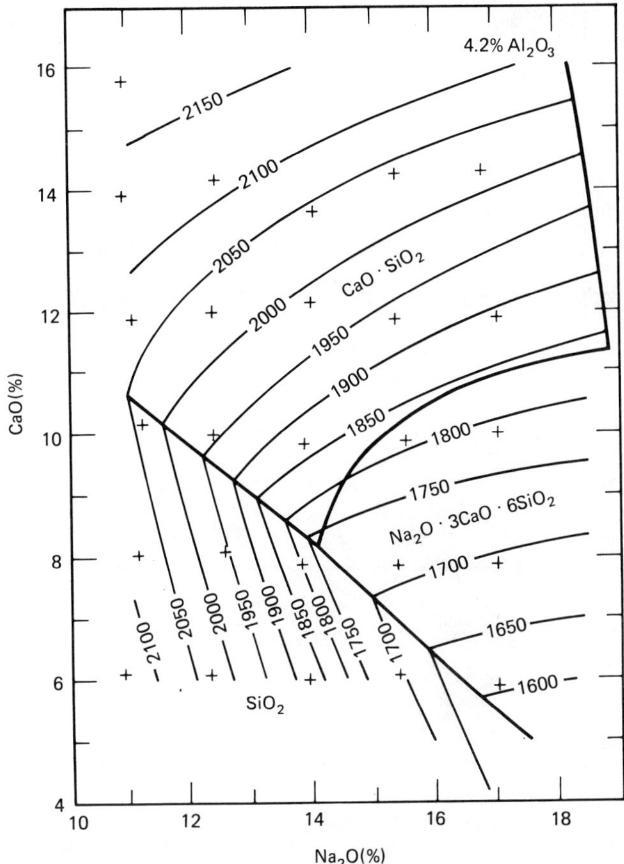

Figure 14.4 Primary phase diagram of 4.2% Al$_2$O$_3$ glasses (liquidus temperature in °F). From Silverman (1939).

ity. The data of Silverman have been expressed in terms of linear least squares equations in the effort to obtain a first approximation to these relationships. Table 14.4 presents results in these terms for glasses in the three phase fields. The equation for the combined group is given, although it is evident that it has no practical uses. Figures on averages and maximum differences between measured and calculated liquidus temperatures indicate an approach to linearity. The equations should be of practical use in surveying this system in terms of magnitudes of liquidus temperatures. However this survey should be followed by liquidus temperature measurements on glasses of particular interest. Representation of pri-

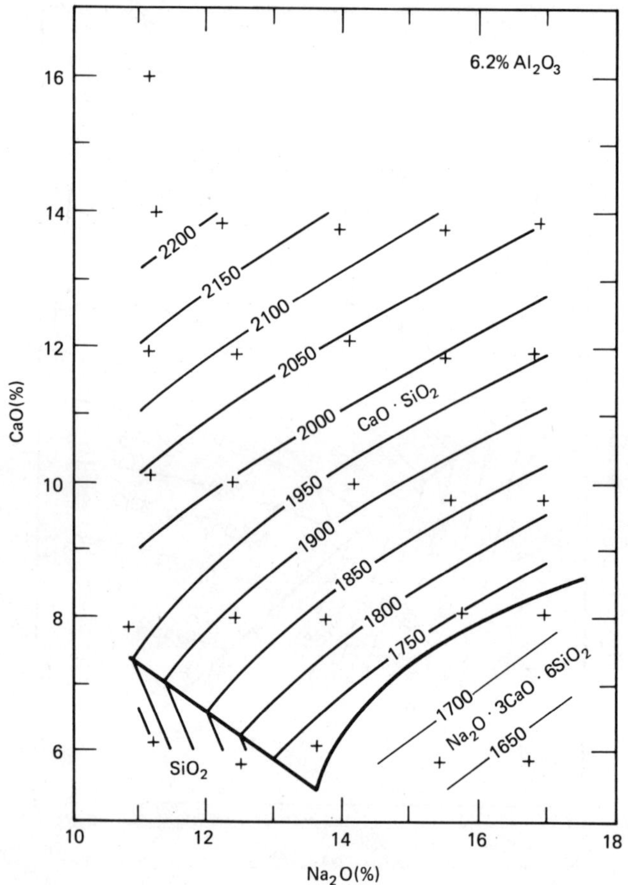

Figure 14.5 Primary phase diagram of 6.2% Al$_2$O$_3$ glasses (liquidus temperatures in °F). From Silverman (1939).

mary phase diagrams by the usual graphical methods is a difficult task when glasses contain four or more oxides. In the glasses under discussion, Silverman chose three-dimensional representation at the five Al$_2$O$_3$ levels. Determination of the phase diagram at intermediate Al$_2$O$_3$ levels, say at 3.5%, would require considerable graphical manipulation. The first approximation equations in Table 14.4 can be used to estimate liquidus temperatures continuously throughout the ranges of all four oxides. The calculation of liquidus temperatures for glasses containing 3.5% Al$_2$O$_3$ is an easy matter after compositions have been converted

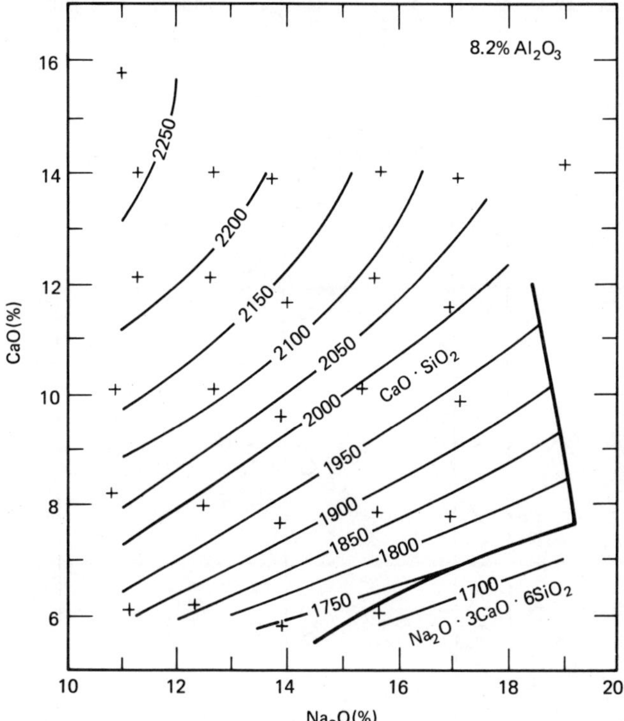

Figure 14.6 Primary phase diagram of 8.2% Al$_2$O$_3$ glasses (liquidus temperatures in °F). From Silverman (1939).

to mole fractions. There is of course no limit to the number of oxides that can
be analytically represented in theis manner.

The author is indebted to Owens-Illinois, Inc., for the use of unpublished data
on viscosity of the above-mentioned glasses at their liquidus temperatures. Table
14.5 allows one to write linear equations representing these data in the three
phase fields and for the combined groups of glasses. Indicated differences between measured and calculated values show reasonabley linear relations between
viscosity at the liquidus temperature and composition. The property-composition relations outlined in Tables 14.3 to 14.5 are described because of the importance in glass manufacturing. These linear relations are used in Part V to give an
overall view of their significance in glassmaking.

TABLE 14.4 Liquidus Temperatures ($^\circ$C) in Na_2O-CaO-Al_2O_3-SiO_2 Glasses

$$^\circ C = A\ SiO_2 + B\ Al_2O_3 + C\ CaO + D\ Na_2O$$

Phase field	Ratio of number of glasses used to number available	A	B	C	D	Differences between measured and calculated temperatures ($^\circ$C)	
						Average	Maximum
Tridymite	27/36	2324.29	-7016.19	-341.67	-4115.46	8	17
Devitrite	20/23	833.33	2093.29	3061.01	68.70	5	15
Wollastonite*	57/75	938.43	3834.26	3306.66	-764.59	7	18
Combined	104/134	1384.25	1459.28	2395.39	-1619.48	48	197

Data accurate to 8°C.

Data source. Silverman (1939).

TABLE 14.5 Log Viscosity at Liquidus Temperatures in Na_2O-CaO-Al_2O_3-SiO_2 Glasses: Equations for Log Viscosity at Liquidus Temperatures

$$\text{log viscosity} = A\ SiO_2 + B\ Al_2O_3 + C\ CaO + D\ Na_2O$$

Phase field	Ratio of number of glasses used to number available	A	B	C	D	Differences between measured and calculated log viscosity	
						Average	Maximum
Tridymite	20/29	-1.8324	66.5116	8.1225	27.0064	0.05	0.14
Devitrite	20/23	7.4109	6.9106	-15.1593	2.7403	0.05	0.15
Wollastonite	58/72	7.1036	-2.1266	-14.2724	3.7279	0.06	0.15
Combined	98/124	5.4521	6.7225	-12.3659	8.1835	0.24	1.19

Data accurate to 0.05.

Experimental data provided by Owens-Illinois, Inc., Toledo, Ohio.

Huff and Call (1973) describe a computer technique involving both linear and nonlinear terms which allows prediction of commercial glass compositions that have arbitrarily desired sets of properties. Their method which includes a computerized retrieval system, was derived from properties and chemical analyses in weights percent of 1400 experimental glasses. The glasses contained ranges of the oxides SiO_2, Na_2O, K_2O, CaO, MgO, BaO, B_2O_3, Al_2O_3, Fe_2O_3, and SO_3. Property coverage consisted of viscosities of 10^2, 10^3, 10^4, and 10^7, fiber softening point, annealing point, liquidus temperature, thermal expansion in the range O to $300°C$, acid and water chemical durabilities, and cooling times for the temperature range between 10^3 and 10^7 viscosities. Property-composition correlations on 48 Na_2O-CaO-Al_2O_3-SiO_2 glasses are of interest here. Their standard deviations between measured and predicted values of 10^3 viscosity temperatures, fiber softening point, and liquidus temperature were 8.3, 3.3, and $10°C$, respectively. Standard deviations for these respective properties in the present study were as follows: tridymite glasses, 6, 6, and $6°C$; devitrite glasses, 2, 2, and $3°C$; wollastonite glasses 10, 7, and $9°C$.

The nonlinear correlation terms in the equations of Huff and Call generally involved the squares of oxide amounts, the products of two oxide amounts, or the ratio of amounts of two oxides. Such nonlinearity between properties and compositions is understandable on the basis of the random network theory and has been discussed in Chaper 4. These relations in glasses containing substantial amounts of B_2O_3 are particularly nonlinear. It is noted that the substructure method can be used to predict the composition of a glass having a specified set of properties. This method was described in Section 6.4 in determining the composition of a glass having specified values of refractive index and Abbe number (Table 6.5). The composition of a glass having specified values of log 3 viscosity temperatures and liquidus temperature can be determined by using appropriate equations from Tables 14.3 and 14.4. Glass compositions having given values of liquidus temperature and log viscosity at the liquidus temperatures can be determined by using appropriate equations from Tables 14.4 and 14.5. Equations from the same primary phase fields must of course be used. These solutions are first approximations: they can be useful in the planning of glass formulations but should be checked by experimental data before being put into production.

15

Relaxation Phenomena

Reactions of glass to imposed static and cyclic mechanical forces have been outlined in Chapters 12 to 14. Glasses simultaneously react as viscous liquids and as elastic solids over wide temperature ranges. This viscoelastic behavior is time dependent and is easily observed in the transformation and annealing temperature ranges. Glass deforms under constant load, with an elastic aftereffect. When the material is subjected to a cyclic stress, the cyclic internal strain lags behind the stress. This chapter deals with relaxation processes of particular interest to glass technologists.

15.1 RATE OF STRESS RELEASE IN GLASS

Measurements on the effect of temperature on stress-optical coefficient by Van Zee and Noritake (1958) were discussed in Chapter 7. The same paper also describes measurements on rate of stress release in commercial plate and container glasses and includes details of the furnace and optical system used. The furnace was capable of being held to within $\pm 0.25°C$ for a period of hours. For measuring the release of mechanically applied stress, the glass sample was loaded as in Figure 15.1 by applying a constant displacement instead of a constant pressure. The displacement was adjusted to apply a load that would give a birefringence of some 600 to 700 millimicrons at the top and bottom of the sample. Readings of birefringence began immediately after release of the load and were taken at time intervals chosen so that changes in stress of 20 to 30 millimicrons occurred between readings. Data could be obtained at temperatures corresponding to the glass viscosity range $10^{12.7}$ to $10^{15.4}$ poises.

It was found that stress release data could be represented by the equation in Table 15.1,

where S = stress level after given soaking time t
            τ_1 = relaxation time for a rapidly decaying mechanism
            τ_2 = relaxation time for a slowly decaying mechanism
$C_1, C_2,$ and C_3 = constants

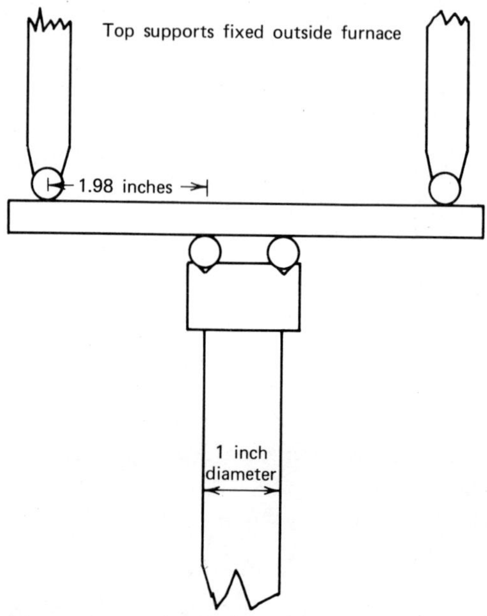

Figure 15.1 Four-point loading mechanism of sample. From Van Zee and Noritake (1958).

TABLE 15.1 Summary of Stress Release Equations

$$S = C_1\, e^{-(t/\tau_1)} + C_2\, e^{-(t/\tau_2)} + C_3$$

Temperature ($^\circ$C)	τ_1 (min.)	τ_2 (min.)	C_1 (lb/in.^{2})	C_2 (lb/in.^{2})	C_3 (lb/in.^{2})
Plate glass					
507	11.84	161.0	390	1120	0
527	7.58	51.8	690	640	0
538	2.35	12.6	547	710	−6.2
549	1.07	4.03	950	775	−5.8
561	0.238	0.978	444	810	−2.5
Normal container glass					
527	10.15	57.1	775	545	0
532	8.70	41.0	985	315	0
538	3.97	17.7	790	560	−13.2
549	0.913	4.38	660	585	−4.6
560	0.291	1.314	1030	605	−6.8

Source. Van Zee and Noritake (1958).

This table also gives values of parameters of equations for plate and container glasses over indicated temperature ranges. Our discussion has distinguished between relaxation times for electrical and mechanical processes and denotes them respectively in terms of τ and λ. The change in nomenclature in Table 15.1 follows Van Zee and Noritake. Mechanical relaxation times for these glasses decrease rapidly with temperature, with indicated differences between those for rapid and slow decay processes. The coefficients C_1, C_2, and C_3 are not independent constants. At time $t = O$ the sum of these three constants equals the original stress. Stress release measurements starting with a different initial stress would give different values of these constants. Values for C_3 in Table 15.1 might suggest the possible existence of a third decay mechanism, but the accuracy of measurements did not justify such an assumption. In terms of their ionic structures, more refined measurements on the glasses might disclose a more complex spectrum of decay times. Reference is made to Table 5.9 on dielectric relaxation times measured by Hakim and Uhlmann (1973) on a $Cs_2O\text{-}SiO_2$ glass that exhibited rapid and slow decay times. Mechanical and electrical phenomena are different, but it appears that their decay times involve at least two decay mecha-

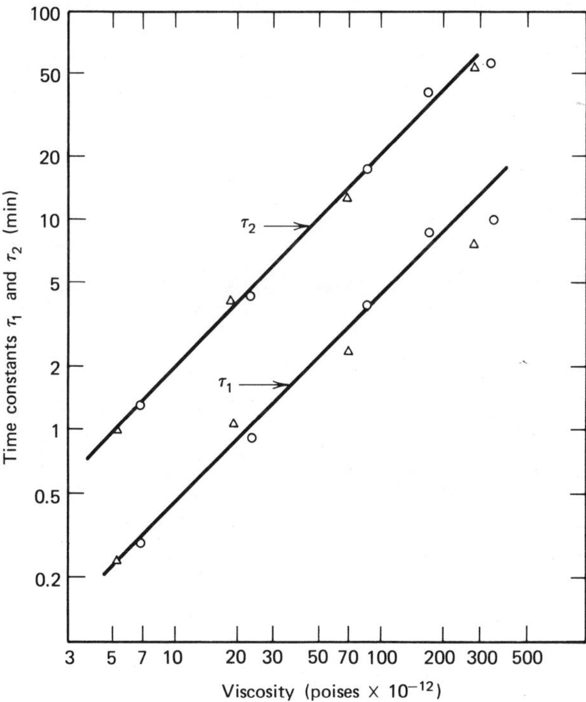

Figure 15.2 Time constants for release of stress as a function of viscosity. Triangles, plate glass; circles, normal container glass. From Van Zee and Noritake (1958).

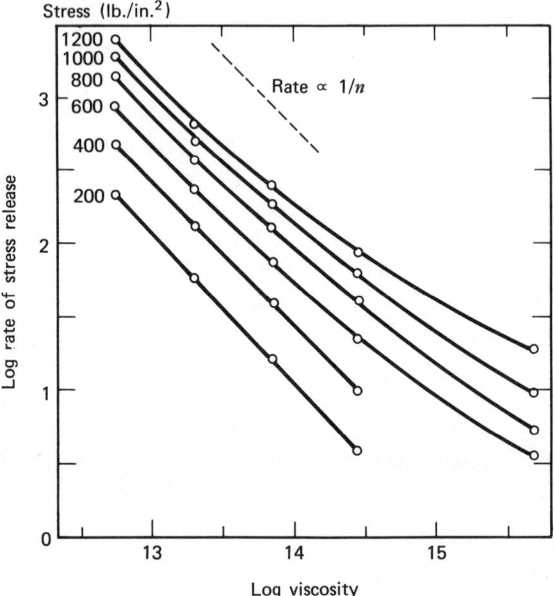

Figure 15.3 Rate of stress release in plate glass versus viscosity. From McGraw and Babcock (1959).

nisms. Figure 15.2 shows that mechanical relaxation times for the two glasses rapidly increase as the glass viscosity increases.

McGraw and Babcock (1959) used the apparatus of Van Zee and Noritake for stress release measurements on two additional types of glass, namely, a potash-barium and a borosilicate. It was found that stress release data on these glasses could be represented by equations of the type provided for soda-lime-silica glasses in Table 15.1. Equilibrium viscosity data were obtained to clarify the role of viscosity in stress release. The data relations on plate glass are described here.

Effect of Viscosity at Constant Stress

Figure 15.3 indicates linear relations between viscosity and rate of stress release at low viscosities (high temperatures) at all stress levels. Thus the stress is released by viscous flow alone in this viscosity range. The rate of stress release dS/dt is proportional to $1/\eta$. At stresses above 400 lb/in.^{2} the rate becomes faster, indicating the presence of an additional stress release mechanism.

Effect of Stress at Constant Viscosity

Figure 15.4 shows that none of the curves is linear. Linearity in this plot indicates that rate of stress release is proportional to S and to $1/\eta$. The curves approach

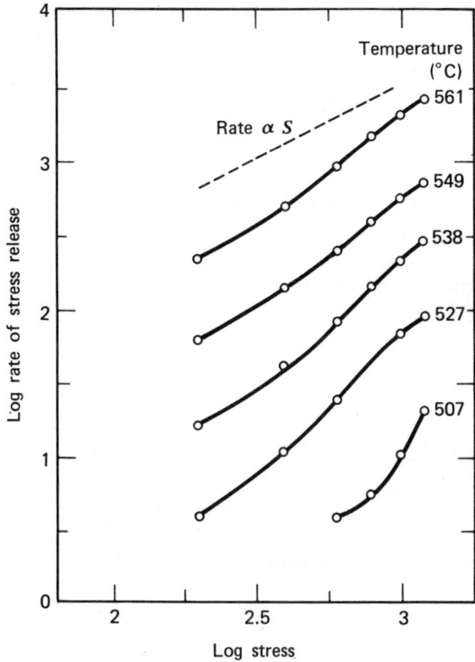

Figure 15.4 Rate of stress release in plate glass versus stress. From McGraw and Babcock (1959).

linearity at low stresses, but at higher stresses and lower temperatures the data yield rates of stress release proportional to S^2 and S^3.

Suggested Mechanisms for Stress Release

The foregoing behavior suggests that both viscous and viscoelastic properties play important roles in stress release in glasses. The data can be reasonably accounted for in terms of the complex actions of viscous and elastic elements represented in Figure 15.5. This model was used by Isard and Douglas (1955) to describe their measurements of stress release in vitreous silica. The data just given on silicate glasses indicate that stress release is dependent on viscosity, delayed elasticity, and stress level. The data were not sufficient to determine the effects of glass composition on these properties.

15.2 DAMPING IN OSCILLATING GLASS FIBERS

Numerical values of internal friction in glasses can be expressed in terms of Q^{-1} measured by the logarithmic decrement of a simple torsion pendulum containing the test specimen of glass as the suspension.

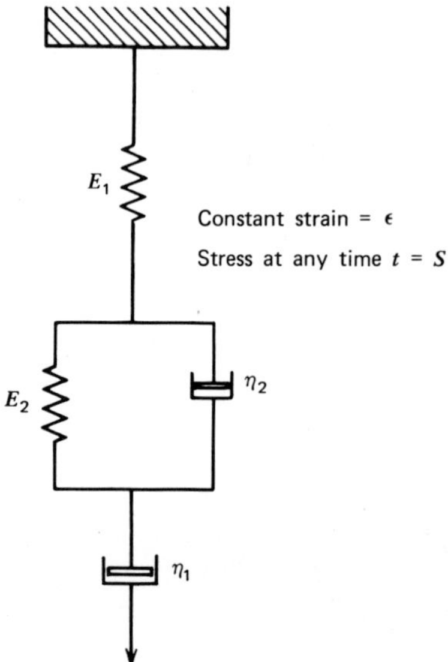

Figure 15.5 Elastic and viscous elements in stress release. From McGraw and Babcock (1959).

$$Q^{-1} = \log \text{ decrement} \Big/ \pi = \pi \tan \delta = \pi G \big/ \omega \eta \qquad (15.1)$$

where $\tan \delta$ = loss angle, G = dynamic shear modulus, ω = frequency (Hz) and η = viscosity coefficient. These relations indicate laminar viscous flow in the fiber. The phase angle between stress and strain in experiments on forced oscillations of glass fibers was shown by de Bast and Gilard (1963) to be quantitatively related to observations on stress release in glasses.

Douglas, Duke, and Mazurin (1968) measured logarithmic damping on vitreous silica and silicate glasses in the temperature range 350 to 450°C and in the frequency range 9.0 to 26.4 Hz. Logarithmic decrements decreased rapidly with temperature for a 0.8 SiO_2-0.2 Na_2O glass showed the well-known low and high- temperature absorption maxima, which appeared superimposed on a background that increased rapidly at high temperatures. Measurements on annealed and unannealed fibers exhibited a very marked effect of thermal history of the samples; decrements of the unannealed fibers being greater at all temperatures. Measurements made on vitreous silica fibers in the same temperature range and at a frequency of 9.0 Hz showed essentially no damping with very low values of decrement that were unaffected by temperature. The very low damping of vitreous silica has long been demonstrated by its use as suspensions in string electrometers. Douglas et al. observed that the damping was related to the lattice or network and to stress-induced diffusion of ions, which gives rise to peaks in the

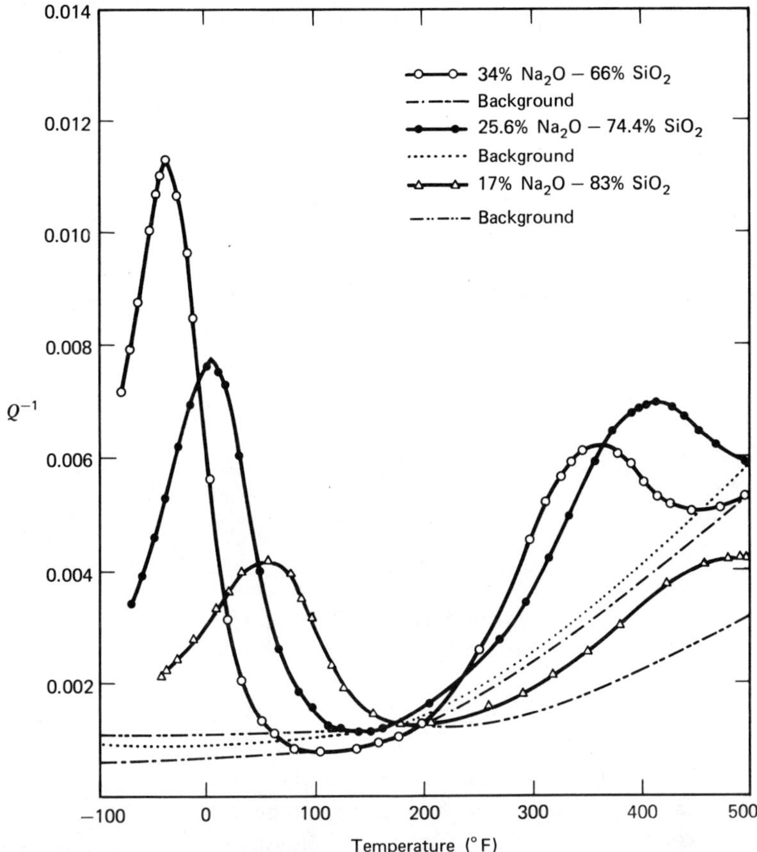

Figure 15.6 Comparison of internal friction versus temperature curves for Na_2O-SiO_2 glasses. From Forry (1957).

curves. The two peaks in this case derive from diffusion of Na ions.

More detailed information on the two Na peaks has been given by Forry (1957), who measured Q^{-1} on three Na_2O-SiO_2 glasses as functions of temperature, composition, and frequency in the range 1 to 3 Hz and from -90 to $500°F$. He found that temperature locations of the peaks were nearly the same for the small range of frequencies used. The Q^{-1} values of unannealed fibers were higher and shifted slightly to lower temperatures as compared to annealed fibers. Figure 15.6 shows compositions of the glasses and relations between Q^{-1} and temperature. Values of Q^{-1} decrease in magnitude and shift to higher temperatures as the SiO_2 contents increase. Forry's measurements also show that dynamic shear modulus of these glasses decrease with temperature and that its magnitude increases with SiO_2 contents. Forry concluded from his measurements that the observed peaks in the curve of internal friction versus temperature were not the result of a single relaxation mechanism; rather, they were due to a distribution of such mechanisms.

The effects of glass composition, frequency, and heat treatment on internal friction have been studied by a number of investigators. Glass compositions have generally been limited to alkali silicates. The data reveal the presence of low and high temperature absorption peaks for all the alkalies from Li^{1+} to Cs^{1+}. Their temperature locations when measured at the same frequency are nearly the same. Jagdt (1960) found that increasing the frequency caused a shift of the maxima to higher temperatures. He found that the maximum in the low temperature peak of an $Na_2O \cdot 2SiO_2$ glass was at $-50°C$ for a frequency of 1 Hz and at $30°C$ for a frequency of 10^3 Hz. The corresponding temperatures for the high temperature peak were 190 and $290°C$, respectively.

Glass structure implications of internal friction data are often given in terms of random network nomenclature (bridging and nonbridging oxygens, etc.). Ratobylskaya and Tarasov (1960) found distinct differences between the internal friction behaviors of SiO_2 and B_2O_3 glasses. They observed that SiO_2 acted like a rigid three-dimensional network, and the structure of vitreous B_2O_3 resembled that of organic polymers. This view is in general agreement with behaviors of these glasses when subjected to high pressures as discussed in Section 13.7. The studies of Coenen and Amrhein (1962) on internal friction and elastic constants are of particular interest because the authors related their data to primary phase diagrams of the glasses. They covered a wide variety of compositions in the temperature range -150 to $1000°C$ and found a pronounced minimum in the value of Young's modulus E at the eutectic composition of 0.24 mole fraction Na_2O in a series of Na_2O-SiO_2 glasses. They also discovered that shear modulus G had a minimum value and Poisson's ratio had a maximum value at this same eutectic. They attributed these minima and maxima to the presence of submicroscopic heterogeneities in the glass. This behavior may be explained in terms of the postulate advanced here of primary phase substructures in silicate glasses. Coenen and Amrhein found that internal friction peaks had maximum values at eutectics where E and G had minimum values and Poisson's ratio was maximum. They describe similar changes in elastic moduli and internal friction for glasses in the systems Na_2O-B_2O_3, $Na_2O \cdot 2SiO_2$-$Na_2O \cdot 2B_2O_3$, and complex aluminoborosilicate glasses. Their research suggests that the substructure postulate may be extended to include borate glasses.

15.3 THERMAL RELAXATION AFTEREFFECTS

Foregoing discussions verify that stress release in glasses occurs in a matter of minutes in the annealing temperature ranges of the respective materials. Annealing processes relate to rapid diffusion of loosely bonded alkali ions as they adjust themselves to positions of minumum potential energy in the glass structure. These processes relate to those occurring in the high temperature, internal friction absorption peaks obtained by Forry and others. Heat treatments of glasses in the annealing range near 10^{13} poises are sufficient to remove internal strain and to stabilize property values to levels that are satisfactory for most glass product uses at room temperature.

When glass thermometers came into universal use during the second half of the last century it became evident that indicated temperatures were dependent on glass composition and heat treatment. Hovestadt (1902) pointed out that if a newly made thermometer is left to itself, the glass, which acts as a containing vessel for the mercury, gradually contracts. The end of the mercury column gives continually higher indications of temperature. Progress of the change can be checked by observing the freezing point in an ice bath. Thermometers are capable of giving correct temperatures only below the transformation point of the given glass. Above this point the thermal expansion increases rapidly and the thermometer will indicate temperatures that are much lower than the actual. Thus when thermometer manufacturers produce glasses, they take into account the transformation points of the glasses as they relate to temperature ranges of intended use. Thermometer glasses usually contain no more than one of the alkali oxides because the combination of Na_2O and K_2O strongly increases the zero constant. These thermal relaxation effects are related to the low temperature absorption peaks obtained in internal friction measurements, which occur near $0°C$. Research of Shelby and Day (1969) on internal friction in single and mixed alkalies clarified understanding of the subject. They studied $R_2O\text{-}SiO_2$ glasses using all the alkalies from Li to Cs. Combinations of two oxides gave new damping peaks whose maxima were considerably higher than peaks of the single alkalies. The lag between manufacturing practices and technological explanations becomes shorter and shorter with advances in the science and technology of glasses. Recent cases can be cited in which this lag has reverse directions, and such instances can be expected to increase in number. Further references to the subject include Volf (1961), Weyl and Marboe (1964), and thermometer manufacturers' catalogs.

Thermal relaxation aftereffects occur in all types of glass. There is current interest in the very slow deformation with time of glass and combinational glass-crystalline astronomical mirror blanks. These small deformations are of much lower magnitude than those which can be caused by improper mounting of the astronomical mirror.

15.4 MICROHARDNESS OF GLASSES

The meaning of hardness of a material was early related to Mohs' mineralogy scale of scratch hardness, which ranged from 1 to 10. On this scale glasses occupy a position intermediate between apatite (5) and quartz (7). Auerbach (1894) determined that the scratching test was not decisive for glasses. He discovered that each one of a series of glasses ranging from the softest flints to the hardest borosilicate crowns scratched every other one. Others have since learned that stress level and its rate of application strongly affect scratch hardness values. Microhardness measurements made by indenting the surface with a pyramidal diamond have become accepted as a more precise method of evaluating the hardness of glass. Both types of measurement show definitely that hardness must be understood in terms of a complex mixture of glass properties and specific test conditions. Hardness can be described following Bridgman's operational scheme in which the final result depends on the glass sample and all the steps involved in the experimental procedure. Experimental data point to the connection between microhardness and glass relaxation phenomena, hence the inclusion of microhardness in this chapter.

The grinding and polishing of glass surfaces is of prime importance to opticians in fabricating precision optical components. Research has indicated that process parameters resulting in wearing away or attrition of glass surfaces is influenced by the glass properties of elasticity, delayed elasticity (relaxation processes), viscosity, and ionic exchange. Hardness plays an important role in the attrition of glass surfaces but its exact nature is not well understood. Holland (1964) comprehensively reviewed optically worked surfaces and optical properties of glass surfaces. The polishing of glass surfaces involves a mixture of art and technology, and Holland states that it "has been the subject of conjecture for at least three hundred years and modern research has not yet produced a qualitative and generally accepted theory of the process."

Studies of microindentation hardness of surface layers have been useful in revealing some of the problems associated with the preparation of glass surfaces. When a diamond pyramid is pressed on to the surface of a glass sample, it leaves an identation whose dimensions can be evaluated in terms of hardness values. Figure 15.7 gives the dimensions of the standard Knoop diamond indenter used by Georoff and Babcock (1973) for studies on a number of experimental glasses. Loads of 15, 25, 50, 100, 200, 300, and 500 g were applied at a constant rate of loading, which was controlled by a dashpot type oil brake. A full load was attained in 17 sec, after which the load was maintained for an additional 10 sec period. The Knoop Hardness Number (KHN) was calculated from

$$KHN = 14{,}230P/L^2 \qquad\qquad (15.2)$$

where KHN is in kilograms per square milimeter, 14,230 is a constant that takes into account the geometry of the indenter, P is the load in grams, and L is the length of the long diagonal of the indentation in microns. The Leitz Durimet hardness tester was calibrated by making a series of 10 indentations in a hardened

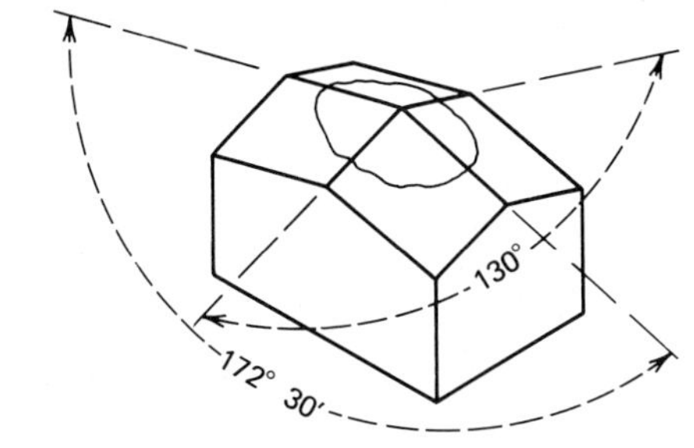

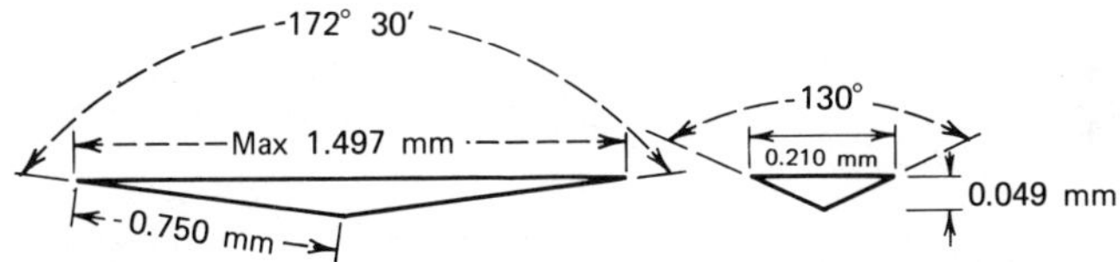

Figure 15.7 Knoop indenter showing maximum usable dimensions. From Georoff and Babcock (1973).

steel block having a certified KHN of 970 ±30 kg/mm^2. The measured value was 979 ±13 kg/mm^2, well within the specified range.

Samples were obtained from sixteen 500 g melts of Na_2O-CaO-SiO_2 glasses and five 2000 g melts of Na_2O-Al_2O_3-SiO_2 glasses. Glasses were melted in platinum crucibles and homogenized by quenching, crushing, and remelting. They were annealed at temperatures corresponding to a viscosity of 10^{13} poises. Glass surfaces were polished on a 100 hour schedule—52 hours using graded sizes of SiC, Al_2O_3, and Barnesite abrasives in water, followed by water polishing with no abrasives on a pitch lap for 48 hours. Samples were stored in a dessicator after polishing; those in the $Na_2O \cdot 2SiO_2$ phase field were hygroscopic and were washed with ethanol before storage.

At least six indentations were measured on each glass at each load level and the measured value of KHN taken as the average at a given load. Figure 15.8 plots KHN versus load for glasses in the devitrite phase field. This plot, which is typical, clearly distinguishes two regions: (1) the load-dependent region from 15 to 100 g, and (2) the load-independent region from 100 to 500 g. Our discussion is confined to data in the load-independent region and their relations with glass compositions. Measured values of KHN given in Table 15.2 are average values obtained with loads of 100, 200, 300, and 500 g. For example, the value 440.2 for glass 70-1 is an average of 24 measurements. The substructure method was

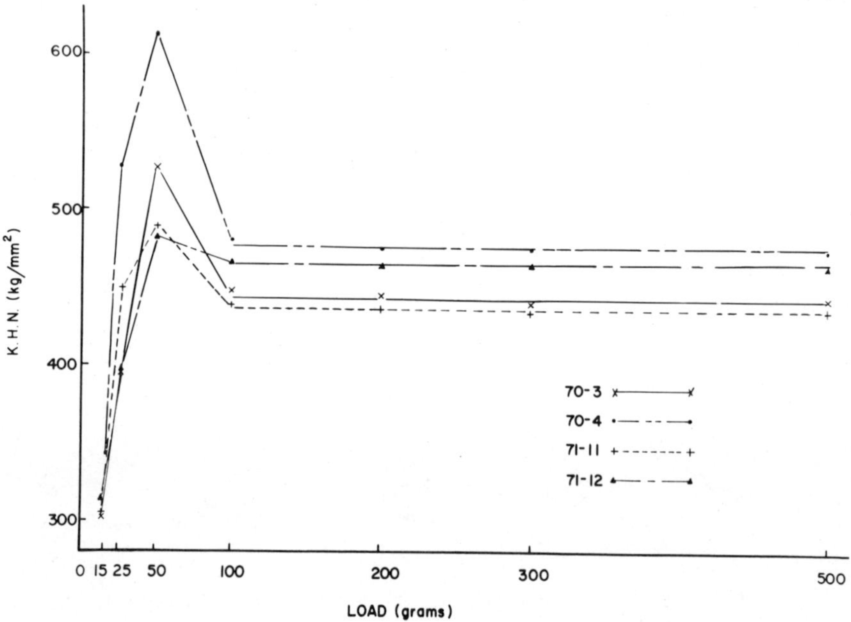

Figure 15.8 KHN versus load for devitrite glasses. From Georoff and Babcock (1973).

used to determine relations between KHN and composition. Separate equations for glasses in the tridymite, devitrite, calcium silicate, and sodium disilicate phase fields were as follows:

$$KHN = 338.55\ SiO_2 + 1285.65\ CaO + 617.57\ Na_2O \qquad (15.3)$$
$$KHN = 525.23\ SiO_2 +\ \ \ 874.02\ CaO +\ \ \ 49.46\ Na_2O \qquad (15.4)$$
$$KHN = 541.79\ SiO_2 +\ \ \ 599.59\ CaO + 615.18\ Na_2O \qquad (15.5)$$
$$KHN = 102.69\ SiO_2 + 2633.04\ CaO + 839.10\ Na_2O \qquad (15.6)$$

All property-compositon data were than combined into one group and the computerized equation found to be

$$KHN = 361.61\ SiO_2 + 1500.83\ CaO + 317.84\ Na_2O \qquad (15.7)$$

Differences between measured and calculated KHN values are quite small when separate equations are used for the different phase fields. Differences are quite large and beyond the precision of the measurements when equation (15.7) for the combined group of 16 glasses is used. Fidelity of data representation using separate equations illustrates the merits of the substructure method as compared to

TABLE 15.2   Knoop Hardness Numbers for Na_2O-CaO-SiO_2 Glasses

Glasses in four phase fields	Mole Fractions			KHN (kg/mm^2)		
	SiO$_2$	CaO	Na$_2$O	Measured	Calculated	Calculated from (15.7)
Tridymite—equation (15.3)						
70-1	0.760	0.050	0.190	440.2	438.9	410.3
70-2	0.760	0.100	0.140	471.0	472.3	469.4
71-9	0.800	0.050	0.150	426.5	427.8	412.0
71-10	0.800	0.100	0.100	462.5	461.2	471.2
Devitrite—equation (15.4)						
70-3	0.700	0.100	0.200	445.1	445.2	466.8
70-4	0.700	0.135	0.165	477.6	477.5	508.2
71-11	0.725	0.075	0.200	436.5	436.5	438.3
71-12	0.720	0.110	0.170	465.8	465.9	479.5
β calcium silicate—equation (15.5)						
70-5	0.700	0.155	0.145	561.5	561.4	531.8
70-6	0.700	0.175	0.125	561.2	561.1	555.5
71-13	0.675	0.175	0.150	562.4	562.9	554.4
71-14	0.650	0.185	0.165	564.8	564.6	565.1
Sodium disilicate—equation (15.6)						
70-7	0.700	0.020	0.280	359.5	359.5	372.1
70-8	0.700	0.040	0.260	395.4	395.4	395.8
71-15	0.650	0.020	0.330	396.3	396.3	370.0
71-16	0.725	0.030	0.245	359.0	359.0	385.1

Primary phase diagram 482 in Levin, Robbins, and McMurdie (1964).
Data source. Georoff and Babcock (1973).

that using equation (15.7), which follows the random network theory. Figure 15.9 plots CaO against SiO_2 content with lines of equal KHN, measured values, and composition locations.

Data on the Na_2O-CaO-SiO_2 glasses were measured with great care and are believed to represent an accurate experimental description of KHN versus glass composition. Interpretation of the data is a difficult matter and must be considered to be an open question. Suggestions of the authors may be helpful in arriving at reasonable explanations. All glass surfaces were subjected to identical polishing schedules, which involved contact with water for 100 hours. Izumitani and Harada (1971) indicated that polishing rate was related to microindentation hardness for dry and oil polishing. Their data also showed that polishing rate was related to both microindentation hardness and chemical durability for polishing

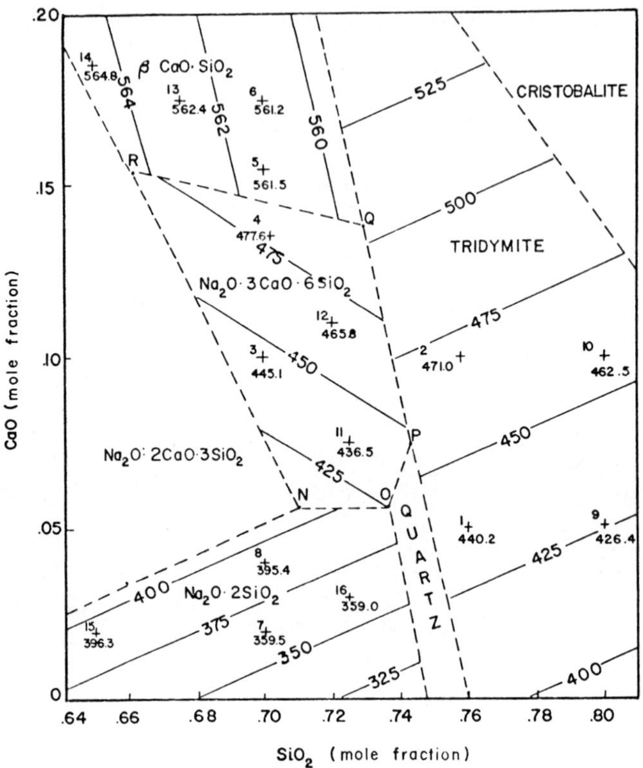

Figure 15.9 KHN data for Na_2O-CaO-SiO_2 glasses. From Georoff and Babcock (1973).

with water and with aqueous acid. Attrition of the glass surfaces in the latter case was caused by mechanical and chemical forces. Long contact with water of the Georoff and Babcock samples undoubtedly involved ion exchange processes in the formation of surface layers. Depth of penetration of these glass-water reactions depended on resistance of a given composition to ionic exchange reactions. Starting with the geometric surface of the sample, it is believed that the composition varies continuously through an altered surface layer of indeterminate depth. Data in Table 15.2 indicate that KHN are independent of depth of penetration for the 100 to 500 g range of loads used. There is no doubt that the most active ionic exchange reactions would occur between Na^{1+} and OH^{1-}. This suggests that glasses in the different phase fields would be altered to different depths and may account for the failure of lines of equal KHN to run continuously across phase boundaries as indicated in Figure 15.9. Thus it may be that KHN measured in this study are strongly influenced by ionic exchange reactions and

TABLE 15.3   Knoop Hardness Numbers for Na_2O-Al_2O_3-SiO_2 Glasses

Oxide	Compositions (mole fraction) of five glasses				
	1	2	3	4	5
SiO_2	0.620	0.620	0.620	0.620	0.620
Al_2O_3	0.100	0.150	0.190	0.205	0.220
Na_2O	0.280	0.230	0.190	0.175	0.160
Load (g)	KHN (kg/mm^2)				
100	505.3	490.0	478.5	493.5	506.3
200	504.4	488.5	480.0	494.0	506.3
300	503.5	491.4	479.2	493.8	504.4
Average	504.4	490.0	479.2	493.8	505.6

Primary phase diagram 501 in Levin, Robbins, and McMurdie (1964).
Data source. Georoff and Babcock (1973).

are at levels characteristic of sample preparation procedures used in this study. These measurements may not be amenable to scientific explanations, but they are quite useful in clarifying polishing procedures, most of which use water as a vehicle. Attrition of glass surfaces in polishing is strongly dependent on the pH level of water or aqueous solutions used in the polishing slurries. Other parameters in these measurements influence the KHN values. Energy imparted at the point of contact of the indenter with the glass will raise the temperature, which in turn produces changes in elastic and viscous deformation. When the indenter is retracted, the indentation (which is subsequently viewed and measured) may be an altered structure that is a good approximation of the structure at the instant of retraction.

Table 15.3 gives data obtained on Na_2O-Al_2O_3-SiO_2 glasses at the indicated loads. Pronounced microfracturing that occurred in these glasses at loads of 500 g prevented measuring KHN. Figure 15.10 is a plot of these data together with least squares lines representing the data. Here again ionic exchange between Na^{1+} and OH^{1-} ions may have altered the true hardness values. Reference is made to previously discussed substructure relations of these glasses for data on refractive index and stress-optical coefficient. It is obvious that these KHN data could not be represented in terms of the random network theory.

Westbrook (1960) studied the effects of temperature on the microindentation hardness of vitreous silica and other types of glass. He found discontinuities in the hardness-temperature curves for vitreous silica and related these to tempera-

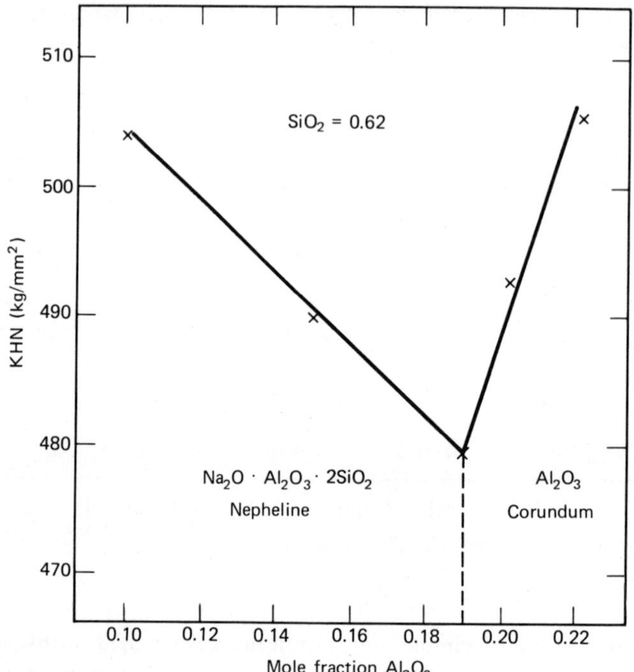

Figure 15.10 Effect of substituting Al_2O_3 for Na_2O on KHN. From Georoff and Babcock (1973).

ture inversions of quartz and cristobalite as noted in Section 4.6. The discontinuity in crystalline quartz was especially pronounced at 573°C, which is the high-low quartz inversion temperature. The hardnesses of silicate glasses decreased monotonically with increases in temperature. The same was true for vitreous GeO_2 glass, which has no high-low inversion in the range 0 to 400°C. An abrupt change in slope of the hardness-temperature curve for BeF_2 glass was related to its high-low temperature inversion at about 220°C. Westbrook's review indicates that microindentation measurements can reveal structural changes in glasses and suggests that further research is needed to evaluate the role of flow in such tests.

When penetrating a sample, a pyramidal diamond indenter gives rise to shear stress because of its geometrical shape. This results in viscous and/or viscoelastic flow as noted by different investigators, including Taylor (1950), Brueche and Schimmel (1954, 1955), and Prod'homme (1968), who used the term "deformation hardness" in describing the tests. Using a Vickers indenter, which leaves a square indentation, Taylor described evidence of flow. The glass flowed to form a raised ridge at the edge of the indenter. These studies and others reviewed by

Holland (1964) demonstrate that glass flow occurs in indentation hardness tests and in glass polishing. Determining the exact nature of this flow poses difficult experimental problems and presently constitutes an open question. Prod'homme found that indentation hardness numbers increased linearly with viscosities for a number of optical glasses. He and others have also observed that hardness numbers of annealed glasses are higher than those obtained when the glasses are quenched.

16
Surface Tension

The term "surface tension" refers to the properties of an interface between two phases of matter, and its value is dependent on the state and nature of the contiguous masses. Unless further qualified, the surface tension of a material usually implies air or vapor of the material as the second phase. In a multicomponent system such as glass, there is a concentration at the interface of those components which reduce the surface tension. This necessitates the diffusion of material both to and from the interface, and the process requires time to reach equilibrium. The surface tension is defined as a force per unit length and can be expressed in dyness per centimeter. The surface energy per unit of surface has the same dimensions and can be expressed in the same units. This chapter covers methods of measurement and discusses surface tension data on selected glasses.

16.1 METHODS OF MEASUREMENT

Morey (1954) and Holland (1964) review methods of measuring surface tension in glasses and some of the published data. Figure 16.1 illustrates methods for measuring surface tension in glasses. The sessile drop method which is suitable for low temperature measurements between glass and solids, has been used to evaluate wetting properties between glass and metals as in glass forming and in making glass-metal seals. The other methods are appropriate for high temperature measurements at glass-air interfaces. The bubble, ring-breaking, and dipping cylinder methods are usually used for property measurements, whereas the others are better suited for control of glass fabrication processes. There is a balance in all methods between surface tension and opposing forces, particularly gravity and viscosity. Usable ranges of the methods depend on viscosity-temperature relations of the glass being measured.

Reactions between liquid glass and solids such as metals and refractories involve three independent surface tensions corresponding to the three interfaces (liquid-air, solid-air, and solid-liquid). Thus the contact angle in the sessile drop method is strongly influenced by constituents of the surrounding air and by ambient oxidation-reduction conditions. Our interest in this chapter is confined to glass-air interfaces at high temperatures, which are covered in terms of effects of

254

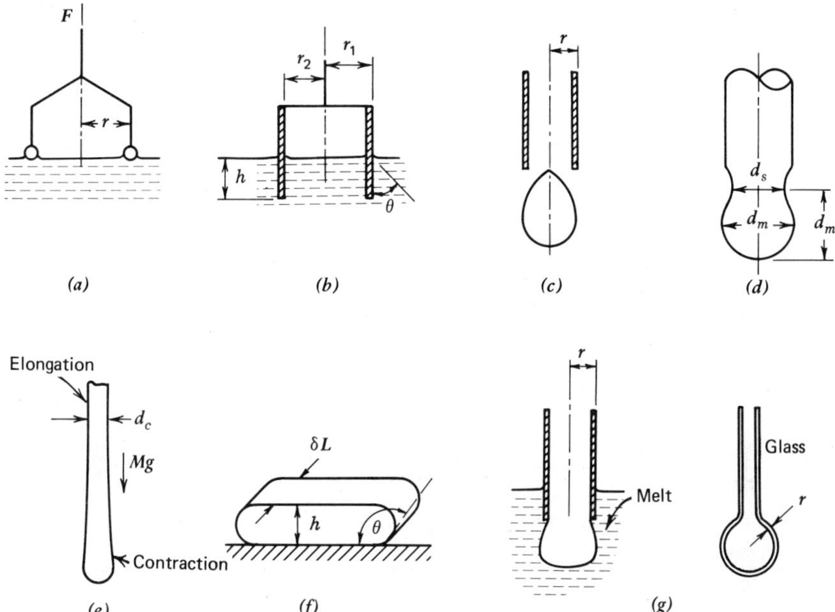

Figure 16.1 Methods for determining the surface tension of glass. *(a)* Ring breaking. *(b)* Dipping plate or cylinder. *(c)* Drop weight. *(d)* Pendant drop. *(e)* Elongated fiber. *(f)* Section of a pessile drop. *(g)* Bubble method. From Holland (1964).

composition and temperature on surface tension.

Jaeger (1917) developed the maximum bubble pressure method and used it to measure changes in molecular free surface energy of a number of liquids in the range from -80 to $650°C$. Subsequent investigators, including Parmelee and Harman (1937), Badger, Parmelee, and Williams (1937), and Keppler (1937), used this method for surface tension measurements on glasses. Harkins and Jordan (1930) described determination of surface and interfacial tension from the maximum pull on a ring. Harrison and Moore (1938) used this approach to measure surface tension on vitreous enamel frits. Babcock (1940) described a modified dipping cylinder method for measurements on a series of glasses in the temperature range from 1100 to $1400°C$. The dipping cylinder was supported from the spring of a Jolly balance. The level of the molten glass was raised slowly until it contacted the bottom rim of the platinum cylinder. The cylinder was immediately pulled down into the glass by surface tension forces; the depth was limited by stops. The movable barrel of the Jolly balance was then slowly raised. Successive lowering of the glass surface and raising of the balance cylinder drew the glass out into a thin film which extended above the glass surface. Equilibrium readings of surface tension were determined in this manner. The following equation was used for this method

$$\text{surface tension (dynes/cm)} = kgE/2L \qquad (16.1)$$

where k = spring constant (1.04 g/cm extension)
      g = acceleration of gravity (980.24 cm/sec^2)
      E = extension of spring (cm)
      L = circumference of cylinder (cm)

The apparatus was calibrated by measuring the surface tensions of distilled water, glycerol, and ethyl alcohol. The data described here were measured with the dipping cylinder and ring methods.

16.2 SURFACE TENSION DATA ON GLASSES

Data on systematically arranged compositions can be used to evaluate the effects of temperature and composition on glasses.

Binary Glass Systems

Investigators at the National Bureau of Standards; Shartsis, Spinner, and Smock (1948), Shartsis and Canga (1949), Shartsis and Spinner (1951), and Shartsis and Capps (1952) measured surface tensions on glasses in the following binary systems:

Li_2O-SiO_2	PbO-SiO_2	K_2O-B_2O_3
Na_2O-SiO_2	Li_2O-B_2O_3	PbO-B_2O_3
K_2O-SiO_2	Na_2O-B_2O_3	ZnO-B_2O_3

Space does not permit detailed discussions of these extensive data, which cover wide composition ranges and temperature ranges from 600 to 1400°C, depending on the glass system studied. Relations between surface tension and composition were usually nonlinear, as were relations between surface tension and temperature.

Some of the data are of special interest. The surface tension of vitreous B_2O_3 was found to be 83 dynes/cm, whereas that of PbO was 135 dynes/cm. Substitution of B_2O_3 for other oxides lowered the surface tensions in all glasses. The substitution of Li_2O for SiO_2 in the binary system increased surface tension in the range 12 to 49 weight percent Li_2O. Below 20% Li_2O, the surface tension decreased with decreasing temperature. Above 20% Li_2O, the surface tension increased with decreasing temperatures. The surface tension did not change with temperature at 20% Li_2O; the average value in the range 1100 to 1400°C was 320.0 +0.2 dynes/cm. Levin, Robbins, and McMurdie (1964, Fig. 182) show that the eutectic between SiO_2 and $Li_2O \cdot 2SiO_2$ is at 20% Li_2O. Thus the data are understandable in terms of the author's substructure postulate. Substitution of Na_2O for SiO_2 resulted in increases in surface tension in the range 20 to 50% Na_2O. Surface tension increased with decreasing temperatures throughout the range. Substitution of K_2O for SiO_2 caused surface tensions to decrease through the range 24 to 44% K_2O; surface tensions increased with decreasing temperatures. Data on Na_2O-SiO_2 and K_2O-SiO_2 glasses cannot be related to their respective primary phase disgrams.

Shartsis, Spinner, and Smock (1948) measured surface tensions on 29 glasses in the PbO-B_2O_3 system with PbO ranging from 0 to 100%. They related their data to the primary phase diagram of this system of Levin, Robbins, and McMurdie (1964, Fig. 201). Surface tensions at 900°C in the two-liquid immiscribility region from 0 to 40% PbO were essentially constant, the average of nine glasses being 79.1 ±0.6 dynes/cm. Further increases in PbO caused surface tensions to increase rapidly to about 162 dynes/cm at the major eutectic near 88% PbO. Further increases in PbO resulted in decreases in surface tension to a value near 135 dynes/cm at 100% PbO. These data support the substructure postulate and suggest that it may be extended to borate glasses.

Ternary Glasses

Shermer (1956) measured surface tensions on 154 glasses in the BaO-B_2O_3-SiO_2 system. This system is important because it serves as the starting point of a large class of barium crown optical glasses. Measurements covered the temperature range 900 to 1300°C. Shermer related his data to the phase diagram of a system discussed by Levin, Robbins, and McMurdie (1964: Fig. 558) in terms of regular one-liquid glasses and two-liquid immiscibility glasses.

Babcock examined data on 28 glasses in the primary phase $3BaO \cdot 3B_2O_3 \cdot 2SiO_2$ of this system to test the validity of the substructure postulate. Compositions were converted to mole fraction, and the data were subjected to the least squares computer program. The following equations for surface tension (ST) give good representation of the data:

$$ST_{1300°C} = 263.99\ SiO_2 + 93.52\ B_2O_3 + 423.12\ BaO \qquad (16.1)$$
$$ST_{1200°C} = 254.82\ SiO_2 + 90.83\ B_2O_3 + 444.45\ BaO \qquad (16.2)$$
$$ST_{1100°C} = 247.76\ SiO_2 + 88.91\ B_2O_3 + 463.57\ BaO \qquad (16.3)$$
$$ST_{1000°C} = 251.30\ SiO_2 + 94.02\ B_2O_3 + 468.92\ BaO \qquad (16.4)$$

Differences between measured and calculated values of surface tension are 1.2, 0.9, 0.9, and 1.2 dynes/cm, respectively. When these constants were used to calculate surface tensions of glasses located in other one-liquid phase fields, it was found that differences between measured and calculated values of surface tension were large and beyond precision of the measurements. Surface tension data at $900°C$ were not evaluated because liquidus temperatures in this phase field are above $900°C$. Surface tensions decrease linearly from 1000 to $1300°C$ within experimental error, although there is a slight concavity in the plots of surface tension versus temperature.

Na_2O-CaO-MgO-Al_2O_3-SiO_2 Glasses

Babcock (1940) measured surface tensions on 24 glasses in this system over the temperature range 1100 to $1400°C$. Measurements covered the following oxide ranges in weights percent

SiO_2	64-74
Al_2O_3	0-4
MgO	0-7
CaO	6-16
Na_2O	11-17

These ranges include the major oxides used in large tonnage production of containers, flat glass, tableware, and so on. The role of surface tension in glass manufacturing is discussed in Part V.

The author has reexamined the published experimental data in terms of the substructure method. Two of the glasses are located in the devitrite (Na_2O·$3CaO$·$6SiO_2$) phase field and could not be evaluated. Twenty-two glasses lie in three primary phase fields—tridymite (SiO_2), wollastonite (CaO·SiO_2), and diopside CaO·MgO·$2SiO_2$), and the numbers are 7, 6, and 9, respectively. Surface tension data at 1400 and $1200°C$ for groups of glasses in the separate phase fields were subjected to least squares analyses. Table 16.1 lists surface tension-composition relations for glasses in the separate phase fields and allows one to find the equations for the 22 glasses when combined into one group. The indicated average and maximum differences between measured and calculated values of surface tension shows the advantage of using the substructure method. Combining all the glasses into one group follows the random network theory. It was found that the temperature coefficient of surface tension was -0.02 dyne/(cm) ($°C$). That is, the surface tension increases 2 dynes/cm for each lowering of the temperature by $100°C$. A linear relation between surface tension and temperature had to be assumed in terms of measurement precision, but extra polation of the data below $1000°C$ or above $1500°C$ is not recommended for estimation purposes. Surface tensions can be closely estimated by multiplying the oxide constants by their respective mole fractions. Surface tensions calculated with constants in Table 16.1 were in good agreement with measurements by other investigators for similar glasses (see Babcock, 1940). Glasses in the wollastonite field do not have constants for MgO; they contained 0.2 weight percent MgO, which was added to their CaO contents. The equations can be used for calcite glasses made from calcite lime and for dolomite glasses made from dolomitic lime.

TABLE 16.1 Surface Tension (ST) Equations for Na_2O-CaO-MgO-Al_2O_3-SiO_2 Glasses

$$ST = A\ SiO_2 + B\ Al_2O_3 + C\ CaO + D\ MgO + E\ Na_2O$$

Phase field	Temperature (°C)	A	B	C	D	E	Differences between measured and calculated temperatures (°C)	
							Average	Maximum
Tridymite	1400	368.18	565.73	343.00	337.06	−5.36	0.6	1.8
Wollastonite	1400	333.52	596.17	363.41	—	204.48	0.5	0.9
Diopside	1400	339.86	594.58	742.93	139.30	74.18	0.8	1.8
Combined	1400	324.64	675.60	446.16	447.49	143.30	1.7	4.5
Tridymite	1200	374.64	620.12	353.00	345.39	−16.95	0.6	1.7
Wollastonite	1200	332.15	567.67	399.00	—	209.63	0.1	0.2
Diopside	1200	342.45	621.54	1063.92	−184.73	82.47	0.6	1.2
Combined	1200	327.68	672.35	458.09	453.40	150.63	1.7	4.5

Data source. Babcock (1940).

IV

Thermal Properties

The equalization of temperature in bodies at rest is effected not by heat conduction alone but also by heat radiation, which is a totally different mode of propagation of heat.

Max Planck (1858-1947)

Heat conduction through a material wherein the temperature gradient does not change with time is analogous to the transfer of electrical and mechanical energy under similar static conditions. The driving forces are, respectively, temperature, electomotive force, and mechanical force. Their mathematical expressions are in the form of Ohm's law. There is one important difference. Electrical and mechanical energy are usually confined within the material, whereas thermal energy is subject to losses caused by thermal radiation. Temperature-dependent thermal radiation results in heat transfer from hot to cold layers within the material and from surface radiation of the hot material to colder surrounding bodies.

Part IV covers heat transfer in glasses in terms of density, specific heat, regular conduction processes, thermal diffusivity, and radiation conductivity. Heat transfer in molten glass involves both steady state and transient conditions. The effects of temperature and composition on thermal expansion and specific heat are outlined.

17

Heat Transfer

Heat may be transferred by three different mechanisms. (1) Conduction occurs within continuous materials in which energy is transmitted directly between adjacent molecular or ionic aggregates without mass motion of the materials. (2) Convection occurs in liquids and gases, the transfer being effected by motion of the fluid from a location at which it receives heat to a location from which it gives up heat. (3) Radiant heat transfer occurs between two bodies that are at different temperatures and between two locations in a continuous material that are at different temperatures.

17.1 DEFINITIONS AND NOMENCLATURE

Heat may be defined as the energy in transition within a material or between two materials under the driving force or potential of a temperature difference. Depending on circumstances, heat energy may be transferred by simultaneous actions of conduction, convection, and radiation.

The thermal condutivity coefficient gives the amount of heat, in gram-calories per second, which will be conducted through a centimeter cube of the material under a temperature difference of $1°C$ between faces. The time rate of heat flow is proportional to the condutivity coefficient, the cross-sectional area, and the temperature gradient. The conductivity coefficients of silicate glasses near room temperature, where glass products are used, are in the range from 0.002 to 0.003 cal/(sec) (cm) ($°C$). This coefficient increases with temperature and is dependent on glass composition. It is difficult to determine experimentally, and its quantitative relations with glass composition are not well defined. Boundary conditions in terms of surface contacts and so on in such tests are difficult to define and control. Once the coefficient of a given glass is known, it is not an easy matter to apply it to practical situations such as heat conduction through a glass container of complex shape. Problems in heat conduction of glass products must be solved uniquely in terms of ambient boundary conditions and the product under consideration. Pursuit of this subject would lead to areas outside our present interests.

Convective heat transfer is of importance in large glass melting tanks because of the viscous liquid characteristics of the molten glass. Although the paths of convection currents in glass tanks can be mapped and controlled to a certain extent, the transfer of heat by this means is difficult to evaluate.

According to Kirchhoff's law, every substance emits as much heat radiation as it receives for any wavelength and temperature. An ideal black body is defined as being perfectly absorbing for all wavelengths. Materials differ in their abilities to absorb electomagnetic energy of different wavelengths and can only approach the characteristics of a black body. Following the Stefan-Boltzmann law, the net transfer of radiant energy between two bodies is known to be proportional to the difference between the fourth powers of their absolute temperatures. It is noted that this law holds for radiant energy transfer between two locations within a material which are at different temperatures.

Glassmaking processes cover the temperature range from melting down through annealing and heat treating. It is convenient to consider heat transfer processes in the temperature range between 1500 and 500°C. Large tonnage glass melting, for containers and flat glass, is carried on by the use of radiant energy from gas and oil flames. To a first approximation, glass melting involves nearly steady state conditions. However from the time the glass leaves the conditioning chamber and is subjected to machine forming the heat transfer processes are transient. These processes are similar in principle for containers, flat glass, tableware, and so on. However the machine forming processes are quite different and

TABLE 17.1   Heat Transfer Data of a Typical Commercial Na_2O-CaO-SiO_2 Glass

Temperature ($^\circ$C)	Density, d (g/cc)	Specific heat [cal/(g)($^\circ$C)]		Thermal diffusivity, D (cm^2/sec)	Effective conductivity, K*[cal/(sec)(cm)($^\circ$C)]
		c_m	c		
500	2.474	0.2445	0.3136	0.007	0.005
700	2.438	0.2695	0.3397	0.015	0.012
900	2.398	0.2882	0.3594	0.030	0.026
1100	2.367	0.3031	0.3751	0.065	0.058
1300	2.344	0.3167	0.3893	0.130	0.119
1400	2.335	0.3264	0.3995	0.200	0.187

*Effective conductivity caclulated from the equation $K = Ddc$.

Data sources: Density from Sawai and Inoue (1940); specific heat from babcock and McGraw (1957); thermal diffusivity from Van Zee and Babcock (1951).

pose individual problems in transient heat transfer. Generally speaking heat transfer by conduction per se occurs up to about 500°C. At higher temperatures "radiation conductivity" increases in importance and predominates over pure conduction mechanisms. This complex mechanism can be designated as "effective conductivity" as defined in Table 17.1, which gives pertinent properties on a typical commercial $Na_2O\text{-}CaO\text{-}SiO_2$ glass. Information in this table forms the basis of our discussion on heat transfer in glass over the temperature range from 500 to 1500°C.

17.2 SPECIFIC HEAT

Specific heat, the thermal capacity of a substance per unit mass, is usually expressed in calories per gram per degrees C. Specific heat is usually understood to be measured at constant pressure. Figure 17.1 illustrates the difference between true specific heat c and mean specific heat c_m. True specific heat equals the slope of the heat content-temperature curve at a given temperature; that is, $c = dH/dT$. It is the quantity of heat required to change the temperature of a unit mass of glass from T to $T \pm 1°$. The mean specific heat c_m is the average

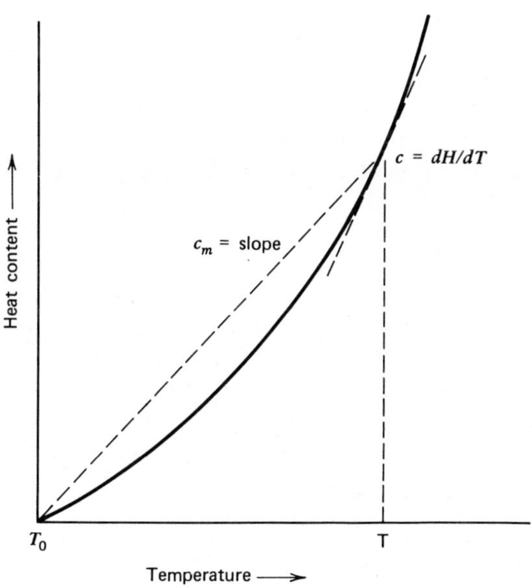

Figure 17.1 True and mean specific heats.

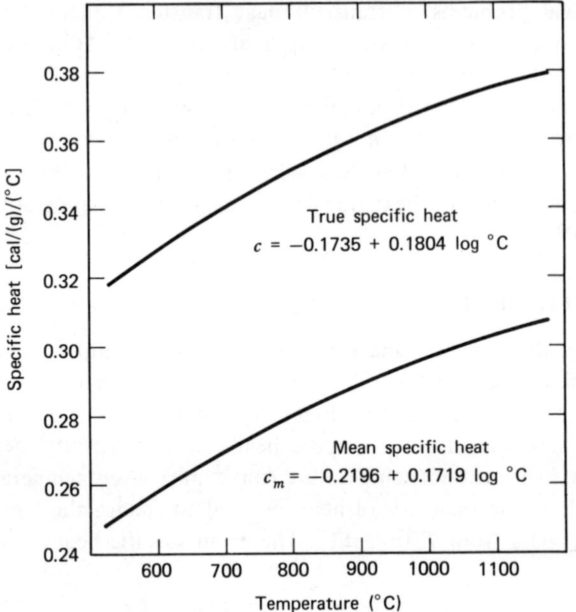

Figure 17.2   Specific heats of Na_2O-CaO-SiO_2 glasses.

quantity of heat, per unit mass per degree, required to raise the temperature from T_0 to T. It is usually determined by the method of mixtures, which involves measuring the difference in heat contents between some elevated temperature and that of the calorimeter. Goranson (1942) describes the method of converting mean specific heat data at constant pressure to true specific heat as a function of temperature.

Figure 17.2 plots specific heat-temperature data given in Table 17.1. These data can be accurately represented by the following equations:

$$c = -0.1735 + 0.1804 \log_{10} (^{\circ}C) \tag{17.1}$$
$$c_m = -0.2196 + 0.1719 \log_{10} (^{\circ}C) \tag{17.2}$$

Review of published data shows that equations of this form may be used to represent specific heat data on all silicate glasses. For example, data on c_m for vitreous silica given by Sosman (1927) can be represented by the equation

$$c_m = -0.013 + 0.09 \log_{10} (^{\circ}C) \tag{17.3}$$

TABLE 17.2   Oxide Factors for Calculation of Mean Specific Heat of Glasses

Oxide	a	c_0
SiO_2	0.000468	0.1657
Al_2O_3	0.000453	0.1765
CaO	0.000410	0.1709
MgO	0.000514	0.2142
K_2O	0.000335	0.2019
Na_2O	0.000829	0.2229
B_2O_3	0.000635	0.198
SO_3	0.00083	0.189
PbO	0.000013	0.049

Data source: Sharp and Ginther (1951).

This equation represents the experimental values of c_m within 0.001 from 400 to 1400°C.

In his review of specific heat of glasses, Morey (1954) notes the lack of studies on systematically arranged compositions. However several investigators have published factors for most glassmaking oxides which can be used to estimate specific heats. The most useful of these studies was performed by Sharp and Ginther (1951) who made a critical review of effects of temperature and composition from published data. They used the following equation to represent mean specific heat:

$$c_m = aT + c_o \big/ 0.00146T + 1 \tag{17.4}$$

Where a and c_0 are constants given in Table 17.2 and T is temperature in °C. They state that accuracy is generally better than 1% in the range from 0 to 1300°C. Following their method, c_m for vitreous silica would be represented by

$$c_m = 0.000468T + 0.1657 \big/ 0.00146T + 1 \tag{17.5}$$

The equation for mean specific heat for a glass containing 75% SiO_2, 10% CaO, and 15% Na_2O would be determined as follows;

$$SiO_2 \quad 0.75 \times 0.000468 = 0.0003510$$
$$CaO \quad 0.10 \times 0.000410 = 0.0000410$$
$$Na_2O \quad 0.15 \times 0.000829 = 0.0001244$$
$$a = 0.0005164$$
$$SiO_2 \quad 0.75 \times 0.1657 = 0.1243$$
$$CaO \quad 0.10 \times 0.1709 = 0.0171$$
$$Na_2O \quad 0.15 \times 0.2229 = 0.0334$$
$$c_o = 0.1748$$

Therefore the equation for mean specific heat of this glass is

$$c_m = 0.0005164T + 0.1748 \,/\, 0.00146T + 1 \qquad (17.6)$$

Sharp and Ginther further describe equations for calculating true specific heats. The interested reader is referred to their paper for details. This method, which considers glasses as mutual solutions of oxides on the basis of the random network theory, is recommended for calculating specific heats of glasses. Examination of values of oxide constants given in Table 17.2 indicates that specific heat is not very sensitive to changes in silicate glass compositions.

17.3 ABSORPTION AND EMISSION OF RADIATION

Spectral absorption of vitreous silica and silicate glasses in the infrared region has been discussed in Section 9.3. Several investigators have studied the effects of wavelength and temperature on absorption, transmission, and reflection in this spectral region. Typical publications are those of McMahon (1951), Genzel (1951), Neuroth (1952, 1953), and Grove and Jellyman (1955). Others have studied applications to glassmaking processes, including Kellet (1952, 1953); Czerny, Genzel, and Heilman (1957), and Gardon (1961).

McMahon (1951) studied thermal radiation characteristics of several commercial glasses in the range 3 to 12 microns and at temperatures from 500 to 1500°F (260 to 816°C). Figure 17.3 shows emissivity, apparent transmissivity, and apparent reflectivity of a soda-lime-silica glass at 1000°F (538°C) for a glass thickness of 1/8 in. The reflection peak at 9.5 microns is associated with the Si-O-Si stretching frequency and corresponds with the decrease in emissivity. This curve is typical of all silicate glasses with the exception of relatively minor differences in the near infrared caused by specific absorptions of FeO and so on. The temperature dependence of emissivity and transmissivity was very small and almost independent of wavelength. Transmissivities of silicate glasses are essentially zero at wavelengths beyond 5 microns. A black body may also be defined as one whose reflectivity is zero for radiation of all frequencies. The high emissivity

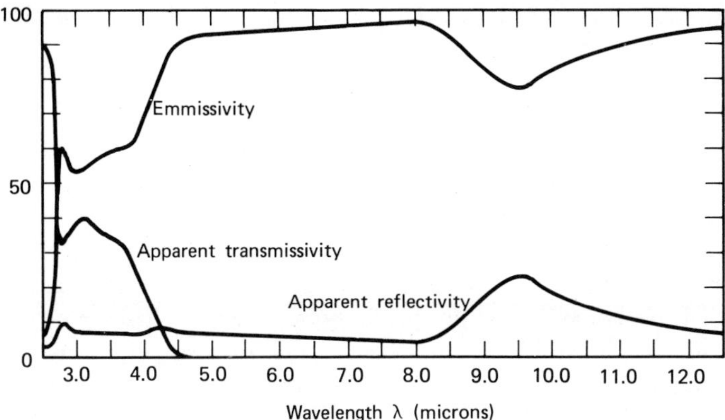

Figure 17.3   Radiation characteristics of Na_2O-CaO-SiO_2 glass. From McMahon (1951).

and low reflectivity of silicate glasses indicate their approach to properties of a black body.

Sorting out the various heat transfer mechanisms that occur in a transparent or semitransparent material like glass is more difficult than dealing with heat transfer in an opaque material. The thickness of the glass sample under consideration is of importance. Energy leaving the glass surface is the sum of several components. Portions arising from various depths within the glass travel to the surface where part is reflected and the rest emitted. Interaction of simultaneous emission and absorption of radiation throughout the volume leads to dependence of emissivity of tranparent glasses on their thickness. Thus heat transfer problems dealing with glass thicknesses of a meter or more in the melting chambers of glass tanks are quite different from those posed by glass thicknesses in millimeters in glass forming processes. An additional difference must be pointed out. Though heat transfer processes in melting chambers approach steady state conditions, those in forming processes are definitely time-dependent or transient. The latter are most difficult to describe analytically.

The effects of emissive or radiating powers of several commercial glasses were related to temperature gradients in glass tanks by Holscher, Rough, and Plummer (1943). They measured temperature gradients in a small experimental furnace using a base glass containing 74% SiO_2, 10% CaO, and 16% Na_2O, to which small amounts of colorants were added to match amounts used in commercial glasses. The added colorants FeO, Fe_2O_3, Cr_2O_3, and MnO increased the radiating powers of the glasses because of their absorption in the very near infrared region. For example, FeO strongly absorbs at a wavelength of 1.05 microns. The radiating powers and temperature gradients increased with amounts of absorbing constituents. The investigators measured temperature gradients in large commercial melting tanks and related them to experimentally determined values, as in Figure 17.4. The glass designated "flint" is an essentially colorless glass.

Burch and Babcock (1938) observed that during the manufacture of bottles of the same size and shape, higher machine speeds could be obtained on colored glasses such as emerald green and amber. A laboratory test was devised to measure the relative cooling rates of several colored and colorless glasses. Amber glass cooled about 5% faster than the "flint" glass, and the cooling rate of emerald green was about 8% faster. Commercial production figures showed that machine speeds followed this relative order and were almost in quantitative order. Results indicated the relations, even though the types compared had identical viscosity characteristics. This test was used in conjunction with viscosity measurements to closely estimate machine speeds with different compositions of glasses, both colored and colorless. These data suggest that heat loss in machine forming principally involves radiative properties of the glasses.

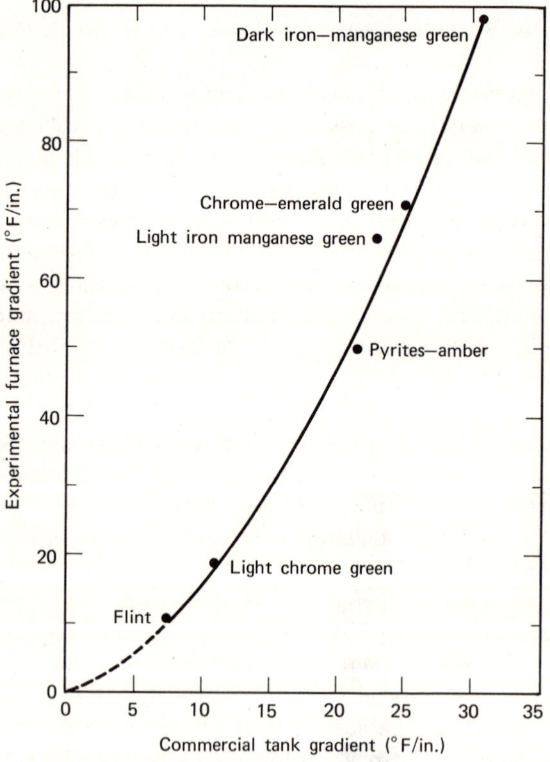

Figure 17.4 Temperature gradients for experimental and commercial furnaces. From Holscher, Rough, Plummer (1943).

17.4 THERMAL DIFFUSIVITY

Thermal diffusivity is equal to the ratio of the thermal conductivity of a substance to the product of the density and true specific heat. It determines the rate at which a temperature wave is propagated in a conducting substance and can be defined by the equation

$$K = Ddc \tag{17.7}$$

where K = thermal conductivity $[\text{cal}/(\text{sec})(\text{cm})(^{\circ}\text{C})]$, D = diffusivity (cm^2/sec), d = density (g/cc), and c = true specific heat $[\text{cal}/(\text{g})(^{\circ}\text{C})]$, as indicated in Table 17.1. Heat transfer at high temperatures covering the glass manufacturing range is effected by conduction per se, augmented by radiation within the glass. Therefore thermal conductivity in terms of equation (17.7) can be considered as the effective conductivity.

Van Zee and Babcock (1951) measured thermal diffusivity data on two typical commercial soda-lime-silica glasses in the range from 700 to 1400°C. The glass sample was contained in a long crucible having cylindrical symmetry. A sinusoidal temperature variation was superimposed on the mean temperature of the sample. Measurements were made on the time lag for the heat wave to travel radially from a point on the circumfrence to the center of the sample; a distance of 3 in. Cycle times from 1 to 4 hours had no effect on diffusivity values, as illustrated in Figure 17.5 for glass number 2. The 1951 paper gives detailed descriptions of furnace construction, method of controlling sinusoidal heat input, methods of measurement, and related mathematical theory.

This work was carried out before any theoretical analysis was available and clearly demonstrated the very rapid increase in effective conductivity with temperature. The diffusivity data on glass number 5 have been used with data on density and true specific heat to calculate values of effective conductivity given in Table 17.1. Examination of the data on thermal diffusivity D and effective or radiative conductivity K reveals that both $\log D$ and $\log K$ are linearly related to temperature.

The work published in 1951 was followed by a number of studies on the application of effective conductivity to practical heat transfer problems in glassmaking. The data of Van Zee and Babcock were determined on a 3 in. thickness of molten glass. Glass thicknesses in melting chambers are of the order of a meter or more, whereas those in glass forming are of the order of millimeters. Charnock (1961) measured diffusivity on plate glass compositions using a similar method, with results appearing in Figure 17.6. He reviewed papers dealing with applications to glassmaking, including those of Kellet (1952), Czerny and Genzel (1952), Geffcken (1952), and Genzel (1953). He concluded that measurements on radiation or effective conductivity could be used for steady state heat transfer as in melting tanks and presented substantiating mathematical theory. He

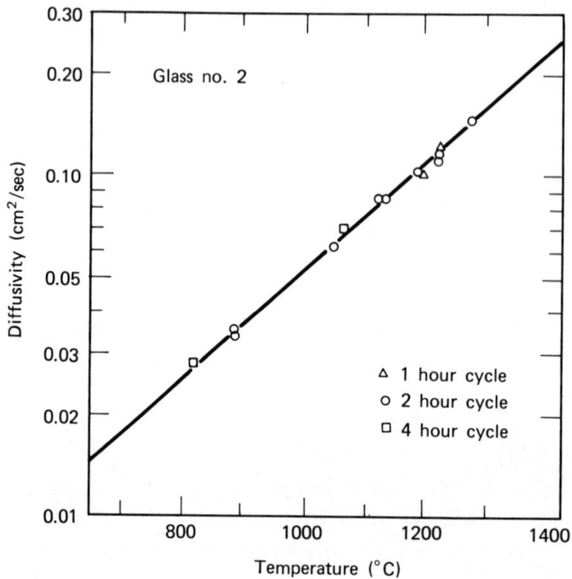

Figure 17.5 Thermal diffusivity of $Na_2O\text{-}CaO\text{-}SiO_2$ glass. From Van Zee and Babcock (1951).

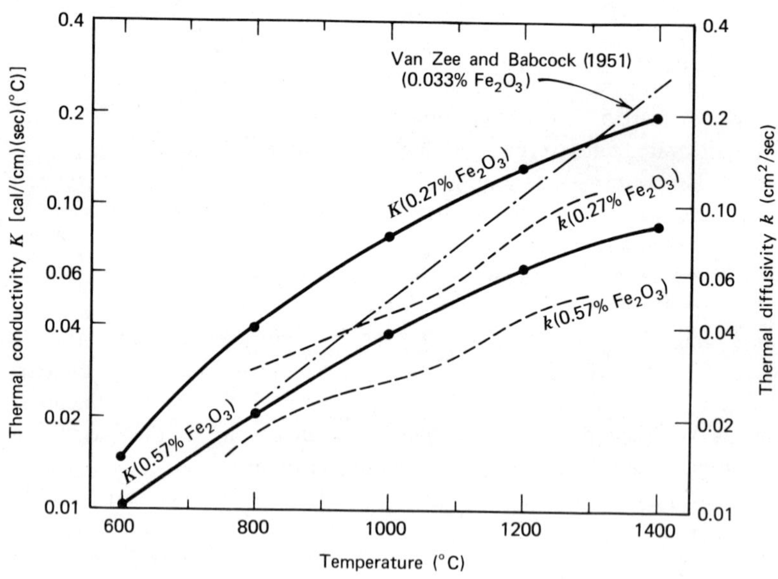

Figure 17.6 Measured thermal diffusivity and calculated effective thermal conductivity of glasses. From Charnock (1961).

272

further outlined a method of applying the data to transient conditions within a large volume of glass.

A number of papers of interest were presented at an international symposium on heat transfer phenomena in glass organized by the author, (Babcock, 1961). The symposium outlined some of the difficulties encountered in applying experimental data and heat theory to practical glass problems. These problems are being continually reviewed by glass manufacturers in terms of their own glassmaking processes. Many of these processes are proprietary and seldom made public. Thus detailed discussion of such applications to glassmaking are outside our present interest, and subsequent discussions deal only in general qualitative terms with heat transfer.

18

Thermal Expansion

The expansion of glass with temperature has long been of interest in the usage of glass products. The change in dimensions of thermometers and changes in volume of glass vessels have led to accurate measurements of coefficients of expansion of glasses. The making of glass-glass seals in bifocal lenses and construction of scientific apparatus involves coefficients of thermal expansion of the glasses being welded together. The making of glass-metal or glass-ceramic seals requires that the joining materials have closely the same coefficients of expansion.

The coefficient of expansion also plays an important role in certain stages of glass manufacturing, especially the machine forming of glass products.

18.1 DEFINITIONS AND NOMENCLATURE

The thermal expansion of a substance may be qualitatively understood in terms of asymmetrical vibrations of atoms or ions about their positions of minimum potential energy. Figure 1.1 indicates that under the application of increased temperature, the outgoing semiamplitude is greater than the ingoing one. This can be assumed to account for the continuous expansion of a substance with continuous increases in temperature.

The coefficient of linear expansion of a substance is the increase in length of a bar of unit length when heated through unit difference of temperature. Data are usually expressed in terms of the mean linear coefficient of expansion α—the increase in length of the sample, divided by the original length, when heated over a given temperature interval. Neglecting terms of higher order than the first, the coefficient of volume expansion is 3α. The rate of increase in length of a glass sample increases slowly with temperature from room temperature up to the transformation region of the given glass, after which the rate increases quite rapidly (Figure 18.1).

Coefficient of expansion measurements may be either volumetric or linear, depending on how the results are to be employed. Volumetric measurements employ a bulb of glass containing mercury in testing thermometer glasses. Linear measurements on glass may be made by the accurate interferometer method described by Peters and Cragoe (1920) or in a simple dilatometer apparatus

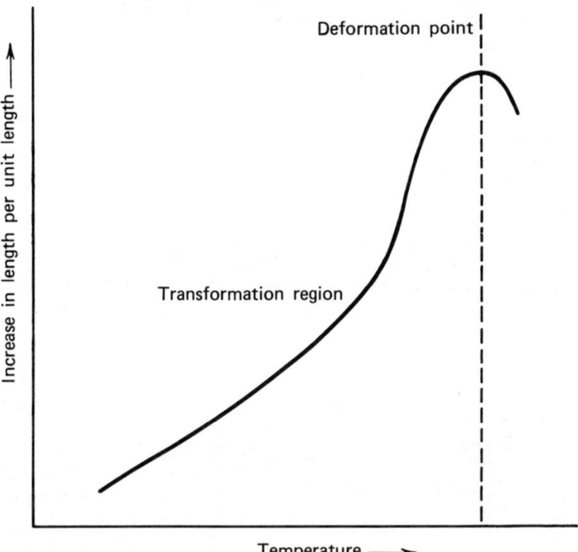

Figure 18.1   Expansion curve of glass.

wherein the increase in length of a glass rod is measured relatively to vitreous silica. Ways of analytically expressing the data are further dependent on needs. In most cases the mean coefficient of expansion is sufficient, but some applications may require accurate representation of the curve in terms of equations requiring three or more constants.

The rate of linear expansion of glass with temperature is dependent on heat treatment carried out in the annealing range and on glass oxide composition. Rapidly cooled glass has a higher coefficient of thermal expansion than annealed glass. Likewise the higher the fictive temperature at which the glass is stabilized, the higher the coefficient of expansion. Treatment of this subject is delayed until we cover annealing, because all glass properties are conditioned by annealing.

18.2   EFFECTS OF GLASS COMPOSITION

The coefficients of thermal expansion of glasses are strongly dependent on glass compositions, as demonstrated in Table 18.1, which lists coefficients for selected commercial glasses. The range of coefficients from vitreous silica at 5.5 to that for an iron sealing glass at 124 covers usual commercial glasses. Coefficients of expansion in the range 0 to 300°C relate specifically to the use of glasses in commercial products. Vitreous silica products are characterized by their very high resistance to thermal shock, as indicated by their low coefficients of ex-

TABLE 18.1   Thermal Expansion Coefficients of Selected Commercial Glasses

Oxide	Compositions (wt %) of six glasses					
	7940	7900	7740	1720	0080	1990
SiO_2	99.9	96.0	81.0	62.0	73.0	41.0
H_2O	0.1					
Al_2O_3		0.3	2.0	17.0	1.0	
B_2O_3		3.0	13.0	5.0		
Li_2O						2.0
Na_2O			4.0	1.0	17.0	5.0
K_2O						12.0
CaO				8.0	5.0	
MgO				7.0	4.0	
PbO						40.0
Mean linear expansion coefficient in the range 0-300°C						
$\alpha \times 10^7/°C$	5.5	8	33	42	92	124

Data source: *H*utchins and Harrington (1966).

pansion. Thermal shock resistance decreases as these coefficients increase. Thermal shock resistance of glass products depends to a lesser extent on the architecture and wall thickness of the items. Generally speaking, glasses with coefficients up to 45 or so are suitable for baking, top-of-the-stove ware, ignition tubes, and other products that are subjected to rapid temperature changes. Glasses of the soda-lime-silica type with coefficients near 90 are not suitable for such applications.

The review of Morey (1954) shows that coefficients of thermal expansion over the range in which the linear expansion is approximately a linear function of temperature are additive in terms of oxide composition. Most published data are in this temperature range and are applicable to glass product usage. Extensive published data are available on thermal expansion and its relations with temperature and composition. Methods of measurement are not uniform with respect to type of apparatus, temperature interval, size of glass sample, heating rate of sample, and precision of measurement. In view of these differences, the separate studies cannot be incorporated in a comprehensive analytical scheme. The best procedure is to consider the data of a given investigator to be correct relative to one another and then, in so far as possible, to compare different sets of data in terms of similarity of temperature interval and so on. Space does not permit dealing with available data in this manner. The selected sets of data reviewed here cover the most used glass oxides in terms of different analytical methods of expressing the data.

Na_2O-CaO-SiO_2 and Na_2O-Al_2O_3-SiO_2 Glasses.

Schmid, Finn, and Young (1934) and Faick, Young, Hubbard, and Finn (1935) have made thermal expansion measurements on these glasses. Both groups of authors found that their data could be represented by equations of the type

$$E = aA + bB + cC + dD \tag{18.1}$$

where E is linear expansion in microns per centimeter, A, B, C, and D are weight percentages of the oxides, and a, b, c, and d are the respective constants of the oxides.

Schmid, Finn, and Young determined that thermal expansion of Na_2O-CaO-SiO_2 in the range 25 to 400°C could be represented by the equation

$$E = a\,SiO_2 + b\,Na_2O + c\,CaO \tag{18.2}$$

where

$$a = 0.00036t - 0.00000036t^2$$
$$b = 0.00245t + 0.0000046t^2$$
$$c = 0.00116t + 0.0000010t^2$$
$$t = temperature - 25°C$$

Equation (18.2) gave good representation of the measured data.

Faick, Young, Hubbard, and Finn found that the following equation gave good representation of their data in Na_2O-Al_2O_3-SiO_2 glasses in the range from 25 to 450°C.

$$(18.3)$$

$$E = aSiO_2 + bNa_2O + dAl_2O_3$$

where

$$a = 0.00034t - 0.0000003t^2$$
$$b = 0.00274t + 0.0000035t^2$$
$$c = 0.00001t - 0.0000003t^2$$
$$t = \text{temperature} - 25°C$$

The constants for SiO_2 and Na_2O in equations (18.2) and (18.3) are nearly the same. The authors point out that calculated values of coefficients of thermal expansion of binary Na_2O-SiO_2 glasses are nearly the same for both sets of constants. Both groups of authors determined the equations by the method of least squares, which was laborious at the time the work was done. Modern objective computer programs can of course make such determinations in a matter of seconds once the program is set up.

The method of representing expansion equations through the temperature range of measurements used by the authors just cited is highly recommended. The mean coefficient of expansion for any temperature range within the interval can then be readily determined. Unfortunately few published studies express the data in this manner. It is noted that accurate measurements made by the foregoing authors employed the interference method described by Merritt (1933) with the interferometer enclosed in a furnace.

Na_2O-CaO-Al_2O_3-SiO_2 Glasses.

Silverman (1940 measured coefficients of expansion on 67 glasses in this series in the temperature interval from 80 to 170°C using the interferometer method. The following oxide ranges in weights percent were covered in this study: SiO_2, 60 to 79; Al_2O_3 contents of 0.2, 2.2, 4.2, 6.2, and 8.2; CaO contents of 6, 8, 12, 14, and 16; and Na_2O contents of 11, 12.5, 14, 15.5, 17, and 19. He developed oxide factors to represent the measured data in terms of weights percent of the oxides.

$$\alpha = 0.28\ SiO_2 + 0.24\ Al_2O_3 + 1.36\ CaO + 3.84\ Na_2O \qquad (18.4)$$

where α is in centimeters per centimeter per °C. The average difference between measured and calculated coefficients is 1.1 and the maximum difference is 3.0.

Thus his constants give good representation of the data.

These data may be represented in terms of the author's substructure postulate. Fifty-seven of the glasses are located in three primary phase fields and are distributed as follows; 16 in the tridymite field, 10 in the devitrite field, and 31 in the wollastonite field. Equations representing separate data in these phase fields appear in Table 18.2. The equation representing the combined group of 57 glasses is also given. Reference to figures on average and maximum differences between measured and calculated values also indicate good representation. From a practical standpoint there is little to choose between the two methods. However the substructure method does indicate that the oxides make different contributions in the different primary phase fields.

Silverman also made measurements on the deformation point, which is the temperature at which viscous flow exactly counteracts thermal expansion. This viscosity in the apparatus used is near $10^{11 \cdot 5}$ poises. Generally speaking glasses cannot support internal strain above the deformation point. Silverman made checks of deformation temperatures and found them to be accurate to within $5^{\circ}C$. Table 18.2 permits determination of deformation temperatures of glasses in separate phase fields and when combined in one group of 57 glasses. The figures on differences between measured and calculated temperatures show good representation of the data in terms of the substructure method and display the merits of this approach over the random network method of combining all the glasses in one group.

Na_2O-SiO_2 Glasses.

Turner and Winks (1930) measured thermal expansion on nine glasses in this series over the temperature intervals 0 to $130^{\circ}C$, 130 to $250^{\circ}C$, and 250 to $350^{\circ}C$, using a differential dilatometer apparatus. These glasses, which lie in the two phase fields of $Na_2O \cdot SiO_2$ and $Na_2O \cdot 2SiO_2$, have been reviewed in terms of the substructure postulate with the results given in Table 18.3. Plots of their data versus temperature demonstrate that these relations are linear within the given phase fields. This is borne out by the linear equations in Table 18.3. According to the figures on average and maximum differences between measured and calculated temperatures, separate equations give better representations than are obtained when data on the nine glasses are combined in one group. The two straight lines in a plot of coefficient of expansion versus composition intersect near the eutectic between the $Na_2O \cdot SiO_2$ and $Na_2O \cdot 2SiO_2$ primary phase fields. Following Morey (1954), this eutectic is at 62.1% or 0.628 mole fraction of SiO_2. This composition can be determined analytically by simultaneous solution of appropriate equations in the $Na_2O \cdot SiO_2$ and $Na_2O \cdot 2SiO_2$ phase fields, together with the equation $SiO_2 + Na_2O = 1$, to be 0.646 SiO_2 for the 0 to

TABLE 18.2   Linear Coefficients of Thermal Expansion α and Deformation Temperatures of Na_2O-CaO-Al_2O_3 Glasses

$$\alpha = A\ SiO_2 + B\ Al_2O_3 + C\ CaO + D\ Na_2O$$
$$\text{Deformation temperature} = A\ SiO_2 + B\ Al_2O_3 + C\ CaO + D\ Na_2O$$

Phase field	A	B	C	D	Differences between measured and calculated temperatures ($^\circ$C)	
					Average	Maximum
Expansion coefficient $[cm/(cm)(^\circ C)]$						
Tridymite	23.533	−94.298	122.629	428.048	0.85	1.6
Devitrite	17.326	−4.082	169.764	414.764	0.62	1.4
Wollastonite	28.658	4.748	129.853	381.169	0.79	1.6
Combined	27.520	−22.503	124.708	398.548	1.84	2.5
Deformation temperature $(^\circ C)$						
Tridymite	607.539	1384.630	951.281	63.960	0.99	2.7
Devitrite	574.417	1291.932	911.774	251.497	1.04	2.2
Wollastonite	643.526	1342.237	877.350	−71.870	1.95	5.3
Combined	629.965	1318.542	875.585	6.865	4.60	8.4

Data source: Silverman (1940).

TABLE 18.3   Linear Coefficients of Thermal Expansion α of Na_2O-SiO_2 Glasses

		$\alpha = A\ SiO_2 + B\ Na_2O$		
Phase			Differences between measured and calculated temperatures ($^\circ$C)	
field	A	B	Average	Maximum
0-130°C				
$Na_2O \cdot SiO_2$	46.44	318.42	1.3	2.5
$Na_2O \cdot 2SiO_2$	24.71	358.00	1.6	2.4
Combined	33.49	336.59	1.7	3.4
130-250°C				
$Na_2O \cdot SiO_2$	74.69	334.20	2.5	3.5
$Na_2O \cdot SiO_2$	10.01	441.32	0.7	1.1
Combined	29.14	394.67	3.2	8.0
250-350°C				
$Na_2O \cdot SiO_2$	40.21	427.58	2.1	2.7
$Na_2O \cdot 2SiO_2$	9.27	477.22	0.8	2.0
Combined	17.37	457.94	2.0	4.9

Data source: Turner and Winks (1930).

130°C interval. In a like manner the compositions of eutectics for the 130 to 250°C and 250 to 350°C interval data are 0.623 and 0.618 SiO_2, respectively, the average of the three being 0.628 SiO_2. Here again the substructure method discloses quite different contributions of the oxides in the two phase fields.

The foregoing examples indicate the feasibility of using the substructure method to accurately represent data on coefficients of thermal expansion.

V

Glassmaking Processes

Still, however, the highest value is set on glass that is nearly colorless or transparent, as nearly as possible resembling crystal. For drinking vessels glass has quite superseded the use of silver and gold.

Pliny the Elder (c. AD 23-79)

Commercial glassmaking deals with the high temperature conversion of raw materials into a homogeneous molten liquid capable of being fabricated into useful glass products. Manufacturing processes for different types of product vary widely in terms of homogeneity and property requirements. Modern large tonnage glassmaking is highly mechanized and strongly dependent on rapid heat exchange processes during all stages of production. Simultaneous rapid changes in glass properties, especially viscosity, exert further controls on fabrication processes.

Part V outlines methods used in control of the melting, glass-forming, and annealing stages of large tonnage glass manufacturing. Information presented in Parts I to IV is related to these manufacturing stages. The methods outlined can be adapted to the making of all types of glass. The processes involved differ only in details of heat exchange processes, glass property changes with temperature, and glass quality.

19

Glass Melting

Glass melting processes vary widely with respect to the kind of product being manufactured. The homogeneity of the glass in terms of properties, bubbles, and stria must be of the highest quality for optical glasses. Such specifications call for exacting choices of melting pot or tank, raw materials, and quality controls. Homogeneity requirements for commonly used glass products are more flexible and specifications less exacting. Glass homogeneity requirements are generally the same all over the world. This has been brought about by international competition and the adoption of more uniform quality standards and nomenclature This effort has been promoted through publications and meetings of the International Commission on Glass, which has headquarters in Brussels. American glass practice is described in a number of publications including those of Phillips (1948), Shand (1958), Tooley (1974), and Scholes (1975). Further reference is made to the *1976 Annual Book of ASTM Standards,* which gives accepted definitions of terms relating to glassmaking and glass products.

19.1 MELTING TANKS AND FURNACES

The type of melting tank depends primarily on the degree of homogeneity required and the amount of glass needed for a given product. Three general types of tank can be distinguished.

1. Clay pots, either open or covered, are used for production of relatively small amounts of special glasses such as optical and colored products. The pots are heated indirectly. Optical glass manufacturers are making increased use of continuous platinum or platinum-lined tanks.

2. The day tank, holding several tons, is so named because it may be charged one day and used on the next. Day tanks are used for limited production runs and are usually heated directly by gas or oil.

3. Most glasses are produced in continuous tanks heated directly by gas or oil. They usually consist of a melting chamber and a cooling and conditioning chamber, from which the glass flows into feeders or forehearths and then to glass machines in the case of container manufacturing. In sheet and plate glassmaking the glass flows from the conditioning chamber to the drawing mechanism.

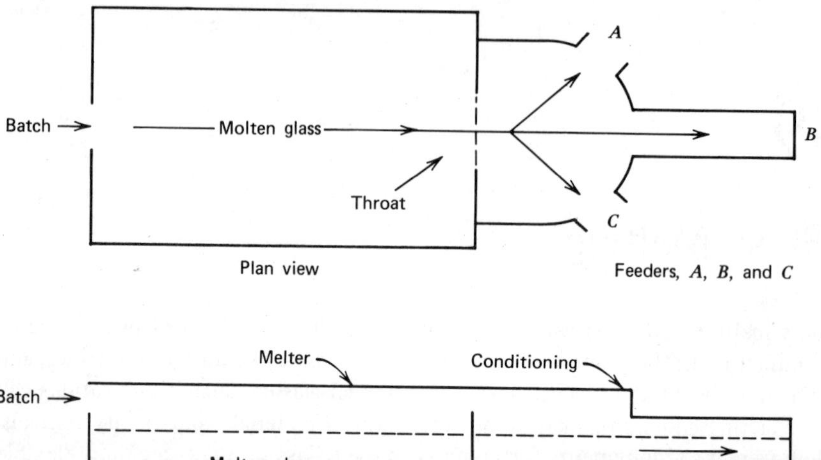

Figure 19.1 Functional sketch of commercial glass furnace.

Figure 19.1 is a functional sketch of a continuous furnace used for production of hollow ware such as containers. These furances have capacities up to 300 tons and the outputs in tons per day may approach the same figure. Continuous furnaces for sheet and plate glass manufacture are of the same general design but much larger. They have capacities of some 1000 tons or so, and outputs per day are about one-fourth their capacities. This low output to capacity ratio is compatible with glass quality requirements of sheet and plate glasses. Continuous furnaces are heated by radiant energy from gas or oil flames. Glass temperatures in the melting chamber are usually maintained near viscosity levels of 10^2 poises of the glasses being melted. The glass flows readily at this viscosity down through the throat on the furnace floor and into the conditioning or cooling chamber, where the viscosity becomes higher. The glass continues to cool as it enters one of the feeders leading to glass machines. The glass as it comes from the feeder and is cut into uniform gobs for machine forming usually has a viscosity level near 10^3 poises.

Melting furnaces vary in design and construction necessary to meet product and glass quality specifications. Detailed information on the subject can be found in Devillers and Waerewyck (1950), Gunther (1958), Spain (1958), and Tooley (1974).

19.2 RAW MATERIALS

The use of raw materials in the formulation of silicate glasses has been covered in previous discussions. Suppliers of raw materials to the glass industry are well equipped to furnish an extensive variety of materials subject to specifications of chemical purity and grain size. Most of the raw materials come from beneficiated natural minerals, all containing minor amounts of ingredients termed "impurities," which vary with glass type. The transition elements are impurities in the making of optical glasses because they produce color. The same elements are not impurities in green and amber hollow ware; in fact their amounts are augmented in some cases. It is most important that impurities in supplied raw materials be held at constant levels for continuous glass production. It is usual for natural minerals to have varying amounts of minor elements even at the same mining locations; amounts at different locations can vary appreciably. Thus raw materials quality controls must be instituted by the supplier and the glass manufacturer.

Continuous production requires the use of materials of uniformly fine grain sizes. Grain sizes must be under continuous control by the supplier and the manufacturer. The ASTM specifies the use of the following U.S. Standard sieves for this purpose:

No. 8	(2.38 mm)	No. 50	(297 microns)
No. 12	(1.68 mm)	No. 70	(210 microns)
No. 16	(1.19 mm)	No. 100	(149 microns)
No. 20	(841 microns)	No. 140	(105 microns)
No. 30	(595 microns)	No. 200	(74 microns)
No. 40	(420 microns)		

Detailed instructions for performing sieve analyses for glassmaking raw materials are given by ASTM.

Raw materials for continuous production are mixed, weighed, and fed automatically into the melting furnace. In many cases these operations are carried out under computer control. The glass manufacturer usually makes periodic checks on chemical analyses and grain sizes in monitoring these operations. Tooley (1974) and others cover raw materials and raw materials handling in detail.

19.3 MELTING PROCESSES

Commercial glass melting deals with the high temperature conversion of raw materials into a homogeneous molten liquid capable of being fabricated into useful glass products. It is a complex process involving chemical and physical reactions between the raw materials being melted, and these are rapidly influenced by properties of the glass being produced. The processes are affected to minor extents by volatilization of some materials and by relatively slow solution of furnace parts and refractories in contact with the glass. For convenience of discussion, the melting process can be described in terms of the melting, fining, and conditioning stages. Glassmaking practice and experimental data show these stages to be quite interdependent.

Melting.

The melting stage implies the complete conversion of crystalline raw materials into molten glass free of crystals. At this point the reader understands that the melting points of quartz sand and most other single batch materials are higher than temperatures attained in glass melting furnaces. The melting points of combinational oxides in glasses in terms of liquidus temperatures are given in the primary phase diagrams of Levin, Robbins, and McMurdie (1964, 1969). Commercial melting temperatures must be distinguished from the melting points of oxide combinations as given by their liquidus temperatures. Commercial melting temperatures must be high enough to attain a low viscosity level needed to carry

TABLE 19.1   Data on a typical Na_2O-CaO-Al_2O_3-SiO_2 Glass*

	Composition	
Oxide	wt%	mole fraction
SiO_2	74.5	0.7508
Al_2O_3	2.1	0.0125
CaO	7.9	0.0853
Na_2O	15.5	0.1514

Glass properties
Liquidus temperature = $1004°C$
Log viscosity at liquidus = 4.15
Log 2 viscosity = $1485°C$
Log 3 viscosity = $1204°C$
Log 5 viscosity = $907°C$
Log 7 viscosity = $752°C$
Surface tenstion $(1500°C)$ = 310 dynes/cm

*This glass is located in the tridymite primary phase field.
The above measured data, supplied through the courtesy of Owens-Illinois, Inc., can be compared to calculated values using tridymite equations from Table 14.3, 14.4, 14.5, and 16.1.

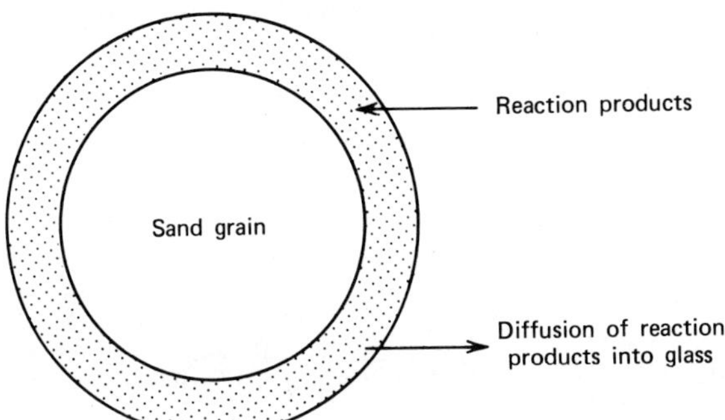

Figure 19.2 Diffusion of reaction products in melting.

on the melting process. Moreover glass temperatures throughout all stages of a continuous operation must be well above the liquidus temperature, to prevent formation of primary phase crystals. Reference to Table 19.1 shows the log 2 viscosity melting temperature to be about 500°C above the liquidus temperature and 200°C above the log 3 viscosity temperature, where the glass normally leaves the melting furnace. The high viscosity of $10^{4.15}$ poises at the liquidus temperature also prevents crystallization during the subsequent forming operations when the glass is cooled very rapidly as it passes through the liquidus temperature. This viscosity can be compared to that at the liquidus temperature of water, which is 10^{-2} poise.

Quartz sand, $CaCO_3$, and Na_2CO_3 batch materials are used in making Na_2O-CaO-SiO_2 glasses. During melting the CO_2 is given off as a gas, leaving only oxides in the mixture. At this stage the reaction products are $CaSiO_3$ and Na_2SiO_3. They react simultaneously with the grains of sand, which is the last material to go into solution. This is illustrated in Figure 19.2 and can be demonstrated in the laboratory by dropping a sand grain into molten glass. The sand grain is soon enveloped by a milky layer of reaction products. These reactions are enhanced by low interface surface tensions of the reacting complexes. The simultaneous diffusion of reaction products into the surrounding glass is dependent on the viscosity of the glass—specifically, the inverse of viscosity or fluidity. These physical and chemical reactions proceed until the sand grains are completely dissolved. The reactions can be speeded up by bubbling dry air through the floor of the melting chamber (Figure 19.1). As the air bubbles rise to the glass surface, they result in mixing of the reaction products and the upward movement of relatively cold glass. Thus radiant heat transfer in the melter is augmented by convection caused by bubblers and by normal flow patterns of the glass.

Refractory materials in contact with the glass give the best service if their Al_2O_3 contents are quite high. Slow solution of these materials into the glass may result in the presence of high Al_2O_3 stria or cords in the glass. Cords may also arise in glasses having high Al_2O_3 contents. In these cases the Al_2O_3 may be the last oxide to go into solution. Silverman (1942) made a study of the effects of surface tension on the persistence of cords in glasses. He drew fibers of glasses containing up to 8% Al_2O_3, placed them on the surfaces of glasses containing much lower amounts of Al_2O_3, and heated the combined sample in a laboratory furnace. In all cases the fiber or simulated cord maintained its identity, even at high temperatures when it sunk below the glass surface. Silverman attributed the results to the high surface tension of Al_2O_3. Refer in this connection to the high constant for Al_2O_3 given in Table 16.1. If the sand grain in Figure 19.2 is replaced by an aluminous cord, the formation of reaction products and their diffusion into the glass would be very slow due to high interfacial surface tensions.

Two different experimental approaches have been made to glass melting problems. These methods have been described in a number of publications. Empirical methods involving grain sizes of raw materials, temperatures, amounts of glass cullet in batches, and glass compositions, have received most attention. The following studies are typical of this approach. Potts (1939) used two methods of investigating rates of melting of glasses made from sodium carbonate, dolomite limestone, and silica sand: (1) batches were melted for a constant period of time, and the temperatures at which they became batch free were measured; (2) batches were melted at constant temperature, and the times at which they became batch free were measured. Potts (1941) measured melting rates of a series of Na_2O-CaO-SiO_2 as influenced by base compositions and minor constituents. Melts were made at $1427°C$ with measurement of times when the batches became batch free. Minor constituents included Na_2SO_4, BaO, B_2O_3, $NaCl$, $NaNO_3$, and CaF_2 in small amounts. These constituents are called refining agents, although they also play a part in melting per se. Review of these studies indicates that batch-free times decrease almost linearly as the viscosity decreases. Complete solution of quartz sand grains was taken as the criterion of complete melting. Potts, Brookover, and Burch (1944) measured melting rates of Na_2O-CaO-Al_2O_3-SiO_2 and Na_2O-CaO-MgO-Al_2O_3-SiO_2 glasses in terms of grain sizes of raw materials and additions of cullet at a temperature of $1427°C$ ($2600°F$). They concluded that the batch-free time is minimized if all batch materials have the same small grain size. The effectiveness of glass cullet additions was not conclusive and varied with glass composition. It is noted that large tonnage commercial batches usually contain some 15% of broken glass or cullet. Most glass manufacturers conduct tests of the type just described and find that the data can be qualitatively related to commercial production.

The second approach to glass melting is that of differential thermal analysis (DTA), which was developed by mineralogists. In this method the temperature of the raw material being examined is compared continuously to the temperature of an inert material while both materials are heated together in a furnace. As the sample undergoes a reaction, additional heat is evolved or absorbed. Thus

DTA curves can be used as fingerprints in identifying raw materials. Thermo-gravimetric methods in which weight changes occur during heating are also widely used. This more quantitative approach is useful for determining evolution or absorption of gases in certain materials. Evolved gases are identified using mass spectrometer methods. Reactions may result in solid reaction products that can be identified by x-ray diffraction methods. High temperature x-ray methods are useful in some cases. The studies of Oldfield (1958) and Wilburn, Thomasson, and Cole (1958) describe typical applications of this method.

Fining.

The fining stage has come to be described in terms of removal of various gases incident to the melting process. Studies by Dalton (1933) and Shadduck and Van Zee (1942) show that most glasses contain O_2, SO_2, H_2O, N_2, and CO_2, depending on glass type. Some of these gases may be dissolved in the finished glass product, and some may be found in small bubbles. Gases in bubbles may contain combinations of gaseous materials. The CO_2 present is due to the use of carbonates in the batch, whereas other gases, like SO_2, come from fining agents added to the batch. The removal and/or solution of these gases in one form or another is qualitatively related to viscosity and surface tension of the glass. Low values of viscous and surface tension forces enhance the fining process.

Conditioning.

Conditioning relates to homogenization of the glass in terms of uniform physical and chemical properties. Conditioning begins in the melting chamber and ends in the feeder. Bubblers through the bottom of the melter, stabilization of gases in the conditioning chamber, and physical mixing by platinum stirrers in the feeder help to accomplish this purpose.

Space does not permit discussion of melting methods of all types of glass. However the outlined physical and chemical principles are essential to all glass melting and are readily adaptable to other kinds of glass.

19.4 GLASS COMPOSITION CONTROLS

Wet chemical analysis methods were used for a number of years as means of glass composition control. These time-consuming techniques were eventually replaced by more rapid measurements of certain physical properties of glasses. In recent years such physical controls have been replaced in some respects by automated instrumental methods of dry chemical analyses. Both physical and chemical methods have their merits, depending on the type of glass under consideration.

Chemical Analyses.

Automatic instruments provide a practical means of composition control in terms of speed and operational costs. X-ray fluorescence is used to determine amounts of elements above atomic number 14. Alkali and alkaline earth ions can be quickly and accurately measured by flame spectroscopy methods. The automatic and quantitative emission spectrograph is also an important analytical tool.

Physical Property Measurements

Most manufacturers periodically measure physical properties as means of glass composition control. The time periods and properties measured vary with the type of product. Properties include density, refractive index, dispersion, spectral absorption, coefficient of thermal expansion, softening point, annealing point, strain point, and glass-metal seal stresses. Measurements of two or more properties are sufficient for good composition control. For example, lines of equal density and equal softening point for Na_2O-CaO-SiO_2 glasses cross at angles great enough to determine the composition from the two measurements. The reader can confirm this by examining oxide constants for specific volume (the inverse of density) and log viscosity = 7 given in previous tables. Optical glass manufacturers can control compositions by measuring refractive index and birefringence on proof samples taken from the melting pots or continuous melters. The practice depends on type of ware being made and customer requirements. Makers of colored television viewing screens, colored hollow ware, and so on, use CIE values of brightness, purity, and dominant wavelength. Several of these routine tests are described in the 1976 *Annual Book of ASTM Standards, Part 17.*

20

Glass Forming

Glass products used for utilitarian purposes were made by hand processes until about the end of the nineteenth century. Glasses for hollow ware were melted in pots and conveyed to simple machines by individual gatherings by hand-operated steel rods or blow pipes. Glasses for artistic purpose are still made in this manner. The invention and development of mechanized methods in the early 1900s revolutionized the making of large tonnage utilitarian glasses. The making of flat glasses underwent similar changes. Flat glasses were at first melted in pots. The molten glass was then poured onto a steel table or plate and rolled to the desired thickness. The flat glass industry is now highly mechanized. Molten glass delivered to the machines should be uniform in composition, temperature, and glass properties.

20.1 DEFINITIONS AND NOMENCLATURE

The wide variety of glass products requires the use of several forming processes, including pressing, blowing, combinational pressing and blowing, casting, rolling, and drawing. Glass manufacturing practice has demonstrated the great influence of viscosity on all glass-forming processes. The report by Cole and Babcock (1970) under auspices of the International Commission on Glass (ICG) reviews relations between viscosity and glassmaking practices. Table 20.1 presents these relations in outline form. Refer to Chapter 14 for a discussion of the effects of glass composition and temperature on viscosity. It is emphasized that glass-forming processes must be maintained under conditions in which the glass flows as a viscous liquid. The significance of this requirement was revealed in Chapters 12 and 15. The closely related properties of elasticity, viscosity, and relaxation time are involved in glass forming. If the machine process is speeded up so greatly that the glass cannot flow, the glass may react as an elastic solid and fracture of the glass will occur. The coefficient of thermal expansion of the glass is also important under such conditions. Glass fracturing under these conditions first appears on the glass surfaces because they cool first and take on the characteristics of an elastic solid. Heat transfer from the glass to the metal mold is of course involved. If the glass is too hot and the heat transfer is not rapid

TABLE 20.1   Viscosity Levels in Manufacturing Processes

Manufacturing process	Log viscosity range	Representative log viscosity points	
Melting and fining	1.7-2.0		2.0
Flow in furnace parts	2.0-2.5		2.5
Forming and afterworking	2.5-8.0		3.0
		W.P.	4.0
			7.0
		S.P.	7.6
Annealing and tempering	12.0-15.0	A.P.	13.0
		St.P	14.5

*W.P. = working point, S.P. = softening point, A.P. = annealing point, St.P. = strain point.

Source. Cole and Babcock (1970).

enough, the glass may stick to the mold, whereupon appropriate adjustments must be made in the operation.

The continuous forming of glass, whatever the process and product, must be compatible with glass composition with respect to liquidus temperature, properties of a viscous liquid, and heat exchange between the glass and its environment during the process. Surface tension and thermal expansion (discussed in Chapters 16 and 18, respectively) also play important roles in glass forming. Under the influence of surface tension forces, the glass tries to assume a shape having a minimum surface for a given volume. These forces tend to smooth out uneven contours in reproducing details in glass finishing molds. In other words they hinder the achievement of desired sharp patterns on the finished surface of the product (e.g., lettering). High coefficients of thermal expansion contribute to surface fracture or checks in the product, particularly in the final stages of forming.

It is apparent from Chapter 17 that heat transfer constitutes a major problem in glassmaking. Molten glass from the melter through the feeder is under approximately steady state conditions. As soon as the glass leaves the feeder (or in case of flat glassmaking, as soon as it leaves the conditioning chamber), it is subject to transient heat transfer until the forming processes are completed.

Glass-forming practice and experimental studies reveal the difficulties of dealing with this type of heat transfer. The difficulties arise from two essential causes: (1) the rapid drop in effective thermal diffusivity as the temperature of the glass sharply decreases, and (2) discontinuities exist in radiative and conductive properties between the glass and its surroundings. In hollow ware production the principal heat exchange occurs between the hot glass and the cooler mold

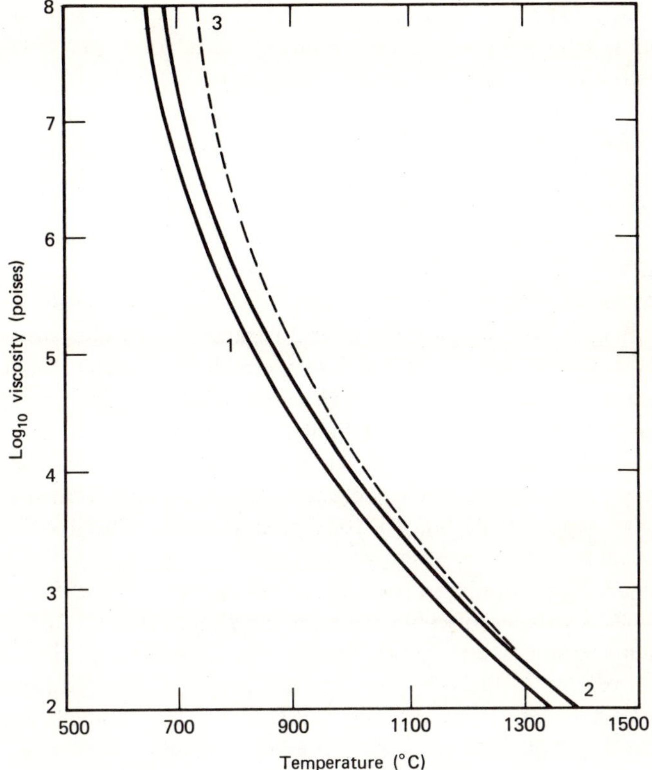

Figure 20.1   Viscosity-temperature curves for glasses.

metals in contact with it. Heat exchange in flat glass production is between the metal rollers, followed by radiative exchange between the glass and its surroundings.

In practice these problems are evaluated and controlled by monitoring of the operation augmented by experimental measurements of thermal properties of the materials involved, time, and temperature. Analytical solutions in terms of potential equations can be helpful as guidelines for some parts of the processes, the potential driving force being temperature. For example, heat losses from simple glass shapes can be calculated when the boundary conditions are specified. The paper by Van Zee and Babcock (1951) gives the general equation for the transfer of heat in a cylinder with radial heat flow.

Table 20.1 lists the viscosity levels in the glass-forming range from log 2.5 to log 8.0. Viscosity working ranges are dependent on the type of product: for example, the working range for machine-produced containers is usually between log 3 and log 7. This means that forming operations are carried out between the corresponding temperatures. Viscosity-temperature relations can be changed by oxide substitutions in the composition as demonstrated in Table 14.3. Consider the relation of the viscosity-temperature curve for glass 1 (Figure 20.1) to the curve for glass 2, which is parallel and has the same temperature difference between log 3 and log 7. The change of viscosity with time is greater for 2 than for 1 because 2 is losing heat at a faster rate. Glass 2 will reach its log 7 temperature before glass 1: it has a faster setting rate through the specified forming range. The rate of heat loss overshadows the slight differences in heat capacity of the two glasses. The setting rate in terms of change of viscosity with time for glass 3 is greater than the rate for glass 2 because it does not have to cool as many degrees and its log 3 and log 7 temperatures are higher. The change of viscosity with time, an important consideration in all forming processes, can be adjusted to a given process by use of known viscosity-temperature relations such as those in Table 14.3. Clearly these adjustments must be compatible with end usage product requirements such as chemical resistance and thermal expansion. Other things being equal, faster setting glasses can be used to obtain faster production rates.

In summary, glass-forming processes are influenced by the following glass properties: heat transfer by conduction and radiation, viscosity, surface tension, liquidus temperature, specific heat, density or specific volume, and coefficient of thermal expansion. Changes in these glass properties in terms of time and temperature caused by heat exchange between the glass and its surroundings exert controlling influences on all glass-forming processes.

The best book on glass forming is the one by Giegerich and Trier (1969), translated from the German by N. J. Kreidl. It reviews all technological aspects of the forming of all commonly used glass products and describes commercial processes also. This book is highly recommended to those engaged in forming practices and to those pursuing research on the subject. Descriptions of commercial processes are given by Holscher (1974), Hutchins and Harrington (1966), and others.

20.2 FORMING OF HOLLOW WARE

Utilitarian glass containers are made on high speed machines, fed continuously from melting tanks of appropriate size. In one type of process the stream of molten glass from the feeder is cut into gobs of uniform size and shape, the gobs fall into a blank mold, where a metal plunger shapes the parison. The parison is then transferred to the finish mold, where the glass is formed into the final shape. The whole operation from the time the glass gob hits the blank mold until the container is discharged from the finish mold is completed in a very short time. Babcock and McGraw (1957), who measured temperature distributions in glass and mold during forming, give representative times of forming of the parison. Their experimental forming machine records show the following time schedule for one cycle of the operation:

Gob in blank	−1.1 sec
Plunder in glass	0
Start of dwell	+0.1 sec
End of dwell	+2.03 sec
Next gob in blank	+7.5 sec

The contact time of the glass with the finish mold, after the glass parison has been transferred, is longer, but it is controlled in terms of seconds. These figures characterize high speed glass production and emphasize the necessity of process controls of heat transfer and glass property changes. After the glass machine has attained a continuous stable operation, the flow of heat in metal mold parts becomes cyclic and the temperature cycles repeat themselves quite closely. This indicates that a heat balance exists between the glass and mold parts under these transient conditions. Babcock and McGraw (1957) made temperature measurements of the glass and metal molds during continuous operation of an experimental forming machine. Their measurements of average glass gob temperatures by specific heat measurements in a calorimeter verified the merits of this method. They used their measurements in drawing up viscosity profiles of the glass in the parison at different times and at different distances from the glass surface.

Giegerich and Trier (1969) review the research just cited in connection with such other relevant aspects of hollow ware forming as measurements of heat content, changes in gob shape, time exposure of the glass, and theoretical studies on flow processes involved. They cover mechanical details of forming machines including different commercial types, characteristics of mold materials, types of mold cooling, and their functions in control of machine forming.

20.3 FORMING OF FLAT GLASS

Flat glass is formed by several modifications of rolling processes. In all cases the glass flows directly from the conditioning chamber and is drawn between metal rollers. Two principal modification are used: (1) the glass may be drawn vertically upward by rollers, after which the sheet continues through the annealing lehr or oven; (2) the glass may be drawn horizontally by rollers at the glass level of the conditioning chamber, after which it proceeds to the annealing lehr. The more recently developed Pilkington float process is of the second type, except that the glass sheet flows from the rollers to a separate furnace containing a bath of molten metal alloy. The sheet then continues into the annealing lehr. The lower surface of the sheet in contact with the molten metal acquires a degree of smoothness that eliminates costly mechanical polishing required by earlier methods. The upper surface of the sheet in contact with air during the molten metal stage also acquires a smoother surface finish as a result of further temperature conditioning. Rights to use of this improved process have been acquired by plate glass manufacturers in several countries.

Flat glass forming processes are usually carried on in the temperature range from the log 4 to log 7.6 viscosity points of the glass. These processes can be used to make pieces measuring about 12 × 12 feet and having thicknesses up to ¾ in. Each forming process presents unique problems of heat exchange and glass properties. Transient heat exchange in making flat glass differs from that which occurs in hollow ware manufacturing and is more amenable to process controls. Flat glass making is comparatively ideal in that it is a continuous process from the melter to the finished product. Window glass production may continue for a year or more with no change in the product being made. In contrast, different sizes and shapes of containers are made from individual feeders on the same furnace, and their production is interrupted from time to time by job changes. Each size and shape presents a different process control problem to the machine operator.

20.4 OTHER GLASS-FORMING PROCESSES

Glass tubing and cane are made using a number of drawing processes, classifiable under three general headings.

1. Glass flows continuously off the surface of an inclined, revolving conical refractory mandrel. Tubing is formed by blowing air through the center of the mandrel. Tubing diameter, wall thickness, and bore are controlled by drawing speed, viscosity, and temperature.

2. Glass is drawn upward from a refractory cone located on the glass surface. Air is blown upward through the cone to form tubing of desired dimensions. This method is used to manufacture large diameter glass pipe (up to 6 in. or more diameter) for transporting certain liquids.

3. Glass flows down through a refractory cone, and the blow air is introduced from above through the cone. Solid glass cane can be fabricated by any of these methods. Viscosity ranges are from log 4 to log 7.6 generally.

Glass fibers are produced using two types of process:

1. Fibers used for heat insulation are made by flowing the glass through a sievelike platinum bushing. The fibers are drawn at a rapid rate by the use of a jet of steam directed in the direction of glass flow.

2. Continuous fibers for technical uses such as battery separators and glass

cloth are made by the melting of glass marbles in an electrically heated platinum bushing. The strands of fibers flow from holes in the bushing and are attached to a revolving drum for continous drawing. The marbles used for textile production are made on a marble machine and are selected for good glass quality before being put into the platinum bushing.

Variations of these general types are used by different glass fiber manufacturers. However in all cases the fibers must be sprayed by binder solutions immediately after formation to prevent abrasion and breakage of the fibers. The viscosity level at the beginning of glass fiber processes is near log 2.5. The working range is very short because the fibers are rapidly quenched into solid glass.

Glass building blocks, television face plates or viewing screens, and similar products are formed by casting large gobs of molten glass directly into the finish molds. These products are then pressed to their final shapes. The viscosity working range is generally from log 4 to log 7.6. Television funnel shapes, which are attached to viewing screens by low melting glasses, are formed by centrifugal casting in the finish mold.

The glass for conversion into zero expansion Cer-Vit for astronomical mirrors is cast directly into the finish mold. In this Owens-Illinois process the final shape of the mirror blank is formed by gravity. In large mirror blanks of about 150 in. diameter, some 25 tons of glass can be cast in about the same number of minutes. The glass is then annealed and given time-temperature treatments that convert it into a combinational glass-crystalline material having a coefficient of expansion near zero.

Incandescent light bulbs, Christmas tree bulbs, and similar glass products are made on ribbon machines. The glass is formed into a ribbon by rollers, and the bulbs are formed by blow heads that come down into contact with the top surface of the ribbon. These thin-walled bulbs from high speed machines go directly to quality control inspection because they require no annealing.

21

Annealing and Tempering

The annealing process reduces internal strains in glasses to levels compatible with given product requirements and stabilizes the microstructures and resultant glass properties at levels that become essentially permanent after the glass is cooled to room temperature.

Product requirements may call for special heat treating or tempering processes, which establish high compressive stresses on the surfaces of the glassware. These processes may be used to effect other desired changes in surface properties.

21.1 DEFINITIONS AND NOMENCLATURE

The foregoing discussion on glass-forming processes make it evident that the surfaces of products cool first and become rigid while the interior layers of glass are still in a viscous liquid condition. The inside layers of glass continue to contract as the glass cools, placing the glass surface under compressive stresses that are balanced by interior tensile stresses. These stresses must be removed or minimized for all types of ware except for glass fibers and thin-walled ware such as light bulbs. Glass is birefringent when stressed, and the birefringence is proportional to the stress (see Chapter 7). Birefringence is measured in terms of stress-optical coefficient, which varies with glass composition. Stress levels are usually given in millimicrons per centimeter and vary with the type of product. A birefringence of 5 mμ/cm is satisfactory for most optical glass uses. Massive astronomical mirror blanks have a required birefringence of about 10 mμ/cm. Upper limits ranging from 50 to 100 mμ/cm cover plate glass and other large tonnage commercial glasses. For glasses of this type a level of 30 mμ/cm corresponds to about 150 lb/in.^{2} .

The establishment of guidelines for annealing processes in terms of temperatures at viscosity levels of log 13 and log 14.5 (annealing and strain points) has served to make these processes more uniform. Figure 21.1 gives an ideal picture of annealing in these terms which can only be approximated in actual practice, where each product must be considered as a unique problem. The solid line represents ideal sequences of the process. The glass enters the annealing lehr at time t_1 in a stressed condition and having an average temperature less than the

300

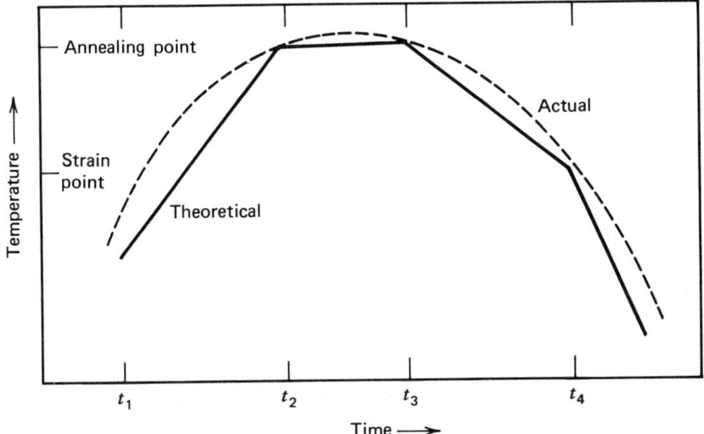

Figure 21.1   Time-temperature annealing curves for glasses.

annealing temperature. After reaching the annealing point at time t_2 it is held until the birefringence is one-half the desired value at room temperature. This is the so-called soaking stage of the process, and its time duration varies with size, weight, and shape of the product. The glass is then cooled from the annealing point at t_3 to the strain point at such a rate that the gain in birefringence is one-half the desired room temperature value. Theoretically the glass can be rapidly cooled to room temperature from the strain point at t_4 without gaining additional birefringence. Thus the glass reaches room temperature with the desired birefringence level.

Laboratory experiments show that the foregoing ideal conditions can be closely approached when single samples of glass are being annealed. In this case birefringence levels in terms of average stress-optical coefficient C, time, and temperature can be closely reproduced.

Morey (1954) covers the technology of annealing in some detail and describes important research studies. References to work performed since 1954 can be found in technical glass journals.

21.2   ANNEALING PRACTICE IN CONTINUOUS PRODUCTION

The annealing stage in continuous production involves transient heat exchanges between the glass product and its immediate surroundings. The problem can be approached in terms of the annealing of plate glass. The continuous ribbon of glass, which may be up to some 12 feet wide, moves through the annealing lehr at a speed of the order of 15 feet per minute. Thermocouples are used to measure ambient temperatures near the moving ribbon throughout its passage through the lehr. The difficulty of equalizing temperatures throughout the ribbon in the soaking stage becomes apparent. Heat exchange from the ribbon to the rollers over which it is moving, radiation of the edges of the ribbon to the side walls of the lehr, and so on, are conducive to temperature gradients in the glass. These conditions during soaking and subsequent cooling stages hinder the following of the ideal annealing curve of Figure 21.1. (Actual curves achieved in continuous production are of the general shape indicated by the broken line.) Forced circulation of air for cooling and electrical elements for heating are used to achieve proper temperature conditions in the annealing furnace.

Similar problems are presented by the continuous annealing of containers placed in order across annealing lehrs several feet wide. Differences in shape, weight, and wall thickness of the ware make the difficulties more serious than those associated with plate glass annealing. Continuous annealing processes are kept under control by the use of empirical methods that are specific for the type of ware being produced. Successful procedures result from a knowledge of heat transfer conditions and pertinent glass properties coupled with good engineering practice. Periodic photoelastic test checks on stress levels and stress patterns in the finished ware are used in monitoring the process. Strength considerations in all types of glassware require that the outside surfaces be under low levels of compressive stresses. Compressive stresses enhance resistance to surface abrasions and mechanical forces to which the ware may be subjected in service. In many cases the ware is sprayed with protecting films that further improve surface properties.

Continuous annealing practices are described by McMaster (1974) and his listed references on the subject. Further reference is made to Phillips (1948), Shand (1958), current glass journals, and files of the United States Patent Office.

21.3 HEAT TREATMENT OF SPECIALTY GLASSWARE

The heat treatments required in the manufacture of optical glass, thermometers, tempered plate, and other special products can properly be termed afterworking processes. These three products can be made by continuous processes and given regular annealing similar to that for ordinary glassware. However in these cases more precise annealing schedules are usually selected. This procedure is followed by further heat treating, which ensures satisfactory service performance of the finished products. Meeting special service specifications requires more precise attention to all phases of manufacture, from raw materials through melting and on to the finished product. Successful manufacture of specialty items calls for closer attention to the mechanical and thermal glass properties discussed in Parts III and IV.

Optical glass, made by either continuous or intermittent processes, necessitates care in all manufacturing stages, plus final annealing to ensure uniformity of refractive index. The refractive index value can be changed in the third decimal by varying the annealing temperature as illustrated in Figure 13.4. Annealing controls and their relations to optical requirements described by Wright (1921) remain essentially the same. However refinements in technique and instrumentation have been steadily improved, permitting manufacturers to meet precision requirements in optical design. As noted in Chapter 10, the visible absorption limits of colored filter glasses containing CdS, CdSe, and similar compounds are dependent on heat treating temperature. Major changes in optical properties result from oxide composition changes. However the relatively minor changes caused by annealing are very important.

Glass thermometers have been in universal use for about 100 years. It was soon discovered that when glasses are cooled they do not at once revert to their original condition. A thermometer whose capillary tube changes in volume with changes in temperature does not quickly revert to its original diameter. During its slow contraction the inside diameter and volume decrease, causing the mercury level to rise. The zero point of melting ice lies above the original marking in this case. This characteristic varies with glass composition and is strongly dependent on heat treatment. Thermometers demonstrate the degree to which the properties of glass are so strongly dependent on its thermal history. We recall at this point that structures and properties of glasses are in metastable conditions at temperatures down to absolute zero. These conditions are attained by most glass products during ordinary annealting. These items are not subjected to recurrent heating during their usage, and their metastable conditions are not disturbed. Recurrent heating in thermometers at temperatures well below the strain point at log viscosity = 14.5 causes volume changes that must be dealt with during manufacture. Although the change of volume with temperature depends primarily on glass composition, its change with heat treatment in terms of recurrent heating is of major importance in thermometers. Manufacturers keep these problems under control by proper choice of glass composition and by artificial aging achieved by heat treatments at low temperatures in the range of intended usage. Studies indicate that use of both Na_2O and K_2O strongly increases the zero constant; single alkali oxides are usually employed. Refer to relaxation phenomena of glasses covered in chapter 15. Further reference is made to Volf (1961).

Plate glass as regularly manufactured has low compressive stresses on its surfaces, and these improve the product's resistance to abrasion and other mechical forces. These strengths can be greatly enhanced by afterworking processes. Two methods can be distinguished: (1) after reheating the plate, both surfaces are rapidly cooled by applying compressed air, and (2) the plate is heated and immersed in a bath of molten salts at a temperature below the annealing zone of the given glass; ionic exchange of alkali ions between the glass and molten salt occurs, changing properties of the plate near its surfaces. It is usual to replace

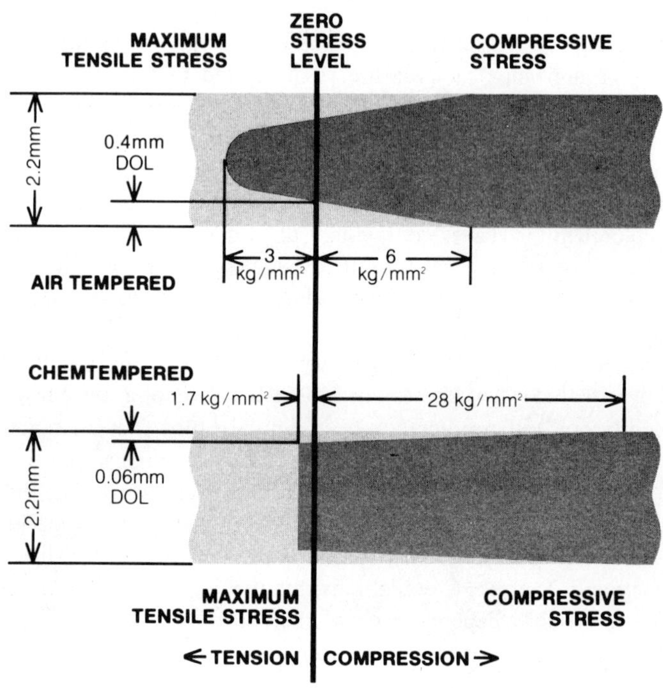

NOTE: 1 kg/mm² = 1,422 pounds per square inch (psi)

Figure 21.2 Typical stress distribution in 2.2 mm thick Chemtempered and Air Tempered lenses. From Corning Glass Works brochure, *Chemtempering Today*.

Na_2O ions in the glass by larger K_2O ions from the molten salt bath. The glass in the plate is in an essentially solid condition during the process. If time and temperature conditions are right, the process leaves the plate with surfaces that are under high compressive stresses. Figure 21.2 compares stress distribution in chemtempered and air-tempered ophthalmic safety lenses 2.2 mm thick. The Corning brochure from which the figure is taken gives the following comparisons of maximum stresses obtained in the two lensmaking processes.

Stress	Chemtempering (kg/mm^2)	Air Tempering (kg/mm^2)
Compressive	28-50	6-14
Tensile	1.7-2.5	3.0-7.0

Tempered products made by both processes are widely used in various applications including the manufacture of bulletproof windows.

All properties of glass surfaces, including refractive index and others, are changed by these tempering processes. Some specialty glasses make commercial use of the property changes that occur.

References

Adams, L. H. and E. D. Williamson (1919), *J. Wash. Acad. Sci.,* **9,** 609.

Ahrens, L. H. (1952), *Geochim. Cosmochim. Acta,* **2,** 168.

American Society of Metals Handbook (1961), Cleveland, Ohio.

Anderson. O. L. and H. E. Bommel (1955), *J. Am. Ceram. Soc.,* **38,** 125.

Anderson, O. L. and D. A. Stuart (1954), *J. Am. Ceram. Soc.,* **37,** 573.

Annual Book of ASTM Standards, Part 17 (1976), American Society for Testing and Materials, Philadelphia.

Appen, A. A. (1956), Preprint from Proceedings of Fourth International Congress on Glass, International Congress on Glass, Paris, France.

Armistead, W. H. and S. D. Stookey (1964), *Science,* **144,** 150.

Auerbach, F. (1894), *Ann. Phys. Chem.,* **53,** 1000.

Baak, T. (1969), Report No. 8744, Systems Engineering Group, USAF Wright-Patterson Air Force Base, Ohio

Babcock, C. L. (1934), *J. Am. Ceram. Soc.,* **17,** 329.

Babcock, C. L. (1940), *J. Am. Ceram. Soc.,* **23,** 12.

Babcock, C. L. (1968), *J. Am. Ceram. Soc.,* **51,** 163.

Babcock, C. L. (1961), *J. Am. Ceram. Soc.,* **44,** 301.

Babcock, C. L. (1969), *J. Am. Ceram. Soc.,* **52,** 151.

Babcock, C. L. (1973), *J. Am. Ceram. Soc.,* **56,** 527.

Babcock, C. L., S. W. Barber, and K. Fajans (1954), *Ind. Eng. Chem.,* **46,** 161.

Babcock, C. L. and D. A. McGraw (1957), *Glass. Ind.,* **38,** 137.

Bacon, F. R. and C. J. Billian (1954), *J. Am. Ceram. Soc.,* **37,** 60.

Badger, A. E., C. W. Parmellee, and A. E. Williams (1937), *J. Am. Ceram. Soc.,* **20,** 325.

Baldwin, G. C. (1969), *An Introduction to Nonlinear Optics,* Plenum Press, New York.

Bamford, C. R. (1960), *Phys. Chem. Glasses,* **1,** 159, 165.

Barron, L. M. (1972), *Am. Mineral.,* **57,** 809.

Barth, T. F. W. and T. Rosenquist (1949), *Am. J. Sci.,* **247,** 316.

Bates, T. (1962), "Ligand Field Theory and Absoption Spectra of Transition-Metal Ions in Glasses," in *Modern Aspects of the Vitreous State,* Vol. 2, J. D. Mackenzie, Ed., Butterworths, London.

Becherer, G., O. Brummer, and G. Herms (1962), *Silikatech.,* **13,** 339.

Belov, V. N. (1961), *Crystal Chemistry of Large-Cation Silicates,* Academy of Science, USSR, Moscow; English translation (1963), Consultants Bureau, New York.

Berezhnoi, A. I. (1970), *Glass-Ceramics and Photo-Sitalls,* Translated from Russian by S. A. Mersol, A. G. Pincus, Translation Ed. Plenum Press, New York.

Berthelot, M. (1901), *Compt. rend.,* **133,** 659.

Billmeyer, F. W., Jr., and M. Saltzman (1966), *Principles of Color Technology,* Wiley, New York.

Bingham, K. and S. Parks (1965), *Phys. Chem. Glasses,* **6,** 224.

Birch, F. (1942), "Elasticity," in *Handbook of Physical Constants,* F. Birch, J. F. Schairer, and H. C. Spicer, Eds., Special Paper No. 36, Waverly Press, Baltimore.

Birch, F. and R. B. Dow (1936), *Bull. Geol. Soc. Am.,* **47,** 1235.

Biscoe, J. (1941), *J. Am. Ceram. Soc.,* **24,** 262.

Biscoe, J., M. A. Druesne, and B. E. Warren (1941), *J. Am. Ceram. Soc.,* **24,** 100.

Biscoe, J., A. G. Pincus, C. S. Smith, and B. E. Warren (1941), *J. Am. Ceram. Soc.,* **24,** 116.

Biscoe, J. and B. E. Warren (1938), *J. Am. Ceram. Soc.,* **21,** 287.

Blau, H. H. (1951), *J. Soc. Glass Technol.,* **35,** 304.

Bloss, D. F. (1961), *Introduction to Methods of Optical Crystallography,* Holt, Rinehart & Winston, New York.

Bockris, J. O., J. L. White, and J. D. Mackenzie (1959), *Physicochemical Measurements at High Temperatures,* Academic Press, New York.

Bogdk'yants, G. O. (1955), *Tr. Soveshch. Str. Stekla, Akad, Nauk SSSR,* pp. 216-218.

Borrelli, N. F. (1971), *Phys. Chem. Glasses,* **12,** 93.

Borrelli, N. F. and R. A. Miller (1968), *Appl. Opt.,* **7,** 745.

Botvinkin, O. K. and N. L. Ananich (1958), *Steklo,* Information Bulletin of Soviet Glass Institute, No. 2.

Bragg, W. L. (1937), *Atomic Structure of Minerals,* Cornell University Press, Ithaca, N. Y.

Bray, P. J. and A. H. Silver (1960), "Nuclear Magnetic Resonance Absorption in Glass," in *Modern Aspects of the Vitreous State,* Vol. 1, J. D. Mackenzie, Ed., Butterworths, London.

Breen, R. J., R. M. Delaney, P. J. Persiani, and A. H. Weber, *Phys. Rev.,* **105,** 517.

Bridgman, P. W. (1936), *The Nature of Physical Theory,* Princeton University Press, Princeton, N. J.

Bridgman, P. W. (1948), *Proc. Am. Acad. Arts Sci.,* **76,** 55.

Bridgman, P. W. (1924), *Am. J. Sci.,* **7,** 81.

Bridgman, P. W. (1925), *Am. J. Sci.,* **10,** 359.

Bridgman, P. W. (1939), *Am. J. Sci.,* **237,** 7.

Bridgman, P. W. and I. Simon (1953), *Appl. Phys.,* **24,** 405.

Bruckner, R. (1970), *J. Non-Cryst. Solids,* **5,** 123; (1971), **4,** 177.

Brueche, E. and G. Schimmel (1954), *Glastech. Ber.,* **27,** 239.

Brueche, E. and G. Schimmel (1955), *Z. Angew. Phys.,* **7,** 378.

Brosset, C. (1962), *Proceedings of the Eighth International Ceramic Congress,* Copenhagen.

Buerger, M. J. (1956), *Elementary Crystallography,* Wiley, New York.

Burch, O. G. and C. L. Babcock (1938), *J. Am. Ceram. Soc.,* **21,** 345.

Buzhinskii, I. M. and N. I. Bodrova (1962), *Opt. Spectr. Osc. USSR,* English transl., **12,** 212.

Cabezas, A. Y. and L. G. DeShazer (1964), *Appl. Phys. Lett.,* **6,** 117.

Caley, E. R. (1962), *Analyses of Ancient Glasses 1790-1957,* Vol. I, Corning Museum of Glass, Corning, New York.

Carlson, D. E., K. W. Hang, and G. F. Stockdale (1972), *J. Am. Ceram. Soc.,* **55,** 337.

Charnock, H. (1961), *J. Am. Ceram. Soc.,* **44,** 313.

Chayes, F. (1968), *Am. Mineral.,* **53,** 359.

Clarke, F. W. and H. S. Washington (1924), "The Composition of the Earth's Crust," U.S. Geological Survey, Professional Paper 127.

Cleek, G. W. (1966), *Appl. Opt.,* **5,** 771.

Cleek, G. W. and C. L. Babcock (1973), "Properties of Glasses in some Ternary Systems Containing BaO and SiO_2," National Bureau of Standards Monograph 135, Washington, D.C., Government Printing Office.

Coenen, M. and E. M. Amrhein (1962), Union Scientifique Continentale du Verre Proceedings, Charleroi, Belgium.

Cole, H. (1950), *J. Soc. Glass Technol.,* **34,** 220.

Cole, H. (1951), *J. Soc. Glass Technol.,* **35,** 5, 25.

Cole, H. and C. L. Babcock (1970), Report on Viscosity-Temperature Relations in Glass, International Commission on Glass, Charleroi, Belgium.

Czerny, M. and L'. Genzel (1952), *Glastech. Ber.,* **25,** 387.

Czerny, M., L. Genzel, and G. Heilmann (1957), *Glastech. Ber.,* **28,** 185.

Dalton, R. H. (1933), *J. Am. Ceram. Soc.,* **16,** 425.

de Bast, J. and P. Gilard (1963), *Phys. Chem. Glasses,* **4,** 117.

Deeg, E. (1957), *Naturwissenschaften,* **44,** 448.

Deer, W. A., R. A. Howie, and J. Zussman (1966), *An Introduction to the Rock-Forming Minerals,* Wiley, New York.

Demkina, L. I. and B. I. Kisin (1972), *Opt. Technol.,* **39,** 414.

Devillers, R. W. and F. E. Vaerewyck (1950), *Glass Tank Furnaces,* Ogden, New York.

Dickson, R. W. and J. B. Wachtmann, Jr. (1971), *J. Res. U. S. Nat. Bur. Stand.,* Sect. A, **75,** 155.

Dietzel, A. (1967), *Glastech. Ber.,* **40,** 378.

Doborzynski, D. (1938), *Bull. Int. Acad. Polon. Sci., Classe Sci. Math. Nat.,* **193A,** 153.

Doremus, R. H. (1973), *Glass Science,* Wiley, New York.

Doremus, R. H. (1974), *J. Am. Ceram. Soc.,* **57,** 478.

Douglas, R. W. and J. O. Isard (1951), *J. Soc. Glass Technol.,* **35,** 206.

Douglas, R. W., P. J. Duke, and O. V. Mazurin (1968), *Phys. Chem. Glasses,* **9,** 1969.

Earnshaw, A. (1968), *Introduction to Magnetochemistry,* Academic Press, New York.

Eden, C. (1961), *Glastech. Ber.,* **34,** 120.

Eitel, W. (1965), *Silicate Science,* Vol. II, Academic Press, New York.

English, S. (1924), *J. Soc. Glass Technol.,* **8,** 205T.

Estermann, I. (1959), "Classical Methods," in *Methods of Experimental Physics,* L. Marton, Ed., Academic Press, New York.

Evans, R. C. (1946), *An Introduction to Crystal Chemistry,* Cambridge University Press, Cambridge, England.

Faick, C. A., J. C. Young, D. Hubbard, and A. N. Finn (1935), *J. Res., (U.S.) Nat. Bur. Stand.*, **14,** 133.

Fajans, K. (1923), *Naturwissenschaften,* **11,** 165.

Fajans, K. and N. J. Kreidl (1948), *J. Am. Ceram. Soc.,* **31,** 105.

Faraday, M. (1825), *Ann. Chim. Phys.,* **25,** 99.

Faraday, M. (1830), *Trans. Roy. Soc. (London),* **120,** II, 1.

Figgis, B. N. (1966), *Introduction to Ligand Fields,* Wiley, New York.

Filon, L. N. G. and F. C. Harris (1923), *Proc. Roy. Soc. (London),* **103A,** 561.

Forry, K. E. (1957), *J. Am. Ceram. Soc.,* **40,** 90.

Friedrich, W., P. Knipping, and M. von Laue (1913), *Radium,* **10,** 47.

Frondel, C. (1962), *Dana's System of Mineralogy,* Vol. 3, *Silica Minerals,* 7th ed., Wiley, New York.

Fulcher, G. S. (1925), *J. Am. Ceram. Soc.,* **8,** 339.

Furakawa, M. (1960), *Sci. Rep. Res. Inst., Tohoku Univ.* **A12,** 150.

Gardon, R. (1961), *J. Am. Ceram. Soc.,* **44,** 305.

Gaskell, P. H. (1966), *Trans. Faraday Soc.,* **62,** 1505.

Geffcken, W. (1952), *Glastech. Ber.,* **25,** 392.

Genzel, L. (1951), *Glastech. Ber.,* **24,** 55.

Genzel, L. (1953), *Glastech. Ber.,* **26,** 69.

Gehlhoff, G. and M. Thomas (1925), *Tech. Phys.,* **6,** 333, 544.

Georoff, A. N. and C. L. Babcock (1973), *J. Am. Ceram. Soc.,* **56,** 97.

Gibbs, J. W. (1906), *Scientific Papers of J. Willard Gibbs,* Longmans Green, New York.

Giegerich, W. and W. Trier, Eds., (1969), *Glass Machines,* Trans. N. J. Kreidl, Springer-Verlag, New York.

Gladstone, J. H. and T. P. Dale (1863), *Trans. Roy. Soc. (London),* **153,** 317, 337.

Glaze, F. W. (1953), *Ceram. Soc. Bull.,* **32,** 282, 313.

Goranson, R. W. (1942), "Heat Capacity; Heat of Fusion," in *Handbook of Physical Constants,* F. W. Birch, J. F. Schairer, and H. C. Spicer, Eds., Geological Society of America, Special Paper No. 36, Waverly Press, Baltimore.

Grove, F. J. and P. E. Jellyman (1955), *J. Soc. Glass Technol.,* **39,** 3T.

Gunther, R. (1958), *Glass-Melting Tank Furnaces,* Society of Glass Technology, Sheffield, England.

Gutzow, I. (1962), *Z. Phys. Chem.,* **221,** 153.

Hakim, R. M. and D. R. Uhlmann (1971), *Phys. Chem. Glasses,* **12,** 132.

Hakim, R. M. and D. R. Uhlmann (1973), *Phys. Chem. Glasses,* **14,** 81.

Hamilton, E. H. and G. W. Cleek (1958), *J. Res. (U.S.) Nat. Bur. Stand.,* **61,** 89.

Han Fu-si (1963), *Sci. Sin.,* **12,** 1365, Institute of Optics and Precision Mechanics of the Chinese Academy of Sciences.

Hardy, A. C. (1936), *Handbook of Colorimetry,* M.I.T. Press, Cambridge, Mass.

Harkins, W. D. and H. F. Jordan (1930), *J. Am. Chem. Soc.,* **52,** 1751.

Hartleif, G. (1938), *Z. Anorg. u Allg. Chem.,* **238,** 353.

Harrap, B. S. and E. Heymann (1955), *Trans. Faraday Soc.,* **51,** 259.

Harrison, W. N. and D. G. Moore (1939), *J. Res. (U.S.) Nat. Bur. Stand.,* **21,** 337.

Hensler, J. R. and E. Lell (1969), Meeting of International Commission on Glass, Toronto, Canada.

Herzfeld, C. M. Ed. (1962), *Temperature, Its Measurement and Control in Science and Industry*, Vol. 3, Reinhold, New York.

Hirayama, C. and D. Berg (1961), *Phys. Chem. Glasses, 2*, 145.

Holland, L. (1964), *The Properties of Glass Surfaces*, Wiley, New York.

Holscher, H. H., R. R. Rough, and J. H. Plummer (1943), *J. Am. Ceram. Soc., 26*, 398.

Holscher, H. H. (1974), "The Processing of Bottles and Other Hollow Ware Articles," in *Handbook of Glass Manufacturing*, F. V. Tooley, Ed., Books for Industry, New York.

Hovestadt, H. (1902), *Jena Glass and Its Scientific and Industrial Applications*, trans. J. D. Everett and A. Everett, Macmillan, New York.

Huff, N. T. and A. D. Call (1973), *J. Am. Ceram. Soc., 56*, 55.

Huggins, M. L. (1940), *J. Opt. Soc. Am., 30*, 420.

Huggins, M. L. and Kuan-han Sun (1943), *J. Am. Ceram. Soc., 26*, 4.

Hutchins, J. R. and R. V. Harrington (1966), "Glass," in *Encyclopedia of Chemical Technology*, Vol. 10, pp. 533-604, Wiley, New York.

Ide, J. M. (1937), *J. Geol., 45*, 689.

Indenbom, V. L. (1953), *Dokl. Akad. Nauk SSSR*, No. 3, 89.

Isard, J. O. (1969), *J. Non-Cryst. Solids, 1*, 235.

Isard, J. O. and R. W. Douglas (1955), *J. Soc. Glass Technol., 39*, 61T.

Izumitani, T. and S. Harada (1971), *Glass Technol., 12*, 131.

Jaegar, F. M. (1917), *Z. Anorg. All. Chem., 101*, 1.

Jagdt, R. (1960), *Glastech. Ber., 33*, 10.

Jellyman, P. E. and J. P. Proctor (1955), *J. Soc. Glass Technol., 39*, T173.

Johnston, R. A. and C. L. Babcock (1975), *J. Am. Ceram. Soc., 58*, 85.

Jones, T. D. (1973), M. S. thesis, University of Arizona, Tucson.

Karapetyan, G. O. (1961), *Izv. Akad. Nauk SSSR, Ser. Fiz., 25*, 539.

Karapetyan, G. O. (1963). *Izv. Akad. Nauk SSSR, Ser. Fiz., 27*, 779.

Kaufman, L. and H. Bernstein (1970), *Computer Calculation of Phase Diagrams*, Academic Press, New York.

Kellett, B. S. (1952), *J. Soc. Glass Technol., 36*, 115T.

Kellett, B. S. (1953), *J. Opt. Soc. Am., 42*, 339.

Keppler, G. (1937), *J. Soc. Glass Technol., 21*, 53.

Kim, K. S. and P. J. Bray (1974), *Phys. Chem. Glasses, 15*, 47.

Kingery, W. D. (1959), *Property Measurements at High Temperatures*, Wiley, New York.

Kozlovskaya, Y. I. (1960), *Transactions of the Third All Union Conference*, Academy of Sciences Press, Moscow-Leingrad, P. 340.

Kracek, F. C. (1939), *J. Am. Chem. Soc., 61*, 2863.

Kriz, H. M., M. J. Park, and P. J. Bray (1971), *Phys. Chem. Glasses, 12*, 45.

Lark-Horovitz, K., and V. A. Johnson, Eds., (1959), *Solid State Physics*, Vol. 6, Parts A and B, Academic Press, New York.

Lange, S. R. (1973), M. S. thesis, University of Arizona, Tucson.

Lange, S. R. and W. H. Turner (1973), *Appl. Opt., 12*, 1733.

Lebedev, A. A. (1921), *Arb. Staatl. Opt. Inst., Leningrad,* **2;** 10.

Levin, E. M., C. R. Robbins, and H. F. McMurdie (1964), *Phase Diagrams for Ceramists,* American Ceramic Society, Columbus, Ohio.

Levin, E. M., C. R. Robbins, and H. F. McMurdie (1969), *Phase Diagrams for Ceramists,* 1969 Supplement, American Ceramic Society, Columbus, Ohio.

Lillie, H. R. (1929), *J. Am. Ceram. Soc.,* **12,** 505.

Linwood, S. H. and W. A. Weyl (1942), *J. Opt. Soc. Am.,* **32,** 443.

Littleton, J. E. and G. W. Morey (1933), *The Electrical Properties of Glass,* Wiley, New York.

Lippincott, E. R., A. Van Valkenburg, C. E. Weir, and E. N. Bunting (1958), *J. Res. (U.S.) Nat. Bur. Stand.,* **61,** 61.

Lorenz, L. (1880), *Ann. Phys. Chem.,* **11,** 70.

Lorentz, H. A. (1880), *Ann. Phys. Chem.,* **9,** 641.

Lucas, A. (1926), *Ancient Egyptian Materials,* Longmans Green, New York.

Mackey, J., H. Smith, and A. Halperin (1966), *J. Phys. Chem. Solids,* **27,** 59.

Malitson, I. H. (1965), *J. Opt. Soc. Am.,* **55,** 1205.

Maloney, F. J. T. (1968), *Glass in the Modern World,* Doubleday, Garden City, New York.

Mallinder, F. P. and B. A. Proctor (1964), *Phys. Chem. Glasses,* **5,** 91.

Manghnani, M. H. (1972), *J. Am. Ceram. Soc.,* **55,** 360.

Marion, J. B. (1971), *Physics and the Physical Universe,* Wiley, New York.

Marinov, M. R. and T. S. Modeva (1965), *Proceedings of the Seventh International Congress on Glass, Brussels, Belgium, p. 301.*

Mason, B. (1960), *Principles of Geochemistry,* Wiley, New York.

Mazurin, O. V. (1965), in *The Structure of Glass,* O. V. Mazurin, Ed., Consultants Bureau, New York.

Mazurin, O. V. and E. S. Borisovskii (1957), *Sov. Phys.–Tech. Phys.,* **2,** 243.

McGraw, D. A. (1952), *J. Am. Ceram. Soc.,* **35,** 22.

McGraw, D. A. and C. L. Babcock (1959), *J. Am. Ceram. Soc.,* **42,** 330.

McMahon, H. O. (1951), *J. Am. Ceram. Soc.,* **34,** 91.

McMaster, H. A. (1974), "Annealing and Tempering," in *Handbook of Glass Manufacturing,* F. V. Tooley, Ed., Books for Industry, New York.

McMillan, P. W. (1964), *Glass Ceramics,* Academic Press, New York.

McSkimin, H. J. (1961), *J. Acoust. Soc. Am.,* **33,** 12.

Meller, F. and M. E. Milberg (1960), *J. Am. Ceram. Soc.,* **43,** 353.

Merritt, G. E. (1933), *J. Res. (U.S.) Nat. Bur. Stand.,* **10,** 59.

Moore, H. and R. C. De Silva (1952), *J. Soc. Glass Technol.,* **36,** 5.

Morey, G. W. (1930), *J. Phys. Chem.,* **34,** 1745.

Morey, G. W. (1939), U. S. Patent 2,150,694.

Morey, G. W. (1954), *The Properties of Glass,* 2nd ed., Reinhold, New York.

Morey, G. W. and N. L. Bowen (1925), *J. Soc. Glass Technol.,* **9,** 226.

Morey, G. W. and H. E. Merwin (1932), *J. Opt. Soc. Am.,* **22,** 632.

Mozzi, R. L. and B. E. Warren (1969), *J. Appl. Crystallogr.,* **2,** 164.

Mueller, H. (1935), *Physics,* **6,** 179.

Murthy, M. K. and F. A. Hummel (1955), *J. Am. Ceram. Soc.*, **38**, 55.

Nath, P., A. Paul, and R. W. Douglas (1965), *Phys. Chem. Glasses*, **6**, 203.

Navias, L. and R. L. Green (1946), *J. Am. Ceram. Soc.*, **29**, 267.

Nelson, W. F., F. T. King, and S. W. Barber (1964), in *Proceedings of the Third Conference on Rare Earths Research*, K. S. Vorres, Ed., Gordon & Breach, New York.

Neumann, F. (1841), *Ann. Phys. Chem.*, **54**, 449.

Neuroth, N. (1952), *Glastech. Ber.*, **25**, 242.

Neuroth, N. (1953), *Glastech. Ber.*, **26**, 66.

Nicoletta, C. A. and E. G. Eubanks (1972), *Appl. Opt.*, **11**, 365.

Nissle, T. R. and C. L. Babcock (1973), *J. Am. Ceram. Soc.*, **56**, 596.

Norton, F. H. (1935), *Glass Ind.*, **16**, 143.

Ohlberg, S. M. and J. M. Parsons (1965), "The Distribution of Sodium Ions in Soda-Lime-Silica Glass," in *Physics of Non-Crystalline Solids*, J. A. Prins, Ed., Wiley, New York.

Okuda, K. (1927), *Mem. Coll. Eng., Kyushu Imp. Univ.*, **4**, 159.

Oldfield, L. F. (1958), *Proceedings of the Symposium on the Melting of Glass*, Union Scientifique Continentale du Verre, Brussels, Belgium.

Owen, A. E. and R. W. Douglas (1959), *J. Soc. Glass Technol.*, **43**, 159.

Owen, A. E. (1963), "Electrical Conduction and Dielectric Relaxation," in *Progress in Ceramic Science*, Vol. 3, J. E. Burke, Ed., Macmillan, New York.

Parke, S. and R. S. Webb (1972), *Phys. Chem. Glasses*, **13**, 157.

Parmelee, C. W. and C. G. Harman (1937), *J. Am. Ceram. Soc.*, **20**, 224.

Peters, C. G. and C. H. Cragoe (1920), U.S. National Bureau of Standards Scientific Papers, 393.

Pauling. L. (1929), *J. Am. Chem. Soc.*, **51**, 1010.

Pauling, L. (1960), *Nature of the Chemical Bond*, Cornell University Press, Ithaca, N.Y.

Pearson, A. D. and G. E. Peterson (1965), *Proceedings of the Seventh International Congress on Glass*, Brussels, Belgium.

Peddle, C. J. (1920), *J. Soc. Glass Technol.*, **4**, 299, 320.

Petrie, W. M. F. (1924-1925), *Trans. Newcomen Soc.*, **5**, 72.

Phillips, C. J. (1948), *Glass the Miracle Maker*, Pitman, New York.

Phillips, C. J. (1964), *Glass Technol.*, **5**, 216.

Pockels, F. (1902), *Ann. Phys.*, **9**, 220.

Pockels, F. (1903), *Ann. Phys.*, **11**, 651.

Poldervaart, A., Ed. (1955), *Crust of the Earth*, Geological Society of America Special Paper 62.

Porai-Koshits, E. A. (1965), "Structure of Glass and Initial Stages in the Formation of Glass-Ceramics," in *The Strucutre of Glass*, Vol. 5, N. A. Toropov and E. A. Porai-Koshits, Eds., Consultants Bureau, New York.

Potts, J. C. (1939), *J. Soc. Glass Technol.*, **23**, 129.

Potts, J. C. (1941), *J. Am. Ceram. Soc.*, **24**, 43.

Potts, J. C., G. Brookover, and O. G. Burch (1944), *J. Am. Ceram. Soc.*, **27**, 225.

Preston, E. and E. Seddon (1937), *J. Soc. Glass Technol.*, **21**, 123.

Prod'homme, L. (1964), *Comp. Rend.*, **258**, 4476.

Prod'homme, L. (1968), *Phys. Chem. Glasses*, **9**, 101.

Ramachandran, G. N. (1958), in *Progress in Crystal Physics,* Vol. 1, R. S. Krishnan, Ed., pp. 139-167, Wiley, New York.

Raman, C. V. (1950), *Proc. Indian Acad. Sci.,* **31,** 141.

Randall, J. T., H. P. Rooksby, and B. S. Cooper (1930), *Z. Kristallogr.,* **75,** 196.

Ratobylskaya, V. A. and V. V. Tarasov (1960), *The Structure of Glass,* Vol. 1, p. 352, Consultants Bureau, New York.

Rawson, H. (1967), *Inorganic Glass-Forming Systems,* Academic Press, New York.

Riebling, E. F. (1967), *Rev. Inst. Hautes Temp. Refract.,* **4,** 65.

Riebling, E. F. (1968), *J. Am. Ceram. Soc.,* **51,** 143.

Rindone, G. E. (1966), "Luminescence in the Glassy State," in *Luminescence in Inorganic Solids,* P. Goldberg, Ed., Academic Press, New York.

Rindone, G. E. and J. F. Sproull (1964), J. F. Sproull's B.S. thesis, Pennsylvania State University, University Park.

Rinne, F. (1914), *Neues Jahrb. Mineral. Geol. Beliage Bl.,* **39,** 388.

Robinson, H. A. (1969), *J. Am. Ceram. Soc.,* **52,** 392.

Rodriguez, A. R., C. W. Parmelee, and A. E. Badger (1943), *J. Am. Ceram. Soc.,* **26,** 137.

Roedder, E. (1951), *Am. J. Sci.,* **249,** 81, 224.

Roedder, E. (1959), "Silicate Melt System," in *Physics and Chemistry of the Earth,* L. H. Ahrens, F. Press, K. Rankama, and S. K. Runcorm, Eds., Pergamon Press, New York.

Sakka, S. and J. D. Mackenzie (1969), *J. Non-Cryst. Solids,* **1,** 107.

Samson, A. R. (1967), *Techniques of Vacuum Ultraviolet Spectroscopy,* Wiley, New York.

Sawai, I. and S. Inoue (1940), *J. Soc. Chem. Ind. (Japan),* **43,** 47B.

Schmid, B. C., A. N. Finn, and J. C. Young (1934), *J. Res. (U.S.) Nat. Bur. Stand.,* **12,** 421.

Schaefer, C. and H. Nassenstein (1953), *Z. Naturforsch.,* **8A,** 90.

Schairer, J. F. and N. L. Bowen (1956), *Am. J. Sci.,* **254,** 129.

Scholes, S. R. (1975), *Modern Glass Practice,* 7th ed. revised and enlarged by C. H. Greene, Cahners, Boston.

Schulman, J. H. and H. W. Etzel (1953), *Science,* **118,** 184.

Schulman, J. H., R. J. Ginther, and L. W. Evans (1950), U.S. Patent 2,524,839.

Schulman, J. H. and W. D. Compton (1962), *Color Centers in Solids,* Macmillan, New York.

Schultz, P. C. (1974), *J. Am. Ceram. Soc.,* **57,** 309.

Selwood, P. W. (1956), *Magnetochemistry,* 2nd ed., Wiley, New York.

Shadduck, H. A. and A. Van Zee (1942), *J. Am. Ceram. Soc.,* **25,** 69.

Shand, E. B. (1958), *Glass Engineering Handbook,* McGraw-Hill, New York.

Sharbaugh, A. H. and S. Roberts (1959), "Dielectric Measurement Procedures," in *Solid State Physics,* Vol. 6, Part B, K. Lark-Horovitz and V. A. Johnson, Eds., Academic Press, New York.

Sharp, D. E. and L. B. Ginther (1951), *J. Am. Ceram. Soc.,* **34,** 1.

Shartsis, L., S. Spinner, and A. W. Smock (1948), *J. Am. Ceram. Soc.,* **31,** 23.

Shartsis, L. and R. Canga (1949), *J. Res. (U.S.) Nat. Bur. Stand.,* **43,** 221.

Shartsis, L. and S. Spinner (1951), *J. Res. (U.S.) Nat. Bur. Stand.,* **46,** 385.

Shartsis, L. and W. Capps (1952), *J. Am. Ceram. Soc.,* **35,** 169.

Shelby, J. E. and D. E. Day (1969), *J. Am. Ceram. Soc.,* **52,** 169.

Shermer, H. F. (1956), *J. Res. (U.S.) Nat. Bur. Stand.*, **56**, 155.

Shionoya, S. and E. Nakazawa (1965), *Appl. Phys. Lett.*, **6**, 117.

Silver, A. H. and P. J. Bray (1958), *J. Chem. Phys.*, **29**, 984.

Silverman, W. B. (1939), *J. Am. Ceram. Soc.*, **22**, 378.

Silverman, W. B. (1940), *J. Soc. Glass Technol.*, **24**, 59.

Silverman, W. B. (1942), *J. Am. Ceram. Soc.*, **25**, 168.

Simon, F. and F. Lange (1926), *Z. Phys.*, **38**, 227.

Simon, I. (1960), "Infrared Studies of Glass," in *Modern Aspects of the Vitreous State,* Vol. 1, J. D. Mackenzie, Ed., Butterworths, London.

Simon, I. and H. O. McMahon (1953a), *J. Chem. Phys.* **21**, 23.

Simon, I. and H. O. McMahon (1953b), *J. Am. Ceram. Soc.*, **36**, 160.

Smith, H. L. and A. J. Cohen (1963), *Phys. Chem. Glasses,* **4**, 173.

Smyth, H. T., J. W. Londeree, and G. E. Lorey (1953), *J. Am. Ceram. Soc.*, **36**, 238.

Smyth, H. T., H. S. Skogen, and W. B. Harsell (1953), *J. Am. Ceram. Soc.*, **36**, 327.

Smyth, H. T. (1955), *J. Am. Ceram. Soc.*, **38**, 140.

Snitzer, E. (1961), *Phys. Rev. Lett.*, **7**, 444.

Sosman, R. B. (1927), *The Properties of Silica,* Reinhold, New York.

Sosman, R. B. (1965), *The Phases of Silica,* Rutgers University Press, New Burnswick, N.J.

Spain, R. W. (1958), *Better Glass Making,* Industrial Publications, Chicago.

Spinner, S. and A. Napolitano (1956), *J. Am. Ceram. Soc.*, **39**, 390.

Spix, W. L. and F. R. Bacon (1953), *J. Am. Ceram. Soc.*, **36**, 377.

Splittgerber, D. K. (1839), *Poggendorffs Ann.*, **47**, 166.

Stanworth, J. E. (1950), *Physical Properties of Glass,* Clarendon Press, London.

Stevels, J. M. (1957), "The Electrical Properties of Glass," in *Handbuch der Physik,* Vol. 20, p. 350, Springer-Verlag, Berlin.

Stookey, S. D. (1949), *Ind. Eng. Chem.*, **41**, 856.

Stong, G. E. (1937), *J. Am. Ceram. Soc.*, **20**, 16.

Strakna, R. E. (1961), *Phys. Rev.*, **123**, 2020.

Swartz, E. L. and L. M. Cook (1965), *Proceedings, Seventh International Congress on Glass,* Brussels, Belgium.

Tashiro, M., S. Sakka, and T. Yamamoto (1967), *Bull. Inst. Chem. Res., Kyoto Univ.*, **75**, 201 (in English).

Taylor, E. W. (1950), *J. Soc. Glass Technol.*, **34**, 69.

Tickle, R. E. (1967), *Phys. Chem. Glasses,* **7**, 101, 113.

Tischer, R. E. and H. G. Drickamer (1962), *J. Chem. Phys.*, **37**, 1554.

Tool, A. Q. (1945), *J. Res. (U.S.) Nat. Bur. Stand.*, **34**, 199.

Tooley, F. V. Ed., (1974), *Handbook of Glass Manufacture,* Books for Industry, New York.

Topping, J. A. and J. O. Isard (1971), *Phys. Chem. Glasses,* **12**, 145.

Toyuki, H. and S. Akagi (1972), *Phys. Chem. Glasses,* **13**, 14.

Toyuki, H. and S. Akagi (1974), *Phys. Chem. Glasses,* **15**, 1.

Turner, W. E. S. and F. Winks (1930), *J. Soc. Glass Technol.*, **14**, 84.

Turner, W. H. (1967), *Symposium on Colored Glasses, Czechoslovak Scientific and Technical Society for the Glass Industry* Prague, Czechoslovakia.

Turner, W. H. (1973), *Appl. Opt.,* **12,** 480.

Turner, W. H. and J. E. Turner (1970), *J. Am. Ceram. Soc.,* **53,** 329.

Turner, W. H. and J. E. Turner (1972), *J. Am. Ceram. Soc.,* **55,** 201.

Van Uitert, L. G. And S. Iida (1962), *J. Chem. Phys.,* **37,** 986.

Van Vlack, L. H. (1970), *Materials Science for Engineers,* Addision-Wesley, Reading, Mass.

Van Zee, A. F. and C. L. Babcock (1951), *J. Am. Ceram. Soc.,* **34,** 244.

Van Zee, A. F. and H. M. Noritake (1958), *J. Am. Ceram. Soc.,* **41,** 164.

Vedam, K., E. D. Schmidt, and R. Roy (1966), *J. Am. Ceram. Soc.,* **49,** 531.

Vogel, W. and K. Gerth (1958), *Glastech. Ber.,* **31,** 15.

Volf, M. B. (1961), *Technical Glasses,* Pitman, London.

Von Hipple, A. R., Ed., (1954a), *Dielectric Materials and Applications,* Wiley, New York.

Von Hipple, A. R. (1954b), *Dielectrics and Waves,* Wiley, New York.

von Laue, M. (1913), *Phys. Zt.,* **14,** 421.

Vukcevich, M. R. (1972), *J. Non-Cryst. Solids,* **11,** 25.

Warren, B. E. and N. S. Gingrich (1934), *Phys. Rev.,* **46,** 368.

Warren B. E., K. Krutter, and O. Morningstar (1936), *J. Am. Ceram. Soc.,* **19,** 202.

Warren, B. E. (1937), *J. Appl. Phys.,* **8,** 645.

Warren, B. E. and J. Biscoe (1938), *J. Am. Ceram. Soc.,* **21,** 259.

Waxler, R. M. and A. Napolitano (1957), *J. Res. (U.S.) Nat. Bur. Stand.,* **59,** 121.

Waxler, R. M. and C. E. Weir (1965), *J. Res. (U.S.) Nat. Bur. Stand.,* **(***Phys. Chem.***),** **69A:** 4, 325.

Waxler, R. M. and G. W. Cleek (1971), *J. Res. (U.S.) Nat. Bur. Stand.,* **75A,** 279.

Waxler, R. M. and G. W. Cleek (1973), *J. Res. (U.S.) Nat. Bur. Stand.,* **77A,** 755.

Waxler, R. M., G. W. Cleek, I. H. Malitson, M. J. Dodge, and T. A. Hahn, (1971), *J. Res. (U.S.) Nat. Bur. Stand.,* **75A,** 163.

Wentorf, R. H., Ed. (1962), *Modern Very High Pressure Techniques,* Butterworths, London.

Westbrook, J. H. (1960), *Phys. Chem. Glasses,* **1,** 321.

Weyl, W. A. (1951), *Coloured Glasses,* Society of Glass Technology, Sheffield, England.

Weyl, W. A. and E. C. Marboe (1962, 1964, 1967), *The Constitution of Glasses,* Vol. I (1962), Vol. II, Part 1 (1964), Vol. II, Part 2 (1967), John Wiley & Sons, New York.

Weyl, W. A., J. H. Schulman, R. J. Ginther, and L. W. Evans (1949), *J. Electrochem. Soc.,* **95,** 70.

White, J. F. and W. B. Silverman (1950), *J. Am. Ceram. Soc.,* **33,** 252.

Wilburn, F. W., C. V. Thomasson, and H. Cole (1958), *Symposium on the Fusion of Glass,* Union Scientifique Continental du verre, Brussels, Belgium.

Wilke, K. T. (1962), *Z. Phys. Chem. (Leipzig),* **219,** 153.

Williams, F. (1966), "Theoretical Basis for Solid-State Luminescence," in *Luminescence of Inorganic Solids,* P. Goldberg, Ed., Academic Press, New York.

Wilmot, G. B. (1954), Ph.D. thesis, Massachusetts Institute of Technology, Cambridge.

Winkelmann, A. and O. Schott (1894), *Ann. Phys. Chem.,* **51,** 697.

Winkler, H. G. F. (1955), *Struktur und Eigenschaften der Kristelle,* Springer-Verlag, Berlin.

Winter-Klein, A. (1959), *Verres Refract.* **13,** 125.

Wright, F. E. (1921), *The Manufacture of Optical Glass and of Optical Systems,* Government Printing Office, Washington, D. C.

Yafaev, N. R. (1966), "Application of the Electron Paramagnetic Resonance of Transition-Metal Ion in the Study of the Structure of Glass," in *The Structure of Glass*, E. A. Porai-Koshits, Ed., Consultants Bureau, New York.

Young, J. C., F. W. Glaze, C. A. Faick, and A. N. Finn (1939), *J. Res. (U.S.) Nat. Bur. Stand.*, **22**, 453.

Zachariasen, W. H. (1932), *J. Am. Chem. Soc.*, **54**, 3841.

Zarzycki, J. (1956), *Proceedings of the Fourth International Congress on Glass*, International Congress on Glass, Paris, France, p. 323.

Zarzycki, J. (1957), *Verres Refract.* **11**, 3.

Zarzycki, J. and R. Mezard (1962), *Phys. Chem. Glasses*, **3**, 163.

Zarzycki, J. (1965), "Coordination Numbers in Vitreous and Crystalline Systems," in *Physics of Non-Crystalline Solids*, J. A. Prins, Ed., Wiley, New York.

Zschimmer, E. (1912), in *Handbuch de Mineralchemie*, C. Dolter, Ed., Vol 1, T. Steinkopff p. 869, Dresden and Leipzig.

Zschokke, W. (1918), *Z. Instrumentenk.*, **38**, 49.

Zsigmondy, R. (1901), *Ann. Phys.*, **4**, 60.

Zwikker, C. (1954), *Physical Properties of Solid Materials*, Pergamon Press, New York.

Appendix

Supplementary Information

BIBLIOGRAPHY

All-Union Conferences on the Structure of Glass (translated from Russian); Vol. 1, 1960; Vol. 2, 1960, Vol. 3, 1964; Vol. 4, 1965; Vol. 5, 1965; Vol. 6, 1966; and Vol. 7, 1966. Consultants Bureau, New York.

Coker, E. G. and L. N. G. Filon, *Treatise on Photoelasticity,* Cambridge University Press, London, 1931.

Cornish, D. C., *Mechanism of Glass Polishing,* British Scientific Instrument Research Association, Kent, England, 1961.

Dubrovo, S. K., *Vitreous Lithium Silicates,* Consultants Bureau, New York, 1964.

Eitel. W., *Silicate Science,* Vol. VII (1975). Vol. VIII (1976), Academic Press, New York, 1975.

Frocht, M. M. *Photoelasticity,* Vol. II, Wiley, New York, 1948.

Goldschmidt, V. M., *Geochemistry,* Alex Muir, Ed., Oxford University Press, New York, 1954.

Hampel, C. A., Ed., *The Encyclopedia of the Chemical Elements,* Van Nostrand Reinhold, New York, 1968.

Haynes, E. B., *Glass Through the Ages,* Penguin Books, Baltimore, 1959.

Hoffman, E., *Dictionary for the Glass Industry,* Springer-Verlag, Berlin, 1963.

Holden, A., *The Nature of Solids,* Columbia University Press, New York, 1965.

Jebsen-Marwedel, H., *Microscopical Examination of Glass Melting,* Verlag der Deutschen Glastechnischen Gesellshaft, Frankfurst, Germany, 1951.

Jones, G. O., *Glass,* Methuen, New York, 1956.

Kingery, W. D., *Introduction to Ceramics,* Wiley, New York, 1960.

Levin, E. M. and H. F. McMurdie, *Phase Diagrams for Ceramists, 1975 Supplement,* M.K. Reser, Ed., American Ceramic Society, Columbus, Ohio, 1975

Lillie, H. R., "Glass", in E. U. Condon and H. Odishaw, Eds., *Handbook Of Physics,* McGraw-Hill, New York, 1958.

Mazelev, L. Y., *Borate Glasses,* Consultants Bureau, New York, 1960.

Proceedings of a Symposium; Nuclear Science and Technology for Ceramists, Government Printing Office, Washington, D. C., 1967.

Ryshkewitch, E., *Oxide Ceramics; Physical Chemistry and Technology,* Academic Press, New York, 1960.

Salmang, H. and Marcus Francis, *Ceramics,* Butterworth, New York, 1961.

Scholze, H., *Glas: Natur, Struktur,* Eigenschaften., Braunschweig, Veiweg, Germany, 1965.

Schott, E., *Contributions to Applied Glass Research,* Wissenschaftliche Verlag, Stuttgart, 1959.

Van Vlack, L. H., *Physical Ceramics for Engineers,* Addison-Wesley, Reading, Mass., 1964.

Wong, J. and C. A. Angell. *Glass Structure by Spectroscopy,* Marcel Dekker, New York, 1976.

Wulff, J., Ed., *Structure and Properties of Materials,* Vol. 1, *Structure,* W. Moffatt et al, 1964; Vol. 2, *Thermodynamics of Structure,* J. H. Brophy et al., 1964; Vol. 3, *Mechanical Behavior,* H. W. Hayden et al., 1965; Vol. 4, *Electronic Properties,* R. M. Rose et al., 1966.

PERIODICALS

American Ceramic Society Bulletin

American Mineralogist

American Society of Testing and Materials, Special Technical Publication

Applied Materials Research

Applied Optics

Applied Spectroscopy

Bell System Technical Journal

Boletin de la Sociedad Espanola de Ceramica y Vidrio (Spain)

Central Glass and Ceramic Research Institute Bulletin (India)

Ceramica y Cristal (Argentina)

Ceramica (Brazil)

Czechoslovak Glass Review (Czechoslovakia)

Egyptian Journal of Chemistry (Egypt)

Glas-Email-Keramo-Technik (Germany)

Glass Industry

Glass and Ceramics (USSR)

Glass Technology (England)

Glastechnische Berichte (Germany)

High Temperature Science

Industrie Ceramique (France)

Industrial and Engineering Chemistry

Jena Review (Germany)

Journal of Applied Physics

Journal of Chemical Physics

Journal of Glass Technology (Sweden)

Journal of Glass Studies (Corning Museum of Glass)

Journal of Materials Science

Journal of Non-Crystalline Solids (Netherlands)

Journal of Physical Chemistry

Journal of Physics and Chemistry of Interfaces (Netherlands)

Journal of Research of the National Bureau of Standards (United States)

Journal of Solid State Chemistry

Journal of the American Ceramic Society
Journal of the American Chemical Society
Journal of the Chemical Society of Japan (Japan)
Journal of the Optical Society of America
Journal of the Physics and Chemistry of Solids
Materials Engineering
Materials Science and Engineering (Switzerland)
Official Gazette of the United States Patent Office
Philips Research Reports
Physical Review, Section B, *Solid State*
Physics and Chemistry of Glasses (England)
Proceedings of the USSR Academy of Sciences, Inorganic Materials (USSR)
Reviews of Modern Physics
Review of Scientific Instruments
Scientific American
Silicates Industriels (Belgium)
Transactions of the Faraday Society (England)
Verres et Refractaires (France)

Index of Properties of Experimental Silicate Glasses

Index of Subjects